高等学校建筑类专业英语规划教材

Environmental Science and Engineering
环境科学与工程专业

石 辉 主编

中国建筑工业出版社

图书在版编目(CIP)数据

高等学校建筑类专业英语规划教材. 环境科学与工程专业/石辉主编. —北京：中国建筑工业出版社，2010.8
ISBN 978-7-112-12301-8

Ⅰ. ①高… Ⅱ. ①石… Ⅲ. ①环境科学-高等学校-教材②环境工程-高等学校-教材 Ⅳ. ①X

中国版本图书馆 CIP 数据核字(2010)第 143574 号

高等学校建筑类专业英语规划教材
Environmental Science and Engineering
环境科学与工程专业
石　辉　主编
*
中国建筑工业出版社出版、发行(北京西郊百万庄)
各地新华书店、建筑书店经销
北 京 天 成 排 版 公 司 制 版
北京建筑工业印刷厂印刷
*
开本：787×1092 毫米　1/16　印张：21½　字数：644 千字
2010 年 8 月第一版　2010 年 8 月第一次印刷
定价：**40.00** 元
ISBN 978-7-112-12301-8
　　　(19562)

版权所有　翻印必究
如有印装质量问题，可寄本社退换
(邮政编码　100037)

本书以城市环境为主线，介绍了五个方面的环境问题。在第一篇中主要介绍了全球环境问题和可持续发展的基本理念，包括全球环境问题、城市环境问题、人居环境问题、城市的可持续发展；第二篇主要介绍了城市的水环境，包括水的利用和有效性、水质参数、安全饮水、饮用水净化、供水系统、水污染、废水的特征、废水处理技术、废水中固体和生物固体的处理、废水的回用等内容；第三篇主要介绍了城市的空气环境，包括空气污染的类型和污染源、空气污染的环境效应、颗粒污染物的控制、气态污染物的控制和大气环境影响评价；第四篇主要介绍了城市的固体废物处理以及声环境的保护，包括城市固体废物的特征、固体废物的产生、有害废物的贮存和处理，噪声对健康的影响、噪声的控制；第五篇以城市的生态环境为主，主要介绍了城市树木绿地的环境效益、绿地减缓大气颗粒物污染、屋顶绿化的结构和功能、生态城市建设。为了保证内容的完整性，还有一部分阅读材料对课程的内容做一补充，以便于根据课时选择使用。为了使该教材能发挥更大的作用，在教材中增加四个附录，一是为如何进行科学写作，二是科技论文写作的一些技巧，三是英文摘要的写作技巧，四是与环境科学相关的一些网站，希望能对学生的学习提供一些帮助。

* * *

责任编辑：张文胜　田启铭
责任设计：李志立
责任校对：赵　颖

本书编委会

主　　　编：石　辉
编写人员名单：（以姓氏笔画为序）
　　　　　　　王怡　田华　石辉
　　　　　　　高湘　曹利

前　言

随着科学技术的发展和学术交流的日益频繁，一方面我们需要学习国外的先进科学技术，另一方面需要将自己的研究成果介绍给世界，这些都离不开专业外语这个平台。经过多年的公共外语学习，现代大学生的外语水平有了较大的提高；但是一旦进入专业文献，学生们经常不知所云；尤其是当他们看到所熟悉的外语单词变成了字同义不同的专业词汇时，会产生既熟悉又陌生的感觉；当他们想用外语表达专业术语时，贫乏的专业词汇、生疏的语法结构又常使他们无从入手。

专业外语具有很强的专业性、针对性和应用性，因而专业外语学习与公共外语学习就有很大的区别。专业外语教学基于外语知识和专业知识于一体，因而专业外语的教学就不同于公共外语教学和专业课教学，它要求教师在教学方法上要有其独特性，不仅要适应学生外语学习的特点，又要适应专业学习的要求，适应学生的接受能力。专业外语是学生已学习了大学英语及一定的专业知识后，进一步熟练掌握和运用专业外语词汇和翻译技术的一门实用性很强的专业课，其任务是使学生掌握外语的特点和翻译方法，准确、流畅地阅读和翻译专业资料和科技论文，并学会用外文书写简短的论文摘要，为以后更好地吸收和交流国内外专业的先进知识和技术打好基础。

专业英语是环境科学与工程专业的专业基础必修课，在教学中起到了承前启后的关键作用。它保证了英语教学的连续性，培养和提高了学生阅读专业英语文献的能力，为以后的毕业论文环节中能顺利检索和阅读英文资料奠定了一定的基础。但现有的环境科学与工程专业外语教材主要关注各种环境污染的治理，如水污染治理、大气污染治理、噪声污染治理、固废污染治理等，极少涉及别的方面。随着环境保护基本原则由末端治理向预防的转变，专业外语的教学也应该适应这种变化，而这恰恰是当前专业外语教材中存在的不足。

考虑到当前面临的环境问题、课堂教学的时间和学生知识结构的调整，本书以城市环境为主线，介绍了5个方面的环境问题。在第一篇中主要介绍了全球环境问题和可持续发展的基本理念，包括全球环境问题、城市环境问题、人居环境问题、城市的可持续发展；第二篇主要介绍了城市的水环境，包括水的利用和有效性、水质参数、安全饮水、饮用水净化、供水系统、水污染、废水的特征、废水处理技术、废水中固体和生物固体的处理、废水的回用等内容；第三篇主要介绍了城市的空气环境，包括空气污染的类型和污染源、空气污染的环境效应、颗粒污染物的控制、气态污染物的控制和大气环境影响评价；第四篇主要介绍了城市的固体废物处理以及声环境的保护，包括城市固体废物的特征、固体废物的产生、有害废物的贮存和处理、噪声对健康的影响、噪声的控制；第五篇以城市的生态环境为主，主要介绍了城市树木绿地的环境效益、绿地减缓大气颗粒物污染、屋顶绿化的结构和功能、生态城市建设。为了保证内容的完整性，还有一部分阅读材料对课程的内容做一补充，以便于根据课时选择使用。为了使该教材能发挥更大的作用，在教材中增加4个附录，一是如何进行科学写作，二是科技论文写作的一些技巧，三是英文摘要的写作技巧，四是与环境科学相关的一些网站，希望能对学生的学习提供帮助。

全书的框架体系和大纲由西安建筑科技大学的石辉提出，其中 Unit1~Unit4 由石辉编写，Unit5~Unit9 由西安建筑科技大学的高湘编写，Unit10~Unit14 由西安建筑科技大学的王怡编写，Unit15~Unit19 由西安建筑科技大学的曹利编写，Unit20~Unit22 由西安科技大学的田华编写，Unit23~Unit24 由王怡编写，Unit25~Unit28 由石辉编写，附录部分的资料由石辉和高湘负责收集整理。

本书涉及到了环境科学的各个方面，除作为专业外语教材使用外，还可以起到环境科学导论的作用，适合于环境科学相关的专业的本科生、硕士研究生使用，包括环境科学、环境工程、给水排水、资源环境与城乡规划、地理科学等。同时，也可以作为相关专业科技人员的参考书。

西安建筑科技大学是以土建类为特色的综合性大学，环境科学和环境工程专业是其重要的组成部分。同时，环境工程是国家的特色专业、国家重点学科，环境科学为陕西省重点学科、国家重点学科的培育学科。为了进一步提高教学质量，凝练专业特色，西安建筑科技大学教务处组织了建筑类专业外语的研讨，在充分讨论的基础上，决定编写以土建类核心专业为基础的特色专业外语系列教材，以满足专业建设的需要。环境科学与工程专业外语是建筑类专业外语规划教材中的一册。在系列教材研讨和编写过程中，西安建筑科技大学以重点教材建设项目给予资助，教务处庞丽娟副处长在教材研讨和编写过程中做了大量的组织管理工作，使得该教材能得以顺利完成。西安建筑科技大学建筑学院、土木学院、环境与市政工程学院、信控学院、艺术学院的教师在研讨中提出了许多有益的建议。在此，向关心和支持教材编写的单位和个人表示衷心的感谢。对于书中所采用文献资料的作者和单位表示诚挚的谢意。

本书在选材上，尽量选取正式出版的外文期刊文献和书籍，力求保证内容的科学性和准确性，但由于编者学术水平所限，在文章的选材和编排方面还存在着不足，恳请读者给予批评指正。

编　者
2010 年 6 月

CONTENS

Part A Global Environmental Problems and Sustainable Development ·············· 1

    Unit 1   Global Environmental Problems ·· 1
    Unit 2   Urban Environment ··· 12
    Unit 3   Housing and Living Environment ·· 24
    Unit 4   Urban Sustainable Development ·· 33

Part B Urban Water Environment ·· 50

    Unit 5   Water Use and Availability ·· 50
    Unit 6   Parameters of Water Quality ··· 65
    Unit 7   Safe Drinking Water Act ··· 80
    Unit 8   Drinking Water Purification ·· 95
    Unit 9   Water Distribution Systems ··· 116
    Unit 10 Water Pollution ··· 134
    Unit 11 Wastewater Properties ·· 140
    Unit 12 Wastewater Treatment ·· 148
    Unit 13 Treatment and Disposal of Solids and Biosolids in the Wastewater ········ 159
    Unit 14 Wastewater Reuse ··· 174

Part C Urban Air Environment ·· 183

    Unit 15 Type and Sources of Air Pollutants ······································ 183
    Unit 16 Effects of Air Pollution ··· 193
    Unit 17 Control of Particulate Pollutants ··· 200
    Unit 18 Control of Gaseous Pollutants ··· 212
    Unit 19 Summary of Environmental Impact Assessment ················· 225

Part D Urban Solid Waste and Sound Environment ························· 231

    Unit 20 Solid Waste Characteristics ·· 231
    Unit 21 Municipal Solid Waste Processing ·· 237
    Unit 22 Treatment, Storage, and Disposal of Hazardous Waste Management ······ 243
    Unit 23 Adverse Health Effects of Noise ··· 253
    Unit 24 Engineering Noise Control ··· 260

Part E Urban Eco-Environment ·· 271

    Unit 25 Benefits of Trees In Urban Areas ·· 271

Unit 26　Urban Woodlands Reducing the Particulate Pollution ……………… 276

Unit 27　Ecological Structure, Functions and Services of Green Roofs ………… 283

Unit 28　The Ecological City: Metaphor versus Metabolism ……………………… 291

Appendix 1　The Science of Scientific Writing ………………………………… 302

Appendix 2　The Skills in Paper Writing ………………………………………… 321

Appendix 3　Skills in Abstracts Writing ………………………………………… 325

Appendix 4　与环境科学相关的网站地址 ……………………………………… 331

Part A Global Environmental Problems and Sustainable Development

Unit 1 Global Environmental Problems

More than a generation ago scientists detected radioactive strontium from atomic tests in reindeer meat and linked DDT to the non-viability of bird eggs. Ever since then, if not before, science has had a central role in shaping what count as environmental problems. During the 1980s environmental scientists and environmentalists called attention, in particular, to analyses of carbon dioxide concentrations in polar ice, measurements of upper atmospheric ozone depletion, remote sensing assessments of tropical deforestation, and, most notably, projections of future temperature and precipitation changes drawn from computation-intensive atmospheric circulation models. This coalition of environmental activism and "planetary science" stimulated a rapid rise in awareness and discussion of global environmental problems. The global environmental problem is a growing concern, and needs to be attended to immediately. Spreading awareness of environmental problems, and responding to them without delay is absolutely necessary to deal with the global problem effectively.

1. Climate Change

Global warming is becoming a critical issue again with the wild weather occurring all over the world. The biggest contributer to global warming is us, who are too reckless in are use of power and making products.

The temperature near the surface of the earth is determined by the balance between solar radiation energy and heat reflected from earth to outer space. Sunlight warms the surface of the earth, which then cools down when the heat (infrared rays) is released from it. If heat exchange were this simple, then the surface of the earth would cool down rapidly as soon as solar radiation stops, with only heat reflection continuing. In reality, however, heat-absorbing atmospheric gases absorb a certain percentage of heat (infrared rays) reflected from the surface of the earth. The atmosphere, thus warmed, radiates infrared rays out toward space and back toward the surface of the earth, warming the latter. In this manner, the earth surface temperature is kept at around 15℃ (global average), realizing an environment suited for the existence of humans, animals, plants and other life forms. The natural process of heating the surface of the earth by sunlight is called the "greenhouse effect", and the infrared-absorbing gases in the atmosphere are called "greenhouse gases". If not for greenhouse gases, the temperature of the earth would drop by over 30℃ to about −18℃.

Since the Industrial Revolution at the end of the 18th century, there has been a marked increase in the emission of greenhouse gases (mainly carbon dioxide), in proportion to industrial expansion. As well, artificial greenhouse gases that did not exist before the Industrial Revolution, such as CFC (chlorofluorocarbon) and sulfur hexafluoride, have been produced, and are being released into the atmosphere. Increased greenhouse gases destroy the heat exchange balance, keeping greater amounts of heat in the atmosphere and on the earth (i.e. intensifying the greenhouse effect) and raising the temperature near the surface of the earth to undesirable levels. This phenomenon is called global warming.

2. Ozone Depletion

The ozone layer, situated in the stratosphere about 15 to 30km above the earth's surface, plays the important role of "space suit" that protects us living beings by absorbing harmful ultraviolet radiation (UVB) from the sun. The ozone layer is currently being destroyed by CFCs and other substances, its depletion progressing globally except in the tropical zone. The ozone layer is disappearing at a particularly high rate in high-latitude areas. In the Antarctic Circle, a large ozone hole has been observed for eight consecutive years from 1989 through 1996. Destruction of the ozone layer increases the amount of harmful ultraviolet radiation (UVB), which in turn can result in increased cases of skin cancer, and visual impediments such as cataract. It can also hinder the growth of plants and negatively impact small living organisms, such as zooplankton, phytoplankton, shrimp larvae and the young of fish.

3. Biodiversity Loss

Biologists most often define "biological diversity" or "biodiversity" as the "totality of genes, species, and ecosystems of a region". An advantage of this definition is that it seems to describe most circumstances and present a unified view of the traditional three levels at which biological variety has been identified. Biodiversity provides many ecosystem services that are often not readily visible. It plays a part in regulating the chemistry of our atmosphere and water supply. Biodiversity is directly involved in water purification, recycling nutrients and providing fertile soils. Experiments with controlled environments have shown that humans cannot easily build ecosystems to support human needs; for example insect pollination cannot be mimicked by human-made construction, and that activity alone represents tens of billions of dollars in ecosystem services per annum to humankind.

During the last century, some studies show that about one eighth of known plant species are threatened with extinction. Some estimates put the loss at up to 140000 species per year (based on Species-area theory) and subject to discussion. Almost all scientists acknowledge that the rate of species loss is greater now than at any time in human history, with extinctions occurring at rates hundreds of times higher than background extinction rates. The factors that threaten biodiversity have been variously categorized, such as habitat destruction, overkill, introduced species, pollution, human over population, and overharvesting.

4. Acid Rain

"Acid rain" is a popular term referring to the deposition of wet (rain, snow, sleet, fog and cloudwater, dew) and dry (acidifying particles and gases) acidic components. A more accurate term is "acid deposition". Distilled water, which contains no carbon dioxide, has a neutral pH of 7. Liquids with a pH less than 7 are acidic, and those with a pH greater than 7 are basic. "Clean" or unpolluted rain has a slightly acidic pH of about 5.2, because carbon dioxide and water in the air react together to form carbonic acid, a weak acid (pH 5.6 in distilled water), but unpolluted rain also contains other chemicals. The extra acidity in rain comes from the reaction of primary air pollutants, primarily sulfur oxides and nitrogen oxides, with water in the air to form strong acids (like sulfuric and nitric acid). The main sources of these pollutants are industrial power-generating plants and vehicles.

The most important gas which leads to acidification is sulfur dioxide. Emissions of nitrogen oxides which are oxidized to form nitric acid are of increasing importance due to stricter controls on emissions of sulfur containing compounds. 70 Tg (S) per year in the form of SO_2 comes from fossil fuel combustion and industry, 2.8 Tg (S) from wildfires and 7~8 Tg (S) per year from volcanoes. The principal cause of acid rain is sulfur and nitrogen compounds from human sources, such as electricity generation, factories, and motor vehicles. Coal power plants are one of the most polluting. The gases can be carried hundreds of kilometres in the atmosphere before they are converted to acids and deposited. In the past, factories had short funnels to let out smoke, but this caused many problems locally; thus, factories now have taller smoke funnels. However, dispersal from these taller stacks causes pollutants to be carried farther, causing widespread ecological damage.

5. Desertification

Desertification (or desertization) is the degradation of land in arid, semi-arid and dry sub-humid areas resulting primarily from human activities and influenced by climatic variations. Current desertification is taking place much faster worldwide than historically and usually arises from the demands of increased populations that settle on the land in order to grow crops and graze animals.

A major impact of desertification is biodiversity loss and loss of productive capacity, for example, by transition from land dominated by shrublands to non-native grasslands. In the semi-arid regions of southern California, many coastal sage shrub and chaparral ecosystems have been replaced by non-native, invasive grasses due to the shortening of fire return intervals. This can create a monoculture of annual grass that can not support the wide range of animals once found in the original ecosystem. In Madagascar's central highland plateau, 10% of the entire country has been lost to desertification due to slash and burn agriculture by indigenous peoples. In Africa, if current trends of soil degradation continue, the continent might be able to feed just 25% of its population by 2025, according to UNU's Ghana-based Institute for Natural Resources in Africa. Desertification is induced by several factors, primarily anthropogenic beginning the Holocene area. The primary reasons for desertification are overgrazing, overcultivation, increased fire frequency, water impoundment, deforestation, overdrafting of groundwater, increased soil salinity, and global climate change.

6. Deforestation

Deforestation is the conversion of forested areas to non-forest land for use such as arable land, pasture, urban use, logged area, or wasteland. Generally, the removal or destruction of significant areas of forest cover has resulted in a degraded environment with reduced biodiversity. In many countries, massive deforestation is ongoing and is shaping climate and geography. Deforestation results from removal of trees without sufficient reforestation, and results in declines in habitat and biodiversity, wood for fuel and industrial use, and quality of life.

From about the mid-1800s, around 1852, the planet has experienced an unprecedented rate of change of destruction of forests worldwide. Forests in Europe are adversely affected by acid rain and very large areas of Siberia have been harvested since the collapse of the Soviet Union. In the last two decades, Afghanistan has lost over 70% of its forests throughout the country. However, it is in the world's great tropical rainforests where the destruction is most pronounced at the current time and where clearcutting is having an adverse effect on biodiversity and contributing to the ongoing Holocene mass extinction.

Generally, the removal or destruction of significant areas of forest cover has resulted in a degraded environment with reduced biodiversity. In many countries, massive deforestation is ongoing and is shaping climate and geography. Deforestation is a substantial contributor to global warming, and although 70% of the oxygen we breathe comes from the photosynthesis of marine green algae and cyanobacteria, the mass destroying of the world's rain forests is not beneficial to our environment. In addition, the incineration and burning of forest plants in order to clear land releases tonnes of CO_2 which increases the impact of global warming.

Deforestation reduces the content of water in the soil and groundwater as well as atmospheric moisture. Deforestation reduces soil cohesion, so that erosion, flooding and landslides often ensue. Forests support considerable biodiversity, providing valuable habitat for wildlife; moreover, forests foster medicinal conservation and the recharge of aquifers. With forest biotopes being a major, irreplaceable source of new drugs (like taxol), deforestation can destroy genetic variations (such as crop resistance) irretrievably.

Shrinking forest cover lessens the landscape's capacity to intercept, retain and transport precipitation. Instead of trapping precipitation, which then percolates to groundwater systems, deforested areas become sources of surface water runoff, which moves much faster than subsurface flows. That quicker transport of surface water can translate into flash flooding and more localized floods than would occur with the forest cover. Deforestation also contributes to decreased evapotranspiration, which lessens atmospheric moisture which in some cases affects precipitation levels down wind from the deforested area, as water is not recycled to downwind forests, but is lost in runoff and returns directly to the oceans. According to one preliminary study, in deforested north and northwest China, the average annual precipitation decreased by one third between the 1950s and the 1980s.

7. Fresh Water Supply

Uses of fresh water can be categorized as consumptive and non-consumptive (sometimes called renewable). A use of water is consumptive if that water is not immediately available for another use. Losses to sub-surface seepage and evaporation are considered consumptive, as is water incorporated into a product (such as farm produce). Water that can be treated and returned as surface water, such as sewage, is generally considered non-consumptive if that water can be put to additional use.

It is estimated that 69% of worldwide water use is for irrigation, with 15%~35% of irrigation withdrawals being unsustainable. As global populations grow, and as demand for food increases in a world with a fixed water supply, there are efforts underway to learn how to produce more food with less water, through improvements in irrigation methods and technologies, agricultural water management, crop types, and water monitoring. It is estimated that 15% of worldwide water use is industrial. Major industrial users include power plants, which use water for cooling or as a power source (i.e. hydroelectric plants), ore and oil refineries, which use water in chemical processes, and manufacturing plants, which use water as a solvent. The portion of industrial water usage that is consumptive varies widely, but as a whole is lower than agricultural use. It is estimated that 15% of worldwide water use is for household purposes. These include drinking water, bathing, cooking, sanitation, and gardening. Basic household water requirements have been estimated at around 50 liters per person per day, excluding water for gardens. Drinking water is water that is of sufficiently high quality so that it can be consumed or used without risk of immediate or long term harm. Such water is commonly called potable water. In most developed countries, the water supplied to households, commerce and industry is all of drinking water standard even though only a very small proportion is actually consumed or used in food preparation.

According to the World Business Council for Sustainable Development, the concept of water stress is applies to situations where there is not enough water for all uses, whether agricultural, industrial or domestic. Defining thresholds for stress in terms of available water per capita is more complex, however, entailing assumptions about water use and its efficiency. Nevertheless, it has been proposed that when annual per capita renewable freshwater availability is less than 1700 cubic meters, countries begin to experience periodic or regular water stress. Below 1000 cubic meters, water scarcity begins to hamper economic development and human health and well-being.

Water pollution is one of the main concerns of the world today. The governments of many countries have striven to find solutions to reduce this problem. Many pollutants threaten water supplies, but the most widespread, especially in underdeveloped countries, is the discharge of raw sewage into natural waters; this method of sewage disposal is the most common method in underdeveloped countries, but also is prevalent in quasi-developed countries. Sewage, sludge, garbage, and even toxic pollutants are all dumped into the water. Even if sewage is treated, problems still arise. Treated sewage forms sludge, which may be placed in landfills, spread out on land, incinerated or dumped at sea. In addition to sewage, nonpoint source pollution such as agricultural runoff is a significant source of pollution in some parts of the world, along with urban stormwater runoff and chemical wastes dumped by industries and governments.

Water stress can also exacerbate conflicts and political tensions which are not directly caused by water. Gradual reductions over time in the quality and/or quantity of fresh water can add to the instability of a region by depleting the health of a population, obstructing economic development, and exacerbating larger conflicts. Conflicts and tensions over water are most likely to arise within national borders, in the downstream areas of distressed river basins. Additionally, certain arid countries which rely heavily on water for irrigation, are particularly at risk of water-related conflicts.

8. Persist Organic Pollution

Persistent organic pollutants (POPs) are organic compounds that are resistant to environmental degradation through chemical, biological, and photolytic processes. Because of this, they have been observed to persist in the environment, to be capable of long-range transport, bioaccumulate in human and animal tissue, biomagnify in food chains, and to have potential significant impacts on human health and the environment.

In May 1995, the United Nations Environment Programme Governing Council (GC) decided to begin investigating POPs, initially beginning with a short list of the following twelve POPs, known as the "dirty dozen": aldrin, chlordane, DDT, dieldrin, endrin, heptachlor, hexachlorobenzene, mirex, polychlorinated biphenyls, polychlorinated dibenzo-p-dioxins, polychlorinated dibenzofurans, and toxaphene. Since then, this list has generally been accepted to include such substances as carcinogenic polycyclic aromatic hydrocarbons (PAHs) and certain brominated flame-retardants, as well as some organometallic compounds such as tributyltin (TBT). The groups of compounds that make up POPs are also classed as PBTs (Persistent, Bioaccumulative and Toxic) or TOMPs (Toxic Organic Micro Pollutants).

Many POPs are currently or were in the past used as pesticides. Others are used in industrial processes and in the production of a range of goods such as solvents, polyvinyl chloride, and pharmaceuticals. Though there are a few natural sources of POPs, most POPs are created by humans in industrial processes, either intentionally or as byproducts.

POPs released to the environment have been shown to travel vast distances from their original source. Due to their chemical properties, many POPs are semi-volatile and insoluble. These compounds are therefore unable to transport directly through the environment. The indirect routes include attachment to particulate matter, and through the food chain. The chemicals' semi-volatility allows them to travel long distances through the atmosphere before being deposited. Thus POPs can be found all over the world, including in areas where they have never been used and remote regions such as the middle of oceans and Antarctica. The chemicals' semi-volatility also means that they tend to volatilize in hot regions and accumulate in cold regions, where they tend to condense and stay. PCBs have been found in precipitation. The ability of POPs to travel great distances is part of the explanation for why countries that banned the use of specific POPs are no longer experiencing a decline in their concentrations; the wind may carry chemicals into the country from places that still use them.

Exposure to POPs can take place through diet, environmental exposure, or accidents. POPs exposure can cause death and illnesses including disruption of the endocrine, reproductive, and immune systems; neurobehavioral disorders; and cancers possibly including breast cancer. The lipid solubility of POPs allows them to bioaccumulate in fatty tissues of animals. Many of the first generation organochlorine insecticides such as DDT were particularly noted for this characteristic.

A study published in 2006 indicated a link between blood serum levels of POPs and diabetes. Individuals with elevated levels of persistent organic pollutants (DDT, dioxins, PCBs and Chlordane, among others) in their body were found to be up to 38 times more likely to be insulin resistant than individuals with low levels of these pollutants, though the study did not demonstrate a cause and effect relationship. As most exposure to POPs is through consumption of animal fats, study participants with high levels of serum POPs are also very likely to be consumers of high amounts of animal fats, and thus the consumption of the fats themselves, or other associated factors may be responsible for the observed increase in insulin resistance. Another possibility is that insulin resistance causes increased accumulation of POPs. Among study participants, obesity was associated with diabetes only in people who tested high for these pollutants. These pollutants are accumulated in animal fats, so minimizing consumption of animal fats may reduce the risk of diabetes. According to the US Department of Veterans Affairs, type 2 diabetes is on the list of presumptive diseases associated with exposure to Agent Orange (which contained the POP dioxin) in the Vietnam War.

9. Over Fishing

Overfishing occurs when fishing activities reduce fish stocks below an acceptable level.

This can occur in any body of water from a pond to the oceans. Ultimately overfishing may lead to resource depletion in cases of subsidised fishing, low biological growth rates and critical low biomass levels (e. g. by critical depensation growth properties). Particularly, overfishing of sharks has led to the upset of entire marine ecosystems. The ability of the fisheries to naturally recover also depends on whether the conditions of the ecosystems are suitable for population growth. Dramatic changes in species composition may establish other equilibrium energy flows that involve other species compositions than had been present before (ecosystem shift). For example, remove nearly all the trout, and the carp might take over and make it nearly impossible for the trout to re-establish a breeding population.

There are three recognized types of overfishing: growth overfishing, recruit overfishing and ecosystem overfishing. A more dynamic definition of economic overfishing may also include a relevant discount rate and present value of flow of resource rent over all future catches.

Sustainable seafood is a movement that has gained momentum as more people become aware about overfishing and environmentally destructive fishing methods. Sustainable seafood is seafood from either fished or farmed sources that can maintain or increase production in the future without jeopardizing the ecosystems from which it was acquired. In general, slow-growing fish that reproduce late in life, such as orange roughy, are vulnerable to overfishing. Seafood species that grow quickly and breed young, such as anchovies and sardines, are much more resistant to overfishing. Several organizations, including the Marine Stewardship Council (MSC), and Friend of the Sea, certify seafood fisheries as sustainable. The MSC has developed an environmental standard for sustainable and well-managed fisheries. Environmentally responsible fisheries management and practices are rewarded with the use of its blue product ecolabel. Consumers concerned about overfishing and its consequences are increasingly able to choose seafood products which have been independently assessed against the MSC's environmental standard and labelled. This enables consumers to play a part in reversing the decline of fish stocks.

10. Megacities

A megacity is usually defined as a metropolitan area with a total population in excess of 10 million people. Some definitions also set a minimum level for population density (at least 2000 persons/km^2). Megacities can be distinguished from global cities by their rapid growth, new forms of spatial density of population, formal and informal economics, as well as poverty, crime, and high levels of social fragmentation. A megacity can be a single metropolitan area or two or more metropolitan areas that converge upon one another. The terms conurbation, metropolis and metroplex are also applied to the latter. The terms megapolis and megalopolis are sometimes used synonymously with megacity.

In 1800, only 3% of the world's population lived in cities, a figure that has risen to 47% by the end of the twentieth century. In 1950, there were 83 cities with populations exceeding one million; by 2007, this had risen to 468 agglomerations of more than one million. If the trend continues, the world's urban population will double every 38 years. The UN forecasts that today's urban population of 3.2 billion will rise to nearly 5 billion by 2030, when three out of five people will live in cities.

The increase will be most dramatic in the least-urbanized continents, Asia and Africa. Surveys and projections indicate that all urban growth over the next 25 years will be in developing countries. One billion people, one-sixth of the world's population, now live in shanty towns, which are seen as "breeding grounds" for social problems such as crime, drug addiction, alcoholism, poverty and unemployment. In many poor countries overpopulated slums exhibit high rates of disease due to unsanitary conditions, malnutrition, and lack of basic health care. By 2030, over 2 billion people in the world will be living in slums. Already over 90% of the urban population of Ethiopia, Malawi and Uganda, three of the world's most rural countries, live in slums.

Global connectedness and local disconnectedness characterize megacities. The level of slums contrasts the global capital building capabilities. This can be viewed as one of the tensions brought about by the globalization of modern cities. In 2000, there were 18 megacities-conurbations such as Tokyo, New York City, Los Angeles, Mexico City, Buenos Aires, Mumbai (then Bombay), São Paulo, Karachi that have populations in excess of 10 million inhabitants. Greater Tokyo already has 35 million, which is greater than the entire population of Canada.

By 2025, according to the Far Eastern Economic Review, Asia alone will have at least 10 megacities, including Jakarta, Indonesia (24.9 million people), Dhaka, Bangladesh (26 million), Karachi, Pakistan (26.5 million), Shanghai (27 million) and Mumbai (33 million). Lagos, Nigeria has grown from 300000 in 1950 to an estimated 15 million today, and the Nigerian government estimates that the city will have expanded to 25 million residents by 2015.

The world's population of "slum" dwellers increases by 25 million every year. The majority of these numbers come from the fringes of urban margins, located in legal and illegal settlements with insufficient housing and sanitation. This has been caused by the massive migration, both internal and transnational, into cities. This has caused growth rates of urban populations and spatial concentrations not seen before in history. These issues raise problems in the political, social, and economic areas. The record-setting populations living in urban slums have little or no access to education, healthcare, or the urban economy.

Vocabulary and Phrase

acidic *adj.* [化] 酸性的
agent Orange 橙剂
agglomerations 群

alcoholism *n.* [医] 酒精中毒
algae *n.* [生] 藻类
aldrin *n.* [化] 艾氏剂

anthropogenic　adj. 有人引起的，人为的
arable land　n. [农] 耕地，农业用地
aquifer　n. [地] 含水层，蓄水层
basic　adj. [化] 碱性的
bioaccumulate　v. 生物累积
biomagnify　v. 生物放大
brominated flame-retardants　溴化阻燃剂
carbonic acid　碳酸
carcinogenic　adj. [医] 致癌的
cataract　n. [医] 白内障
chlordane dieldrin　n. [化] 氯丹狄氏剂
chlorofluorocarbon　n. [化] 氟氯化碳，俗名氟利昂，缩写 CFC
clearcutting　n. [农] 皆发，不分大小全部砍伐
conurbation
crop resistance　作物阻力
cyanobacteria　n. [生] 蓝藻
deforestation　n. [农] 开垦，滥伐
desertification　n. [地] 荒漠化，沙漠化，同 desertization
diabetes　n. [医] 糖尿病
dioxins　n. [化] 二恶英
distilled water　蒸馏水
discount rate　折现率
drug addiction　吸毒
ecolabel　生态标签，绿色标签
ecosystem services　生态系统服务功能
endocrine　n. [医] 内分泌
endrin　n. [化] 异狄氏剂
evaporation　n. 蒸发
evapotranspiration　n. [生] 蒸腾作用
flash flooding　山洪暴发
food chain　食物链
fossil fuel　化石燃料
gene　n. [生] 基因
genetic variations　基因变异
graze　v. 放牧
greenhouse effect　温室效应

greenhouse gases　温室气体
heat reflected　热反射
heptachlor hexachlorobenzene　[化] 七氯六氯苯
Holocene　n. [地] 全新世
Holocene mass extinction　全新世大灭绝
hydroelectric plant　水电站
immune systems　免疫系统
infrared rays　红外线
insulin　n. [医] 胰岛素
invasive grass
larvae　n. [生] 幼虫
megacity　特大城市
metropolitan area　大都市区
mimick　v. 模仿
mirex　n. [化] 灭蚁灵
monoculture　n. [农] 单作，单一种植
native　adj. 本地的，乡土的
nitric acid　硝酸
nitrogen oxides　二氧化氮
nonpoint source pollution　非点源污染
neurobehavioral disorders　神经紊乱
obesity　n. [医] 肥胖
organochlorine insecticides　有机氯杀虫剂
overgrazing　n. [农] 过度放牧
overpopulated　人口过剩
ozone　n. [化] 臭氧
particulate matter　颗粒物
persistent organic pollutants (POPs)　持久性有机污染物
pesticides　n. [农] 杀虫剂
pharmaceuticals　n. [医] 药品
photolytic　n. [化] 光解
photosynthesis　n. [生] 光合作用
phytoplankton　n. [生] 浮游植物
population density　人口密度
pollination　n. [生] 授粉
polychlorinated biphenyls　多氯联苯
polychlorinated dibenzofurans　[化] 多氯

二苯并呋喃
polychlorinated dibenzo-p-dioxins   多氯代二苯丙二恶英
polycyclic aromatic hydrocarbons   [化](PAHs) 多环芳烃
polyvinyl chloride   [化] 聚氯乙烯
productive capacity   生产力
reforestation   n. [农] 植树造林
reindeer   n. 驯鹿
sage   n. [生] 鼠尾草
seepage   n. [地] 渗漏，渗流
semi-arid   n. [地] 半干旱
serum   n. [生] 血清
sewage   n. 污水
shanty towns   棚户区
shrublands   n. [农] 灌木林地
shrimp   n. [生] 虾
slash and burn   刀耕火种
slum   贫民窟
solar radiation   太阳辐射
solvent   n. [化] 溶剂
Soviet Union   苏联
specie   n. [生] 物种
sub-humid   n. [地] 半湿润
sulfur hexafluoride   n. [化] 六氟化硫
sulfuric acid   硫酸
taxol   n. [生] 紫杉醇
tissue   n. [医] 组织
toxaphene   毒杀芬
tributyltin   三丁基锡(TBT)
tropical rainforests   热带雨林
urban margin   城市边缘
visual impediments   视觉障碍
water stress   水分胁迫
zooplankton   n. [生] 浮游动物

Notes

Species-area theory   在岛屿生态学理论中，物种数目随着岛屿面积的增大而增加的关系
background extinction rates   在没有人为干扰下的自然灭绝速率
World Business Council for Sustainable Development   世界可持续发展工商理事会

Question and Exercises

1. which problem are you most concerned about among the global environmental issues? Why?
2. What problems are there in your home town?
3. What do you think which reason is the main for environmental crisis?

Unit 2 Urban Environment

The density and population of today's cities necessitates the equitable distribution of resources that are needed for its various activities. It is therefore necessary to understand the effects of an urban area not only within its immediate boundaries, but also on the region and country it is positioned, due to the large amount of resources necessary to sustain it.

1. Introduction to Urban Environment

Take any of today's environmental problems faced by the inhabitants of Earth, and its causes and pressures can easily be traced back, directly or indirectly, to urban areas. The forces and processes that constitute "urban activity" have far-reaching and long-term effects not only on its immediate boundaries, but also on the entire region in which it is positioned. In a very broad sense, the urban environment consists of resources, human and other; processes, that convert these resources into various other useable products and services; and effects of these processes, which may be negative or positive.

The relationship among resources, processes and effects   Table 2-1

Resources	Processes	Effects	
		Negative Effects	Positive Effects
Human Resources	Manufacture	Air Pollution	Products
Sunlight	Transportation	Water Pollution	Value-addition
Land	Construction	Noise	Knowledgebase
Water	Migration	Waste	Education
Minerals	Population Growth	Garbage	Better services
Electricity	Community Services	Sewage	
Fuels	Education	Congestion	
Finance	Health	Overcrowding	
Intermediary products			
Recyclable materials			

With the inevitable danger of overlap and generalization, three dimensions in urban environments are identified natural, built and socio-economic environments. The natural environments include resources, processes and effects related to flora and fauna, human beings, minerals, water, land, air, etc. The built environments are resources, processes and effects related to buildings, housing, roads, railways, electricity, water supply, gas etc. The resources, processes and effects related to human activities, education, health, arts and culture, economic and business activities, heritage-urban lifestyles in general is considered as socio-economic environments (Table 2-1).

It is the intersection and overlay of these three dimensions that constitutes an "urban environment". Taking any one dimension at the exclusion of the other two poses the inevitable danger of missing the forest for the trees, the interdependency and interdisciplinarity of the three dimensions have to be fully understood in the development of coherent and sustainable policies and programmes for the urban environment. This is particularly true with the multiplicity of actors and activities-there has been a growing realization that state agencies and activities are, but one part of a spectrum of agencies and activities that are involved in the urban environment.

In order to develop an effective response to the range of challenges facing urban environments today, a framework has been developed to tackle these and related issues in the formulation of policy and practice of urban environmental management. The basic aim of this framework is three-fold: (1) to develop awareness and educate on issues related to urban environments; (2) to assist in policy and programme development; and (3) to facilitate monitoring and evaluation. The target audience of this framework is kept broad to increase its utility value-government agencies, planners and planning bodies, NGOs, donor agencies, community groups, academics etc.

2. Urban Environmental Indicators

The approach used for constructing urban environmental indicators is the Driving-Pressure-State-Impact-Response (DPSIR). This approach brings out the cause-effect relations of a given environment and the activities realised by individuals and the society on it. In order to represent the main conditions of the environment at urban level, Istat (2000) elaborates data collected by the mentioned survey and, on the basis of availability and comparability criteria, disseminates over 70 urban environmental indicators.

In the following for each analysed theme, some of the selected indicators are reported showing the DPSIR typology. We underline that the DPSIR scheme for urban environmental indicators has not yet defined at European level for all themes considered by Istat so the classification reported represents the first proposal.

As shown in Table 2-2 an indicator can represent more than one DPSIR typology, according to the context of analysis. For example, the density of waste containers can be a driving, a pressure and a response indicator. In fact, it represents a driving indicator as it can generate pressures (emissions in the atmosphere); it is also a pressure indicator as it exerts a pressure on the soil; moreover it represents a response indicator as it is a proxy for the existence of the service and for the accessibility of inhabitants to municipal waste collection system. Also the indicator related to green areas (density of public green areas) can be considered both as a state indicator and as a response indicator. In fact, the surface of green area of a municipality gives an overview of the state of green areas in that municipality; at the same time the green areas strongly depend on the activity of the local government in defining them, so representing a response indicator. It is clear that the theme we are analysing defines a different classification of the indicator: in the case of waste containers we consider land use, urban environment and air pollution.

Some selected indicators for themes and their DPSIR typology　　　　Table 2-2

Theme	Indicator	DPSIR typology
Population and territory	Number of inhabitants (per km² of municipal territory)	D
Air	Number of air fixed monitoring stations (stations per 100.000 inhabitants and per 100km² of municipal territory)	R
Air	Number of days overcoming of levels defined by legislation for sulphur dioxide, particles, nitrogen dioxide, carbon monoxide and ozone	S
Energy	Implementation of the Environment Energy Plan (Yes or, if No, the level of implementation)	R
Energy	Electricity (in KWh) and gas consumption (m³) per capita	D
Green Areas	Implementation of Green Urban Plan (Yes or, if No, the level of implementation)	R
Green Areas	Public green areas (m² per capita and m² per 100km² of municipal territory)	S/R
Noise	Number of fixed noise monitoring stations (stations per 100km² of municipal territory)	R
Noise	Implemented noise barriers (km of noise barriers per 100km² of municipal territory)	R
Noise	Number of requests of interventions for noise disturbance by typology of noise sources (industrial activities, traffic, recreation activities,...) (per 100.000 inhabitants)	S/R
Transport	Implementation of Urban Traffic Plan (Yes or, if No, the level of implementation)	R
Transport	Density of urban lines transport by typology (km lines per 100km² of municipal territory)	D
Transport	Availability and density of pedestrian areas (m² pedestrian areas per 100 inhabitants and per 100km² of municipal territory)	R
Transport	Density of restricted traffic zones (km² of restricted traffic zones per 100km² of municipal territory)	R
Transport	Number of paying car parks on roads (per 1.000 cars and per 100 inhabitants)	R
Transport	Park fleet (vehicles per 1.000 inhabitants and vehicles per km²)	D

Continued

Theme	Indicator	DPSIR typology
Waste	Collection of municipal waste (kg per capita)	P
	Separate collection of municipal waste by typology (paper, glass, plastic, aluminium, iron material, wood material, organic waste, etc) (kg per capita)	P
	Percentage of separate collection with reference to total municipal of municipal territory	R
	Density of waste containers (number per km^2)	D/P/R
Water	Water consumption per capita (m^3 per capita)	D
	Measures of water supply rationing (Yes/No)	R

Indicators selected by Istat (2000) cover the main environmental issues at cities level and represent the trends above all in respect to the pressures, the driving forces underlying these trends and the policy responses. Even though the selected indicators can not give a complete picture of all environmental issues involved, they are essential for identifying key sectoral trends and represent a first step in building up indicator sets able to measure progress towards environmental sustainability at local level.

3. Urban Environmental Problems

The cities are currently passing through the stage of risk transition in which modern stresses such as chemicals, heavy metals and noise combine with the traditional ones such as bacteria and disease vectors. A wide spectrum of environmental problems from unsatisfactory water supply and sanitation due to lack of basic infrastructure to release of man-made chemicals and hazardous wastes from the industrial and technological processes into water, air and soil.

Water Pollution

Water pollution in urban results from the uncontrolled release of domestic, agriculture and industrial effluents. There are three main sources of pollution: bacterial and organic liquids and solids from domestic sewage; toxic metals, organic loads, acids, and other less-toxic but still polluting substances from industrial discharges; and chemical pollution in the form of pesticide and fertiliser run-off from agricultural lands. All three cause the contamination of both surface and ground-water supplies and both directly and through food chain can cause serious damage to human health.

The industries creating environmental hazards are primarily the manufactures of chemicals (including pesticides), textiles, pharmaceuticals, cement, electrical and electronic equipment, glass and ceramics, pulp and paper board; leather tanning; food processing; electroplating; and petroleum refining. For all practical purposes, industries do not manage their waste water effluents through process controls, waste recycling, or end-of-pipe treatment. For example, the various chemical industries, tanneries, textile plants, steel re-rolling mills, and other operations discharge effluents containing hydrochloric acid and high levels of organic matter directly into streams and canals. Many industries discharge high levels of solids, heavy metals, aromatic dyes, inorganic salts, and organic materials directly into the municipal sewers without any pretreatment, polluting nearby lands. Contamination of shallow ground waters in urban areas from industrial wastes has also been reported.

Air Pollution

The combined emissions of air pollutants from industry, power generation, transportation, domestic activities (particularly energy use), agriculture, and commercial institutions are growing rapidly and air pollution has been estimated to cost more than billion dollars. No doubt the classic source of air pollution is the factory including brick kiln smoke stacks. Such stationary, point-source emissions are highly visible and represent a significant threat to those living nearby. By volume, however, they represent less of a threat to the overall health than do mobile sources such as automobile and other vehicles. During the last few years, traffic in urban areas particularly in the metropolitan cities has increased tremendously.

The truly dangerous pollutants to human health—those that can cause bronchial irritation, hasten asthma attacks, and irritate the eyes—arise primarily from automobiles. Motor vehicle emissions alone account for approximately 90 percent of the total annual emissions of hydrocarbons, aldehydes, and carbon monoxide, and for smaller but still the largest proportion of the emissions of sulphur dioxide and nitrogen oxides. As such, air pollution along busy roads and narrow streets of the main cities should have an order of magnitude much greater than would be predicted from the number of vehicles on the road.

Land Pollution

The main cause of land pollution is industrial effluents (such as those of tanneries) and solid waste from both municipal and industrial sources. In cities, waste disposal typically accounts for 20~50 percent of municipal expenditures. Solid domestic waste is typically dumped into low-lying land without applying even the rudimentary modern sanitary landfill methods. The result is unsightly and unsanitary conditions at and around dump sites, waste of land that could be turned to more productive purposes, and the loss of potentially valuable recyclable materials.

Of considerable concern is the likelihood that quantities of toxic industrial wastes such as discharged batteries have been dumped in municipal disposal areas or are being dumped directly onto lands adjacent to factories with no record of their location, quantity, or toxic composition. There are thousands of small electroplating and other units which are dumping heavy metals on the land and into water without realizing the risks they are causing in the environment. Disposal of medical waste on land is another hazard. In Pakistan, around 250000 tones of hazardous medical waste is produced annually. The hazardous medical wastes can be categorized into infectious chemical and radio-active wastes. Around 10~15 percent of medical wastes is considered infectious, commonly associated to which are the two most common and dangerous diseases-Hepatitis B and Acquired Immune-Deficiency Syndrome (AIDS). Chemical and radio-active wastes also have high risk potential for causing cancer. Unfortunately, these wastes are also usually dumped, along with others, they may remain at these collection points, normally for a few days with great potential health hazards.

Urban Environmental Problems Viewed in Spatial Scale

Viewed at a spatial scale urban environmental problems vary in their impact and intensity. It shows the characteristic problems and the related infrastructure and services needed to address these on a spatial scale. The following points are notable: (1) health impacts are greater and more immediate at the household or community level and tend to diminish in intensity as the spatial scale increases; (2) equity issues arise in relation to (a) the provision of basic services at the household or community scale and (b) intertemporal externalities at the national scale, particularly the intergenerational impacts implicit in unsustainable resource use and release of pollutants; (3) levels of responsibility and decision-making correspond to the scale of impact, but existing jurisdictional arrangements often violate this principle.

4. Urban Environmental Management

In order to understand the context of Urban Environmental Management, the following urban dimensions are described in further detail.

Physical and Functional Dimensions in Urban Centers

A first factor that determines the urban context is the scale or size of the settlement. According to this factor, in Latin America there are: "small cities" that can have more than 50000 inhabitants, intermediate cities with around 500000 inhabitants, big cities with around 2 million inhabitants, and finally, the metropolis or metropolitan areas that can have several million of inhabitants. Generally, the administration of such metropolis covers several municipalities and jurisdictions.

A second characteristic of urban centers is their economic activity or function within its territory of influence, be it local, regional or national. In most cases, urban centers offer a great variety of activities and hold diverse functions that may determine the nature and magnitude of the environmental problems that they may face. For instance, in a city where industry is the main sector, the regulation of air quality and industrial buildings may be a local environmental priority. Likewise, in a city that provides plenty of services, areas such as mobility and transportation and their relationship to the quality of life of its citizens will be a high priority in the environmental agenda. Another example can be illustrated by a port city, whose influence can reach both regional and national levels. Such city will have to prioritize areas like coastal zone management, land use planning, water and air pollution and solid waster management. An adequate management of these environmental sectors may have positive economic impacts both at the local and regional levels.

Moreover, and according to their scale and function, urban centers have a close relationship with the ecosystems surrounding them as well as the overall territorial periphery. Such relationship is determined by the demand of natural resources as well as by the impacts from urban activities affecting them. Because many cities depend on the supply of water from river basins, both the quantity and quality of this resource is essential to sustain the urban ecosystem, which includes both the urban areas and its rural surroundings. Therefore, for cities it is imperative to protect the sources of water that serve them, in particular sources of water in their periphery by establishing effective pollution controls and economic incentives that foster environmentally sustainable activities.

Administrative and Financial Determinants for Urban Environmental Management

Urban centers have administrative structures and financial instruments that enable the development, operation and maintenance of basic local infrastructure (e. g. water and sanitation, solid waste management, transportation, housing, etc.) and the management of natural resources and the environment. In most countries in the region, municipal governments are responsible for different sectors linked to urban environmental management. Nonetheless, this situation may vary depending on the progress of the decentralization processes in each particular country. In this context, municipal responsibilities may include regional planning and development, land use planning and regulation, property registration, provision of basic services and natural resources conservation, among others.

As municipal and local governments continue strengthening their capacities to administer these tasks, urban centers will have a better performance when facing the challenges of rapid urban growth and hence, they will enable a better quality of life for their inhabitants. Moreover, the municipal administration is responsible to ensure the operation, growth and financial sustainability of public services through the adequate pricing and collection of fees, revenue generation, and access to finance both from national and international markets.

Social Dimension of Urban Growth and Environmental Management

Urban environmental management is closely linked to social unrest and poverty. It is estimated that almost 50% of urban citizens in Latin America and the Caribbean live in poverty. In the region, there is direct relationship between environmental degradation and the quality of life, in particular in areas where urban poverty resides. Generally, the poor are located in illegal neighborhoods or marginal settlements that lack an adequate coverage and quality of basic services. As a result, people living in these areas are more vulnerable to environmental and health risks and hazards; in some cities, these conditions are already critical.

As mentioned before, many of the settlements where the poor live are located in risky areas for urban growth including areas with instable soils, steep slopes and watersheds. Location in such areas increases the vulnerability to natural hazards and events such as slides, floods and earthquakes. In addition, location of settlements in these risky areas hinders the efficient progress of road infrastructure, transportation, water and sanitation systems, public lighting and other public services.

Urban Environmental Challenges

In Latin America rapid urbanization and population growth in the cities has increased significantly since the 80s. This phenomenon has come along an escalating demand of natural resources (e. g. water, land, energy, air) affecting negatively the environmental quality of urban centers and their surroundings. Increases in demand of resources by the transport, industry and construction sectors have also put pressure on the availability of natural resources and their carrying capacity. Migration and displacement trends have also accelerated the growth of settlements that do not have the basic infrastructure to adequately secure the quality of life and health conditions that are essential to human development.

As a result of the pressures on the environment aforementioned, urban centers constantly present high levels of air pollution and water contamination, slides, and yet, deforestation and loss of soil for agriculture in the peripheries of the city. Because urban and demographic growth is expected in the following decades, greater demands over natural resources will accompany this process. Therefore, it is important that the diverse set of actors engage in urban development will consider as priorities the challenges that are described below. These are challenges that are closely relate to the field of urban environmental management:

(1) Land-use and Planning

There is an increasing demand over land in most urban centers of Latin America due to expansion of urban infrastructure, population growth and the related demand of services and physical space that these two cause. The continuous growth of illegal settlements, which generally occurs in the edges and limits of the city, has impacted the relationship between the urban center and its periphery. Urban growth has contributed to deforestation and land degradation, in addition to the pressure over the land exerted by agriculture and animal farming. Also, urban expansion has altered the hydrological ecosystems while increasing the vulnerability and risks to natural disasters. As a result, the sum of these forces impacts the availability and sustainability of natural resources that serve not only the urban centers but also regional and national ecosystems.

(2) Water and Sanitation Accessibility and Coverage

In the area of water and sanitation in Latin America there have been remarkable achievements after decades of investments to secure access of urban communities to these services. In the region, institutional reforms have prompted decentralization and privatization processes including that of water and sanitation services. In some countries, the supply of drinking water covers most of the urban population although the deficit in the sanitation services remains high. This situation is worsened for those communities living in the periphery or illegal settlements of the city.

(3) Solid Waste Management

In Latin-American cities solid waste collection services not only have limited coverage, particular in low-income neighborhoods, but they also lack for waste disposal processes. Most of the cities have disregarded the impacts that industrial and toxic wastes represent to the citizens and the environment.

In smaller cities, deficiencies in solid waste management services are even more critical. Most of these cities do not have controlled and regulated land waste sites (landfills) and consequently, most of the waste is disposed in opened lands. The inadequate handling of solid waste has significant environmental impacts including: superficial water pollution, soil and aquifers contamination, air pollution (due to unregulated waste burning) and landscape deterioration. Additionally, deficiencies in solid waste management trigger respiratory and gastrointestinal diseases among the urban population.

(4) Disaster Risks

In Latin America, the riskiest areas in urban centers are generally marginal settlements where lower income groups live in constructions that are not properly designed to resist a natural hazard or disaster risk. Both social and economic forces are triggers of the development and expansion of these illegal neighborhoods that are vulnerable to flooding, slides and quakes. Additionally, the lack of land-use regulations and the weak institutional framework to execute and monitor compliance with norms intensify the vulnerability to disasters. Deforestation and river basin degradation in the peripheries of cities are other factors that increase risks to natural disasters. Therefore, it is important to highlight the economic, social and environmental impacts of disaster vulnerabilities and risks due to the potential loss of infrastructure, social capital and fiscal implications that recovery programs could have.

(5) Air Pollution

In Latin America, the main polluting source of air in cities is urban transport. During the last decades there has been a significant growth of both private and public vehicle fleets resulting in greater levels of air contaminants mainly CO, PST, PM10, HC, NO_2, SO_2 and Ozone. In many cities, the meteorological and geographical conditions of cities are determining factors in air pollution. Examples of this situation in the region are Mexico City, Santiago de Chile and Bogot, which are three of the cities with the highest levels of air pollution.

Air contaminants produced by automobiles cause respiratory and cardiovascular diseases that threaten the lives of the urban population, particularly that of children and elderly people. Today, in many capitals of the region, health problems caused by the high levels of concentration of air contaminants have been registered as public health matter that have broader social and economic implications.

Air pollution also derives from both private and public industry emissions that intensify the environmental health impacts to urban citizens. Examples of these heavy polluting industries include large-scale thermoelectric sites, metal manufacturing and refineries that produce contaminants such as SO_2, PST and CO_2. Emissions from micro-enterprises activities, and biomass and waste burning add to the problem because they are not regulated in most of the cities. Other air contaminants may come from street dust, particles coming from construction and road sites, and fuels used in homes.

(6) Transport Management

The accessibility to transport systems and the mobility of citizens in urban centers is closely related not only to economic productivity but also to urban quality of life. In Latin American cities, public transport is fundamental to their functioning; it offers a more egalitarian mobility alternative to those who do not own a vehicle. Another factor that highlights the relevance of transport management is the decline of air quality in many cities and its public health impacts. City governments should strive to offer transport systems that are economically efficient, financially feasible and environmentally favorable and consequently, they would improve the quality of life in cities.

(7) Institutional Development and Capacity

The development and strengthening of institutions managing and involved with the environment are cross-sectoral factors that need to be articulated with those mentioned above. In the region, despite the progress of decentralization, municipal governments are not completely autonomous and do not always work in close coordination with central governments to manage environmental challenges. In most cases, the municipal-central government relationship is affected by the rigidity of public sector institutions that maintain sectoral and hierarchical organizational structures. Many public administration processes and decisions are fragmented and oriented by a one-sided force: from central to local governments. As a consequence, local governments are unlikely to promote interagency coordination mechanisms and good communication practices. On the other hand, lack of effective coordination can spur jurisdictional conflicts that may intensify the loss and degradation of natural resources.

In addition to the vague definition of national and local competences for urban environmental management, most municipal governments face technical and financial limitations. Nevertheless, during the decentralization process that took place in the 80s and 90s, municipal governments were assigned with responsibilities for planning, financing and implementing environment and development programs and projects in urban centers. Efficient public participation and communication channels in decision-making processes are additional aspects that need continuous support, despite the increasing role that civic organizations are playing in public policy processes. In sum, institutions in charge of urban environmental management can continue promoting efforts to strengthen their capacities (e.g. legal, technical, administrative, financial and public participation) to address more effectively environmental challenges in cities.

Vocabulary and Phrase

Agenda　n. 议程
aldehydes　n. [化] 醛
aromatic dye　芳香燃料
asthma　n. [医] 哮喘
bronchial irritation　支气管发炎
cardiovascular　n. [医] 心血管疾病
cause-effect　因果关系
ceramic　n. 陶瓷，adj. 陶瓷的
cement　n. 水泥
criteria　n. 标准
decentralization processes　权力下放过程
demographic　adj. 人口的
disseminate　v. 散布，传播
domestic sewage　生活废水
egalitarian　n. 平均主义
end-of-pipe treatment　末端处理
escalate　v. 升级
flora　n. [生] 植物区系
fauna　n. [生] 动物区系
framework　n. 框架，大纲
gastrointestinal diseases　胃肠疾病
Hepatitis B　乙型肝炎
hydrochloric acid　盐酸
infectious　n. [医] 传染病，adj. 传染的
infrastructure　n. 基础设施，建筑物
inhabitant　n. 居民
intermediary products　中间产品
interdependency　n. 相互依存
interdisciplinarity　n. 交叉学科
jurisdiction　n. 辖区，行政区
leather tanning　皮革鞣制
manufacture　v. 制造，n. 制造品、企业、生产
meteorological　adj. [地] 气象的，气候的
migration　n. 移民
negative effects　负效应，与之相对的是
positive effects
noise barrier　隔音屏障
periphery　n. 周边，周围
pharmaceutical　n. [医] 制药，药方
proxy　n. 代理，代理人
pulp　n. 纸浆，v. 将……制成浆
petroleum refining　炼油
respiratory　n. [医] 呼吸道
sanitation　n. [医] 卫生，预防
sanitary landfill method　卫生填埋法
sectoral　adj. 行业的，部门的
social unrest and poverty　社会动乱和贫穷
steel re-rolling mill　轧钢厂
tannery　n. 制革
value-addition　增值
vulnerability　n. 脆弱性
waste container　垃圾容器

Notes

Taking any one dimension at the exclusion of the other two poses the inevitable danger of missing the forest for the trees, the interdependency and interdisciplinarity of the three dimensions have to be fully understood in the development of coherent and sustainable policies and programmes for the urban environment.

不考虑其他两个方面的影响而采取单一的措施，就会犯"只见树木，不见森林"的危险，只有多学科的研究才能充分认识城市环境发展的科学性和连贯性。

NGO (non-government organization)　非政府组织

Driving-Pressure-State-Impact-Response (DPSIR)　驱动力—压力—状态—影响—响应，是一个包括城市各个过程和影响因子的系统模型

Acquired Immune-Deficiency Syndrome (AIDS)　获得性免疫缺陷综合症，俗称艾滋病

Question and Exercises

1. Which is the best indicator for urban environment?
2. What do you think which controls are the main for environmental management?
3. If you are a mayor, please give a proposal about urban environmental management framework.

Unit 3 Housing and Living Environment

Urbanization has spread rapidly over the past century, creating major changes in several aspects of human life such as economics, education, housing and public health. In developing countries, a sizable proportion of this urbanization has happened in informal settlements, where low-income dwellers have, since the 1960s, secured shelter in violation of urban regulations and sometimes property rights, and where the quality of housing and services is often markedly below state-sanctioned standards. While the physical and living conditions in these neighborhoods were expected to improve gradually through the self-help efforts of their dwellers, there is widespread evidence that physical consolidation only occurred in particular contexts; in many cases, living conditions have actually deteriorated.

1. Housing Developments and Environment: Interactions

Large-scale housing developments are often associated with changes in land use, large volume of energy and resources consumption for housing construction, movement of populations to the completed projects, provision of infrastructure and services to sustain the populations, and timeless generation of waste and foul emissions from the residences. A wide range of environmental consequences result from these processes: urban design and restructuring of land uses, energy consumption and wastage, environmental pollution, environmental and life cycle impact of materials extraction, processing and their use, changes of interior environmental conditions, human comfort and productivity, to mention a few.

It is clear that housing developments have a major impact on critical social and environmental issues like population, resources, eco-system, social cohesion, public services and transport. The interactions are further complicated by the individual and collective roles of state planners, developers, design professionals, infrastructure providers, builders and consumers. According to statistical data released by the State Administration of Building Material Industry, the annual value-added construction in China is currently 550 billion RMB (US $ 67 billion), reflecting approximately 1.4 billion m^2 of housing under construction and 600 million m^2 completed each year. Given this large volume of construction and the expectation of improving living standard for the huge population of China, the potential environmental effects of post-reform housing boom cannot be ignored within and outside China, in particular, which can increase pressures on future energy consumption and greenhouse gas emission of the world. Currently China is the world's second largest energy consumer, claiming 8.4% of global energy consumption, and "with 11% of total carbon emission in the world, China is the second largest emitter of carbon dioxide after the United States". During the last decade, initiatives taken to modernize the industries and utilities have resulted in environmental improvements in these areas. However, the fastest growing residential construction sector has not kept pace with such improvements. For example, according to Glicksman et al., between 1997 and 1998 residential electricity consumption in China grew at 6% when industrial electricity consumption grew at only 0.1%. Furthermore, China's building-related energy and resources consumption are about three times higher comparing to the developed countries with similar climate conditions.

It is therefore imperative to rethink the existing housing development model in China, understand the interactions between post-reform housing development and environmental problems in Chinese cities, and establish guidelines for sustainable planning. There are various environmental challenges of post-reform housing boom by asking the following key questions:

(1) What is the current situation of China's residential sector, in terms of land-use, energy andresources consumption and related environmental externalities, like air pollution and waste generation?

(2) What are the key factors in the process of housing planning and design that play important roles in shaping these environmental challenges?

(3) How would China integrate its housing development and environment policies in shaping a more sustainable future?

2. Housing Quality and Ill Health

Given the environmental challenges associated with urbanization, and building on the assumption that healthier shelter and environmental infrastructure can improve health factors in low-income settlements, health researchers have turned their attention to densely populated cities in both the developed and developing world where they seek to understand and address the rising health challenges. A number of studies have assessed the direct relationship between ill health and inadequate urban housing quality, namely factors relating to infrastructure and services, housing conditions and overcrowding. Access to basic infrastructure and services includes items such as adequate water and sanitation, reliable electricity supply, and proper disposal of solid wastes. The lack of safe drinking water and the inadequate collection/disposal of solid wastes have contributed to public health outbreaks. Inadequate housing conditions, including the presence of humidity, pest infestation, the absence of amain source of heating and inadequate lighting, have been associated with negative health effects. Crowded, cramped conditions are also linked with acute respiratory infections, poor mental health among children, and household burns and accidents. Some of these factors, such as crowding, depend on social and cultural perceptions of the household, as well as the characteristics of residential units (i.e. individual units or a high-rise unit). Other factors, such as outdoor air pollutants, insecure tenure and poverty, have been linked with urban housing quality.

Housing Quality Indices

Housing quality was evaluated by asking the respondents whether or not they experienced problems with a series of items that ensure a minimum level of cleanliness, comfort and safety. This is a subjective measure at best, and it probably underestimates the problems a person's perception of difficulty or inadequacy may fade with time and habit, or with the appearance of temporary informal solutions in the form of the various self-help projects that frequently evolve in these communities. Nevertheless, these indicators measure the state of well being of the individuals, and provide insights into their lack of comfort and needs. The Hay el Sellom as a community suffers from problems with infrastructure and services, including a deficiency in the provision of municipal potable water, frequent failures of the electricity supply and inadequate sewage disposal methods. Ironically, since public drinking water is unavailable in the neighbourhood and households have to purchase water from private sellers, they all reported having access to adequate water services, and none of the households indicated that this was a problem in this community. Having "self-help" access to the service, in this case, allowed them some control over the quality of the water and hence better safety. However, in such a low-income community, having to buy water continuously is an added financial burden that could be avoided. Similarly, although most households (98%) complained of extended breaks in the electricity supply, only 31% seemed to have a problem with the quality of the power. Once again, this can be explained by the emergence of privately controlled generators that supplement the community with electricity for a fee. A large number of households did not report problems with electrical power since they resort to illegal hook-ups and/or private generators to access those services, again reducing health challenges. Therefore, although most households have access to electricity, it is at an additional cost to the family. This study found that the majority of households are connected to a sewage system despite the absence of official municipal provision of the service. It is well known that the sewage system was historically installed by dwellers as a self-help effort.

The majority of respondents in Hay el Sellom were dissatisfied with the sewage disposal service, but the problems with sewage disposal do not appear to be related to connection to a sewage network. Instead, water overflow from the floor drains seems to be the major problem. The capacity of the sewage system in Hay el Sellom may have been exceeded with the increased size of the neighbourhood, and sewage disposal problems will cause additional concerns in the coming years. Sewage disposal is probably the best indicator of the precarious conditions in this community, as it is the most difficult and probably the most costly problem to deal with individually without relying on state services. As for the problems with garbage disposal, they appeared to be more logistical in nature, dealing with the method and timing rather than the lack of service.

These problems with infrastructure and services are coupled with moderate to high crowding as well as poor housing conditions pertaining to ventilation, cockroach infestation, humidity, seepage and lighting. However, 89% of the households of Hay el Sellom complain of poor ventilation, 75% of cockroach infestation and approximately half have problems with humidity, seepage or lighting. This is especially striking in a nation that has traditionally, but even more so post war, shown a marked concern with projecting an image of beauty and serenity.

Crowding Conditions

These relationships did not show a positive association between household crowding and ill health. This unexpected finding may be due to the complex association between crowding and health which may be affected by the interplay of factors such as personal hygiene and access to healthcare services. Other explanations include the choice of outcome and crowding measures used. Ill health, measured in terms of mental disorders and infectious diseases such as diarrhoea and respiratory problems, has been positively associated with overcrowding in the literature.

The present study focused on "chronic illnesses"; a rather general outcome. Researchers have used various indices/cut-off points to measure crowding. The ratio of dwellers to the number of rooms in the household was generally adopted. Although widely used, the interpretation of this index has varied in different contexts. For instance, in a US study, "overcrowding" referred to more than one person per room, while "crowding" referred to two or more persons per room in Ethiopia. A crowding index of three categories was used in a study in Lebanon, where one person per room implied "undercrowded" households, two to three persons per room implied "crowded" households, and four or more persons per room implied "overcrowded" households.

Infrastructure and Services

No significant association was found between the infrastructure and services index and the presence of illness in the households. This could be explained by how the study investigated perceptions of the quality of services, especially water and sewerage, rather than physical indicators. It could, however, also be because the main health hazard generally identified in the literature is drinking water and, in this community, lack of an adequate source of drinking water is not perceived as a problem since water is systematically purchased from private sellers.

Housing Conditions

The poor housing conditions were strongly associated with the presence of chronic illness among individuals. The results are consistent with work that has demonstrated associations between ill health and precarious poor housing conditions in urban areas, mainly those relating to poor ventilation, cockroach infestation, humid conditions, seepage in walls and ceilings, lack of adequate natural light, cracks in walls, and water overflow from floor drains.

Conditions in Hay el Sellom are similar to those found in other impoverished countries in developing areas, and represent a different set of concerns than those dealt with in more developed countries. Communities like Hay el Sellom are still struggling to obtain, often at additional cost to their financial, psychological and physical health. The government needs to exhibit an active interest in the community and promote social dialogue to find adequate neighbourhood and housing solutions that could alleviate the negative health impact. These should primarily include a number of physical upgrading interventions in the neighbourhood that would improve the level of its urban services, such as the quality of water, the sewage system, solid waste collection, access to electricity, and reduction of ground-level pollution that emerges from heavy traffic in narrow and congested streets as well as small-scale industries scattered around the area. As a first step, housing solutions at household level can also be encouraged to achieve a healthier built environment; these include aerating the houses by opening windows at night when the surrounding levels of ambient air pollution are lowest, promoting hygiene and frequent and proper garbage disposal, and keeping foods in closed containers to avoid cockroach infestation.

3. The Effect of Overcrowded Housing on Children's Performance at School

Children from poor families do not do as well and leave school earlier than children from rich families. These are well-known facts that no longer need to be validated. The interpretation of these facts, however, is still the subject of great controversy. Consequently, public policies that could help reduce inequalities in educational portunities remain poorly defined.

One basic issue is whether increasing financial aid to the poorest families represents a good means for improving their children's performance at school. A number of studies argue that parental income, as such, does not have any impact on children's performance at school. According to these studies, the link between poverty and academic failure is not one of cause and effect. They stress that increasing financial aid to poor families would have no effect on the inequalities between children from rich and poor families. Another important issue concerns the impact of targeted aid, aimed to directly improve the living conditions of poor children. Even if financially assisting the parents of the poorest families would not have any effect on their children's schooling, aid aimed at pacifically improving children's access to medical care or quality of housing could have a very important and positive effect on children's development and performances at school.

The Effects of Overcrowded Housing

The sociological and social psychological literature has long been interested in the problems caused by overcrowded housing. Empirically, the degree of overcrowding is measured by the number of persons per room. Theoretically, the problems caused by lack of living space are conceptualized as the consequences (a) of an excess of interactions, stimulations and demands from the people living in the immediate area, and (b) of a lack of intimacy and the possibility of being alone. People who live in overcrowded housing suffer from not being able to control outside demands. It is impossible for them to have the necessary minimum amount of quiet time they need for their personal development. One of the most convincing sociological studies on this subject is perhaps that of Gove et al. (1979). Using American data, the authors establish the existence of a very clear correlation between the number of persons per room and individuals' mental and physical health.

Medical literature has also shown great interest in the health of people living in overcrowded conditions, i. e. , in houses and/or apartments that are too small for their families. It has been well established that individuals living or having lived in such conditions are sick more often than others, particularly due to respiratory insufficiency and pulmonary problems. In general, people who grow up in overcrowded housing die at a younger age than others, most notably of cancer. The medical literature gives many reasons for these health problems and their persistence. Living in an overcrowded space is a source of stress and favors illnesses linked to anxiety. The members of a family living in a crowded space also transmit their infections to one another more easily, weakening their immune systems. Living in an overcrowded space puts people at greater risk to problems linked to poor ventilation and hygiene conditions, such as poisoning caused by the smoking of one or more family members.

With overcrowded housing, occupants' health at greater risk and their capacity for intellectual concentration being decreased, it is clear that a lack of space is a potentially unfavorable factor for children's success at school. To our knowledge, however, no study that analyzes the nature and intensity of the links between available living space and children's success at school exists in the economic literature. The work published in the sociological and medical literature corresponds essentially to the analysis of statistical correlations. Given that housing and health problems probably share common unmeasured determinants, these statistical correlations do not necessarily correspond to relations of cause and effect. The meaning of the results obtained from this literature is unclear.

Single Room and Diploma

The sample of this retrospective survey consists of about 1000 individuals, representative of the French male population, aged 20~40. The respondents describe their schooling career as well as their housing conditions during childhood. Specifically, they indicate, (1) whether they dropped out of school before earning a diploma, and (2) whether they had their own room at the age of 11. The advantage of this survey is that it gives more direct information on respondents' housing conditions during their childhood and makes it possible to identify the potential longterm effects on educational achievement. The disadvantage of this survey is that it is much smaller than the Labor Force Surveys and does not allow for as precise an identification of the structural parameters. When we restrict the analysis to the individuals who had at least one brother or sister, the sample only contains a little over 600 individuals.

Table 3-1 presents the distribution of the respondents from the 1997 retrospective survey according to family size, year of birth, father's occupation and housing conditions during childhood. The table also describes the variations in the probability of leaving school without a diploma according to the same criteria. These simple tabulations confirm that the probability of dropping out from school without a diploma is greater for older generations than for recent generations, for large families than for those with only one or two children, and finally, for blue-collar families than for white-collar families. The correlation is also very clear between the housing conditions during childhood and the probability of dropping out of school before earning a diploma. Close to 56% of the respondents said that they did not grow up having their own room. One-third of these individuals dropped out of school before earning a diploma, meaning a rate of academic failure twice that of other children.

The impact of overcrowding on the probability of dropping out of school without a diploma

Table 3-1

	No. of observations	Fraction without diploma	Net effects
Family size:			
Three or more children	359	32.6	0.45(0.19)
1 or 2	276	18.9	Ref
Date of birth:			
Born after 1964	357	23.5	0.76(0.19)
Born before 1964	258	31.8	Ref
Father's occupation:			
Manual worker	279	33.8	0.54(0.20)
Nonmanual worker	356	21.6	Ref
Overcrowding:			
Own room at 11	274	18.9	−0.58(0.20)
No own room at 11	341	33.4	Ref

Source: Survey on Educational and Occupational Career, 1997, INSEE.
Note: The third column provides the results of a probit regression where the dependent variable is "To have dropped from school without a diploma" and where the independent variables are an intercept, a dummy for "own room at age 11", a dummy for "Three of more children", a dummy for "Father manual worker" and date of birth.

A multivariate regression confirms that individuals who have their own room during childhood had ceteris paribus a much smaller probability than the others of dropping out of school before earning a diploma, even after controlling for the father's occupation and the number of siblings (Table 3-1, third column). Generally speaking, these supplementary investigations tend to confirm the diagnosis obtained using the Labor Force Surveys.

4. Indoor Environmental Quality

Many building products contain chemicals that evaporate or "off-gas" for several days or weeks after installation. If large quantities of these products are used inside a building, or products with particularly strong emissions are used, they pollute the indoor air. Other products readily trap dust and odors and release them over time. Building materials can also support growth of moulds and bacteria, particularly if they become damp, potentially causing allergic reactions, respiratory problems and persistent odours-symptoms of "sick-building syndrome".

In recent years, several law suits with large damage awards have been won by building occupants suffering from health problems linked to chemicals off-gassed from building materials, setting legal precedents across North America. This has prompted many insurance companies to examine their policies and their clients' design and construction methods. Following a rigorous selection procedure for construction materials, aimed at minimizing occupant chemical exposure, is an effective way to reduce health risks-and exposure to liability by building developers, designers, contractors and operators. Considerations for Indoor Environmental Quality include:

Thermal Comfort

Employee health and productivity are greatly influenced by the quality of the indoor environment. Poor air quality and lighting levels, off-gassing of chemicals from building materials, and the growth of moulds and bacteria can adversely affect building occupants. Sustainable design supports the well-being of building occupants by reducing indoor air pollution through the selection of materials with low off-gassing potential and ventilation strategies, providing access to daylight and views, and controlling lighting, humidity, and temperature levels for optimum comfort.

Indoor Air Quality

Employee health and productivity are greatly influenced by the quality of the indoor environment. Poor air quality and lighting levels, off-gassing of chemicals from building materials, and the growth of moulds and bacteria can adversely affect building occupants. Sustainable design supports the well-being of building occupants by reducing indoor air pollution through the selection of materials with low off-gassing potential and ventilation strategies, providing access to daylight and views, and controlling lighting, humidity, and temperature levels for optimum comfort.

Daylighting

The intention is to provide a connection between indoor spaces and the outdoors for the building occupants through the introduction of daylight and views into the regularly occupied areas of the building.

Vocabulary and Phrase

allergic   *n*. [医] 过敏
chronic   *n*. [医] 慢性病
cockroach   *n*. [生] 蟑螂
diagnosis   *n*. [医] 诊断
diarrhoea   *n*. [医] 腹泻
ceiling   *n*. 天花板，顶棚
disposal of solid wastes   固废处理
healthcare services   医疗服务
healthier shelter   健康住房
hygiene   *n*. [医] 卫生
immune systems   免疫系统
ironically   *adv*. 具有讽刺意味的
mental health   心理健康
off-gas   烟气
paribus   *n*. 不变
psychological   *adj*. [医] 心理的
retrospective   *adj*. 怀旧的，*n*. 追溯
secured shelter   住房担保
sick-building syndrome   病态建筑综合症
social cohesion   社会凝聚力
state-sanctioned standards   国家标准
symptoms   *n*. [医] 症状
urbanization   *n*. 城市化
ventilation   *n*. 通风
violation   *n*. 违反

Notes

State Administration of Building Material Industry   国家建材局

Question and Exercises

1. How do you analyze your living and housing environment?
2. Which is the most that you concerned in living and housing environment?

Unit 4 Urban Sustainable Development

1. Looking Outside the City for Sustainability

   Running parallel with academic and political concern about the continuing globalisation of economic trading systems is an increased awareness of the potential of urban resource demands and waste streams to exert truly global impacts: the "ecological footprint" of cities has rapidly extended itself. One outcome of the stretching tentacles of urban influence is that formerly close ties between cities and their immediate hinterlands have been severed, and a city can potentially over-ride its damaging influence on the local environment by importing resources and exporting wastes further afield. Attempts to improve the local environment without considering the external impacts of urban behaviour, including global issues such as ozone layer depletion and global warming, are not sufficient to address the true imperatives of sustainable development. The sustainable urban development has to be seen as an integral ingredient of a broader goal: achieving global sustainable development, with its wide-ranging agenda of environmental stewardship, inter-generational equity, social justice and geographical equity. The sustainable city, therefore, needs to be seen in its global context, involving a thorough examination of the external impacts that cities generate. Charting where urban impacts are felt can work as a potential tool in devising systems that ensure that the polluter pays appropriately for the full environmental impact of damaging behaviour. For instance, cities could impose directly-linked reparations between polluters and polluted regions. In this paper, four alternative perceptions of the nature of urban environmental problems and the differing policy emphases that have arisen from them. In addressing different urban development models are focused, it is helpful to recall the general debates on sustainable development that contrast "deep green" ecocentric stances, which are antipathetic to major economic expansion, and more "light green", anthropocentric stances, with various options in between. Light green approaches typically see greater possibilities for balancing environmental and economic development imperatives, including the substitution of natural resources with those created by human technologies; wealth creation is seen as essential to making policies politically acceptable and, in principle, to enabling wealth to be spread to currently impoverished groups. Extremes can vary between banning personal motor cars in cities (deep green) and developing energy-efficient or zero-emission motor vehicles (light green). The urban development models outlined here vary from the deep green, self-reliant city, the political-technical fix of redesigning the city, and light green attempts to rely on reforming market mechanisms to bring about changes in support of sustainable development.

   Whilst the primary focus of attention here is on the inequitable impacts that urban consumption habits exercise on outlying areas, the processes of creating intra-urban inequities are clearly enmeshed with external inequities, often mediated through the same management and planning systems and inequitable trading mechanisms. External impacts are the least talked about or understood aspects of the sustainable urban development debate at the moment, and yet arguably the very ability to use urban (and other political administrative) boundaries to avoid accepting responsibility for external impacts helps fuel our current patterns of non-sustainable behaviour, as we transfer the costs of our consumption preferences to other people, other species, and other areas. We need to reform not just the city, but the way in which the city interacts with the rest of the global economy and environment.

2. Four Approaches to Sustainable Urban Development

Whilst the dominant approach to sustainable development is still to use the phrase as a legitimising mantra for maintaining "business as usual" or minimal change in urban development policies, it is also possible to identify four sets of more radical ideas for sustainable urban development. These perspectives reflect different sets of values and judgments about both environmental and urban development: although they frequently indicate similar policies, most notably in advocating measures to limit the damage done to the environment by cars, in respect to land use their approaches can differ markedly. In each of the models, the boundaries of the city-region are potentially drawn differently.

Self-reliant cities: intensive internalisation of economic and environmental activities, circular metabolism, bioregionalism and urban autarky. The self-reliant city model has steadily grown more popular with environmental activists since the early 1980s. Indeed, as Roseland highlights, the overall model outlined here subsumes a wide variety of alternative approaches. The selfreliant city model used here heavily emphasizes sorting out a city's problems from within, in particular by building local economies which are more self-reliant, meeting local needs through local businesses and cooperatives, and so on. This economic self-reliance in turn requires greater use of local environmental resources, and attention to minimising and redirecting waste flows so that they can be absorbed either productively or with minimum ecosystem disturbance. A bioregional emphasis is an important ingredient in most self-reliant city analyses: although defining a bioregion is invariably problematic, it is usually upheld as a natural unit for addressing environmental concerns. Typically, a bioregion is a river basin, a valley, or some similarly distinctive ecosystem that provides natural boundaries for political and administrative units.

In the new bioregional politics, ecological integrity requires a radical shift away from development in which humans dominate and control nature in favour of ways of working in harmony with it. The new mission for settlements is integration with nature at the bioregional level, rather than the eradication of considerable parts of natural ecosystems. Linked to this, the self-reliant city movement has a considerable political agenda: to change moral values away from anthropocentric views of nature and towards more decentralised and cooperative forms of human endeavour. From this perspective, strategies for sustainable cities are part of a far more radical project than merely achieving a weakly defined view of sustainable development. In land use terms, the overall policy is one of settlement decentralization (more smaller towns, fewer large cities), combined with greater compaction and diversity (more houses, neighbourhood-scale employers, shops, and so on) than in the typical low-density US suburb. In addition, in the bioregionalist's eco-city, nature would be restored to the urban parts of bioregions, with more open spaces, roof gardens, and so on, not least as part of a strategy to raise inhabitants' spiritual awareness of their links with nature.

In this vision of the self-reliant city, it is through an intensive internalisation of local economies and resource usage systems that the problems of uneven external exchange are addressed. In essence, the solution to non-sustainable patterns of external dependence is to reduce, rather than reshape, its levels. In effect, the decentralised bioregionalist system would fulfill its global duties by organising itself internally along more sustainable and ethical lines.

Girardet's (1992, 1993) models of urban linear and circular metabolism illustrate some of the resource management implications of the self-reliance approach. In the linear metabolism model, urban development is fuelled by inputs that are sought from huge hinterlands, whilst wastes are discarded with little thought for the consequences. Resource inputs are, to a large degree, divorced from concern over outputs: for instance, trees are felled without concern for replanting and fossil fuels burnt with little regard to the consequences for global warming. It is in this sense that urban metabolism is characterised as linear: what goes into the system is not linked to what goes out. This encourages resource profligacy and a lack of attention to minimising both resource use and waste streams through recycling, reuse, and so on. The alternative to urban linear metabolism is a self-replenishing, more self-reliant system of circular metabolism, where long-term ecological viability requires that cities "learn from nature's own circular metabolism where all wastes end up sustaining and renewing life". Through circular metabolism, the inputs and outputs of the city are connected, as waste products, for instance, are recycled rather than exported.

Most bioregionalist analyses fail to address the problematic question of what degree of self reliance is desirable and what a bioregionalist-organised economy and society should be prepared to compromise given acute geographical resource imbalances (for example, in mineral oil, clay, fisheries and forests). By contrast, Galtung (1986) acknowledges that some inter-regional trading will be necessary and indeed desirable in raising some regions to the life quality standards of others, and proposes a way of doing this in an minimally disruptive fashion. He suggests that trading, be conducted within nested hierarchies of exchange, where each area should aim towards selfreliance and in trading should also seek to privilege neighbouring regions or areas (especially Third World countries) with trade on equal and fair terms, which would help to bring their living standards closer to those of the most prosperous areas.

Although a minority of deep green bioregionalist commentators overemphasizes improving spirituality through communing better with nature, for the most part the bioregionalist emphasis on self-reliance brings a welcome element of radical critique to the principles of organising urban living. Urban self-reliance is for the most part "autarky within limits" rather than total self-enclosure. Certainly, urban self-reliance makes sense-alternatively, for cities to cut themselves off economically and environmentally, from the rest of the world would fundamentally damage much of the basis of modern urban life, not least the exchange of ideas and culture and the spread of information. But even moving closer towards greater urban self-reliance requires such a massive change in political will, and in community engagement with the processes of change, that it remains difficult to see major advances in the immediate future.

Redesigning cities and their regions: planning for compact, energy efficient city regions.

Interest has burgeoned among planners, architects, and others in the possibilities of achieving massive energy savings through more compact city forms, with higher residential density and a reversion to greater mixed land uses. A key assumption is that such changes in the urban fabric would reduce the need to travel long distances while supporting an extensive and viable public transport system, encouraging people to travel less by private transport, and thus reducing energy consumption. Underpinning this concern is the belief that existing patterns of urban settlement are resource-profligate, using a set of environmentally inefficient technologies designed with the assumption that cheap energy, land, water and waste disposal would remain abundant. In effect, the "designers of machines, the designers of buildings, and the designers of cities could ignore the (environmental) efficiencies of these systems and their waste products". However, cities are not inherently profligate in resource usage, and redesigning them could play a central role in reducing resource consumption and waste streams. These reductions would improve the local environment and deter further detrimental impacts on external environmental systems.

Rather than seeking to subordinate nature to every want of human residents, the new designers of cities and technologies that support urban living argue that it is essential to work with nature, and to seek to alter the environmentally damaging ways in which people behave within the city. Although many policy approaches to redesigning cities have much in common with the self-reliance school, the "redesigning the city" approach is more anthropocentric and less nature centred. Rather than attempting to assimilate settlements with nature, this approach more frequently celebrates the "urbanness" of cities, raising residential densities and re-zoning areas for mixed uses in an attempt to break away from the perceived sterility of residential-only suburbs. Compact cities may have less nature in them, as more spare land is allocated to development, but possibly they will generate less negative external impact, not least reduced rural land take, whilst building their own cultural assets, particularly as the human urban fabric of buildings and parks. The other dividing aspect is that the "redesigning cities" approach recognizes the need for fundamental changes in political systems and in environmental ethics, but tends to see these as less transformative. Instead, there is a pragmatic concern with devising systems to alter human behaviour by changing, through a variety of incentives and regulatory controls, the options open to individuals and businesses.

Redesigning the city from within requires a broadly constituted approach to altering the urban environment, from improving building design (through better solar energy capture, improved insulation, use of less energy-intensive and recycled materials, etc) to aiming to create urban settlement forms that encourage a greater conservation of resources. In most redesigning approaches, changing regulatory regimes and standards play a critical role, although they are not necessarily divorced from market-based approaches such as charging residential developers full (rather than subsidised) infrastructure costs for water, electricity, roads and schools.

Rather than revisit all the arguments for and against compact city form, the message here is that "redesigning the city" debates have an internal focus that neglects detailed consideration of external impacts, which are treated as undifferentiated environmental problems. So whilst the general goals of reducing excessive urban resource consumption and minimising waste generation are addressed, much less attention is focused on where resources come from and where wastes go. This reflects not so much a lack of awareness of external impact, but an implicit acceptance that the impact is well-charted, allowing localist parameters to be chosen when examining urban environmental problems and policy solutions in detail. Provided that resource imports and waste exports are halted, this would not be a problem. But as altering city form is usually about minimizing resource use and pollution, rather than eliminating them, there remains a need to understand how negative external environmental, economic and social impacts are distributed.

Externally dependent cities: excessive externalisation of environmental costs, open systems, linear metabolism, and buying-in additional "carrying capacity".

Rather than directly re-regulating the urban environment, the more light green, market-centred approach to sustainable urban development emphasises the benefits of reforming market mechanisms to work more effectively towards environmental goals, in particular by addressing the issue of externalities. A central element of such approaches is that most cities have benefited greatly from externalising some of the environmental costs associated with growth and day-to-day maintenance. For instance, water imported from distant sources may be disrupting riparian ecosystems upstream of a city, whilst urban water pollution may have major effects on downstream river quality. These represent major urban externalities, the environmental costs of urban consumption that are not captured by market pricing mechanisms, because resources and waste streams are currently either not commodified or not properly valued. Since many of the social and environmental impacts of the human use of environmental assets do not get picked up by the market pricing mechanism, the central solution to reducing environmental impact derived from cities is to improve the market system, that is, to make the polluters pay for the full environmental costs of their actions.

The conventional economic approach to solving urban environmental problems by addressing externalities is well-illustrated in a World Bank policy paper, Urban Policy and Economic Development: an agenda for the 1990s. This includes a diagnosis of urban environmental problems and a policy approach centred on the need to resolve market inefficiencies. Whilst acknowledging the importance of issues such as population growth, it argues that many urban environmental problems can be traced to market failings; in particular, inadequate preventative action through economic policy and management measures such as (1) inappropriate economic policies (eg, underpricing of water and other services), leading to resource depletion and higher levels of pollution and (2) inadequate land use control or inappropriate land tenure systems that hinder effective land use or lead to overregulation of land markets.

This diagnosis of urban dysfunctionality, based around inadequate markets and inappropriate subsidies of certain resources, inevitably leads to suggestions for creating and improving markets (eg, land reform, including land registers), and creating incentives to alter behavior patterns. It is argued that through "pricing resources and services at cost, excessive resource use can be discouraged and costly investments postponed". Aspects of this analysis would attract agreement from many environmentalists; however, it lacks attention to the distributive impacts of the proposed market reforms. The document is not blind to social issues or the need for improved regulatory standards in environmental policing; indeed, both are mentioned. The problem lies more in placing market efficiency criteria over considerations of social equity, in regulating for an efficient market without due regard to those whose activities lie outside the formal market such as the workless and the homeless.

The distributional impacts of economic measures are frequently overlooked in neo-classical economics-based analyses of market-led policies. Many cost-benefit analyses, for instance, tend to underestimate the uneven social and geographical nature of many environmental influences—who benefits most, and where; who suffers most, and where. By not applying the means of problem resolution to those most environmentally disadvantaged, overall society and the economy may gain but those locally impacted upon might not. For example, a gasoline tax would not necessarily bring direct relief to those communities most affected by vehicle fumes and noise. However, market reforms are not necessarily geographically neutral in their design and implementation. The introduction of pollution trading permits in the US has been accompanied by the practice of "off-setting" in areas with inadequate air standards. Unless improvements are achieved in the overall ambient air quality, off-setting places local limits on factory expansions and openings, with factories able to trade polluting rights as reduced emissions in one factory allow increased emissions in another. Market reform is essential if sustainable development is to be achieved, but this reform must be geographically sensitised, as well as linked to strong social justice programmes and environmental standards setting to ensure that both local and global environmental carrying capacities are respected.

fair shares cities: balancing needs and rights equitably, with regulated flows of environmental value and compensatory systems.

This final version of the sustainable city draws on some of the most useful aspects of the previous models, incorporating them with an explicit concern for the debates over environmental and social equity.

Although the temptation may be strong to withdraw from all external trading relationships, as in the local economic self-reliance model, it is probably not the best way of helping every area improve its living standards. Some sharing of environmentally benign innovations and technologies is always desirable, whilst geographically uneven resource allocation and efficiency in resource nurturing (eg, crop growing) is likely to make some forms of trading also desirable, even to the most ardent adherent of self-reliance. What is needed is much more detailed consideration of the political, economic, social and environmental conditions under which resources are traded and waste streams sent out to other areas. Examining the environmental value of the resource and pollutant streams that enter and leave city systems is one of the most difficult, yet valuable, reforms required in contemporary regional resource management. It is possible to gain some insights into the complexities of this task by examining two pieces of work centred on the disaggregation of urban externalities. Ravetz (1994) and White and Whitney (1992) argue that non-sustainable urban development involves external exchanges, where cities appropriate the carrying capacity of external areas (in terms of both resource capture and natural assimilative properties in respect of waste streams) without adequate compensation. According to White and Whitney, who clearly attempt to insert a stronger environmental flavour to long-standing unequal global development debates, in many cases this unequal exchange causes hinterland areas to atrophy economically as well as environmentally. White and Whitney state that when the limits of a city's bioregional carrying capacity are reached, agreements could be reached between city and hinterland areas with surplus carrying capacity, provided that no environmental damage is done in the process. Effectively, a region with surplus carrying capacity can 'export' some of this to areas experiencing problems. In a formal relationship, the city would pay compensation costs to the area with surplus carrying capacity. If environmental damage took place, there would also need to be additional reparations. Under the White and Whitney proposals, compensation might vary from financial payments to more favourable terms of trade or relaxed rules on emigration to the richer areas.

The great advantages of this model are its attention to reforming both the terms of trading of environmental assets and the emphasis on assessing regional carrying capacity as the starting point for exchanges of both resources and pollutants. Its great disadvantage is the difficulty of actually implementing it in policy terms. The main problem is the number of flows involved which would need regulating and compensation mechanisms built into them, either separately or in aggregate. Not only would each resource flow and pollution stream need to be assessed, so would its individual components; eg, how do we differentiate between wood imported by different companies, from different countries, or between different types of wood, some harvested sustainably, some not. Even more difficulties arise with pollutant streams, given the complex reactions in air, water and land that result from elements emitted from different sources, and possibly different areas, but which combine toxically in one area. Conceptually, the idea is very attractive, but given the difficulties of gathering, analyzing and interpreting the large amounts of data required, for the immediate future it can function as nothing but a blunt tool.

This demonstrates the difficulties of imposing an approach for using market-type tools to address complex issues. Important though this policy direction is, it will need to be supplemented by other approaches for reducing and re-regulating the negative external impacts of urban activities. In reconsidering the earlier models of creating sustainable development, clearly one of the most powerful ways to realign trading terms between regions is to reform the market mechanism, incorporating the full environmental and social costs of resource capture and remedying damage done by waste streams. However, to achieve a 'fair shares' city, considerations of carrying capacity must be central, effectively creating a pre-condition for trade to take place. Even if we accept (as we must) that resource and pollution flows should not take place where they threaten to breach the local carrying capacity of any ecosystem, problems remain. As natural systems can cleanse wastes, provide sustainable yields of some resources, and withstand the withdrawal of some non-renewable products without major environmental damage, the question arises of who has the priority, for instance, to use the assimilative capacities of global commons, such as oceans or the atmosphere, or more localised capacities, such as rivers and aquifers? If we cannot properly attribute these rights, then attempts to regulate resource usage become exceptionally problematical.

Attempts to reduce the scale of environmental demands, which are central to both the self-reliance and the redesigning the city models, emerge as relatively unproblematic and as a powerful tool for behavioural change in the short-to medium-term, with their emphases on not simply redirecting resource demands but also reducing them. Reducing external impacts—through conservation measures, reuse, recycling, repair, and so on—needs to be made a priority consideration in resource management. Inter-regional exchanges of environmental value will still take place, but these should be fundamentally altered from those of the present: they should not damage the carrying capacity of external areas, they should be conducted on equal terms based on full costings, and they should meet the real needs of urban consumers, not the inefficiencies and profligacy of urban consumption habits. Furthermore, urban consumers have rights of access to external environmental capacities. Just because people are concentrated in one area for economic and social reasons is not a reason for denying them access to resources outside their own bioregion. This is too precious a view of the rights of people living in hinterland regions. Their rights are ones of engaging in exchange on an equal, fair and open basis, without damaging the integrity and sustainability of regional ecosystems, economies and societies. A system of rights of access to environmental assets and a broadly constituted notion of well-being would work far better than a simple system of property rights as the basis for a market-based system of reform.

3. Re-assessing Models of Sustainable Urban Development

It is possible to interpret these four models through a series of alternative analytical avenues. At one level, the different policy approaches represent competing models amongst academic disciplines and also those key decision-making organisations already responsible for resource management issues. Conservative UK central governments in the 1990s espoused a free market economic philosophy of externally dependent cities, which compete in an unfettered global marketplace. Parts of the Department of Trade and Industry, plus aspects of the work on inner city regeneration within the Department of Environment, have pursued a competitiveness agenda that prioritised success in deregulated markets over environmental regulation. Alternatively, the main UK environmental regulator for water, the Environment Agency, promotes a form of bioregionalism through its emphasis on catchment planning. Likewise local authority land use planning departments, many architects, and some officials in the Department of Environment strongly back 'redesigning the city' approaches to sustainable urban development, in particular high-density residential and mixed-use development. A 'Fair Shares' model would find sympathy with the Town and Country Planning Association and its work on the Manchester Sustainable City Region project.

A parallel view might be that the models relate to historical trajectories of urban development. Until recent years, free-market capitalism was fuelled by widespread environmental (and social) cost transference, as a means of underpinning business profitability. In which case, what we are currently witnessing may well be a fundamental ideological battle between those who advocate neo-liberal deregulatory trade reforms to bring about global competitiveness and others who argue for environmental re-regulation in the name of the ecological transformation of capitalism. The outcome of these epochal challenges may determine not just the future of capitalism, but also the functioning of future cities and indeed the very sustainability of the global environment.

Although one might see these models as representing some kinds of sequential trajectory of development approaches, I am increasingly drawn to the notion that aspects of each have considerable merits in their own right. Each model has its own value, not least suggested policy directions: the models lose value only when they inadverantly set in place professional or political blinkers that prevent consideration of a wider range of policy approaches. The Fair Shares Cities model, which I have tried to develop here in a way that integrates the better aspects of the other three models, allied with a greater concern for social justice and geographical equity concerns, provides one possible amalgam of approaches. But it is not the only one: the tair shares approach too is perhaps best seen as complementary to the others, not as their logical end-point.

Despite some experiments, it remains true that urban areas are desensitised to the wider impact of environmental damage caused by residents and businesses. There are various reasons for this: the level of information needed to raise awareness of the nature of transferred impacts is generally not available, whilst systems for either reducing transferred costs or for introducing compensation remain underdeveloped. There is also the problem of legal-jurisdictional boundaries and responsibilities in environmental management, where boundaries too often insulate politicians and other decision-makers from concern over the external impacts of their decision-making.

In consequence, sustainable urban development will require governance, market and regulatory changes not only for cities and nations, but also for environmental hinterlands. As these are in many instances global, it is global reforms in trade and in environmental standards that will force the shift towards sustainable urban development. In addition, for consumers, businesses, and politicians to make more environmentally sensitive decisions, all will need greater knowledge of the damaging impacts of their actions and a greater awareness of the alternatives available to them. The sustainable city's citizen will need to be better informed, embracing practical ethical considerations for everyday decisions in ways that are currently not the norm.

Vocabulary and Phrase

adherent n. 支持者，信徒
ambient adj. 周围的，环境的；n. 环境
anthropocentric stance 人类中心论观点
antipathetic adj. 厌恶的
ardent adj. 热心的
asset n. 财富，资产
assimilate v. 同化，吸收
atrophy n. 萎缩
autarky n. 自给自足
bioregion n. [地]生物地理单元
burgeon n. 萌芽，嫩芽；v. 萌发，发展
commodify v. 商品化
circular metabolism 循环代谢
detrimental impacts 不利影响
dysfunctionality n. 功能失调
ecological integrity 生态完整性

ecosystem disturbance 生态系统扰动
emigration n. 移民
eradication n. 根除，肃清
gasoline n. 汽油
globalization n. 全球化
hinterland n. 腹地
innovation n. 创新
mantra n. 咒语
market reform 市场改革
neo-classical adj. 新古典的，新经典的
nurture v. 培育
pragmatic n. 爱管闲事；adj. 务实的，实用的
profligacy n. 挥霍，浪费
priority n. 优先
radical critique 激进的批判

regulatory　　*n.* 监管，管理
riparian ecosystem　　河岸生态系统
sterility　　*n.* 条件不好，发育不良
stewardship　　*n.* 管理
undifferentiated　　*adj.* 分化的

underpin　　*v.* 支撑，支持
unfetter　　*v.* 束缚
urban fabric　　城市结构
vehicle fume　　汽车尾气

Notes

　　ecological footprint　生态足迹，也称"生态占用"，是在现有技术条件下，指定的人口单位内（一个人、一个城市、一个国家或全人类）需要多少具备生物生产力的土地（biological productive land）和水域，来生产所需资源和吸纳所衍生的废物。生态足迹通过测定现今人类为了维持自身生存而利用自然的量来评估人类对生态系统的影响。通过测量人类对自然生态服务的需求与自然所能提供的生态服务之间的差距，就可以知道人类对生态系统的利用状况，可以在地区、国家和全球尺度上比较人类对自然的消费量与自然资本的承载量。生态足迹的意义在于探讨人类持续依赖自然以及要怎么做才能保障地球的承受力，进而支持人类未来的生存。

　　ecocentric stance　生态中心论观点，这个观点的核心是人类应当把道德关怀的重点和伦理价值的范畴从生命的个体扩展到自然界的整个生态系统。

　　carrying capacity　承载能力，指一个区域或生态系统，或者某种资源所能支撑的人数。

　　environmental ethic　环境伦理，主要认为人类和生存环境系统之间的矛盾——环境污染、破坏和恶化等问题，是人类行为的结果，是一个社会问题。对于这个问题的最终解决，必须提到行为主体——人类环境伦理道德高度去认识和对待才有可能。

Question and Exercises

1. Which sustainable urban development model do you think is best in China?
2. What is the relationship between urban environment and development?

Further Reading (1)

Greenhouse Gases

　　Carbon dioxide is a greenhouse gas that has been increasing remarkably, mostly as a result of the combustion of fossil fuels such as coal, petroleum and natural gas. The atmospheric concentration of carbon dioxide, which was about 280 ppmv (ppmv is a unit for ratio of volume in parts per million) before the Industrial Revolution, has reached 358 ppmv (about 1.3 times) in 1994. If no other measures than those taken at present are introduced, the figure is expected to reach 500 ppmv in 2050 and 700 ppmv in 2100, and to continue to increase for many centuries to come (Figure 4-1).

　　The atmospheric concentration of carbon dioxide can be stabilized by controlling, on a global scale, artificial carbon dioxide emissions to a level sufficiently below the present level.

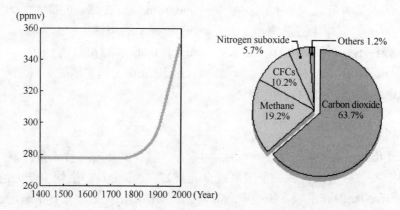

Figure 4-1 Atmospheric Concentration of Carbon Dioxide
and Contribution of Greenhouse Gases to Global Warming

Methane forms naturally in swamps, lakes and marches, and is also generated as a result of human activities such as livestock farming, paddy field cultivation and underground disposal of waste materials. The atmospheric concentration of methane increased from about 700 ppbv (ppbv is a unit for ratio of volume in parts per billion) before the Industrial Revolution to some 1720 ppbv (about 2.5 times) in 1994, as human activities have expanded. Although low in concentration, methane accounts for about 20% of contribution to global warming by greenhouse gases, in consideration of the 20-year total sum of its intensity of greenhouse effect (global warming index), based on the fact that its atmospheric residence time is 56 times longer than that of carbon dioxide.

CFCs are artificial gases that did not exist in nature before the Industrial Revolution. They have been in wide use as coolant gases in air conditioners and refrigerators, and as industrial cleansing agents. In recent years, CFCs have come to be known as an ozone layer destroyer, but they are also greenhouse gases. CFCs known as "alternative CFCs," which deplete the ozone layer less, or not at all, are, however, greenhouse gases. For example, the greenhouse effect intensity of an alternative CFC called HFC 134a is about 3400 times that of carbon dioxide in terms of 20-year total effect, based on its atmospheric residence time. For this reason, despite their small volume, such gases account for about 10% of contribution to global warming by all greenhouse gases.

Nitrogen suboxide is a byproduct of the combustion of organic substances and nitrogen fertilizers. The atmospheric concentration of nitrogen suboxide, about 275 ppbv before the Industrial Revolution, had increased to 312 ppbv by 1994 (about 1.1 times). The intensity of its greenhouse effect (global warming index), in terms of the 100-year total effect based on its atmospheric residence time, is 310 times that of carbon dioxide, but its contribution to global warming is only about 6% because of its low concentration.

Since CFCs are chemically stable, when released into the atmosphere they pass through the troposphere at about 20 km altitude without decomposing, reaching the stratosphere, where they are chemically decomposed by short-wavelength ultraviolet rays, releasing chlorine atoms. In a chain reaction, the chlorine atoms destroy the ozone layer in the stratosphere (Figure 4-2).

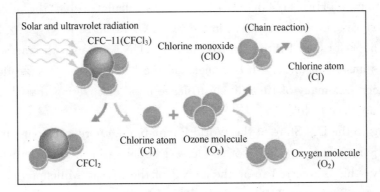

Figure 4-2　Mechanism of Ozone Layer Destruction by CFC in Stratosphere

The 1994 general report of the scientific, environmental impact and technological economic assessment panel of the UNEP predicted that the amount of chlorine and bromine would peak in 1994 in the troposphere, while in the stratosphere it would reach peak 3～5 years later and begin to decrease, provided that all the signatory countries observe the revised Montreal Protocol of 1992. From this, it is predicted that the global ozone decrease will continue during the remaining years of the 20th century, but that if other factors remain unchanged, the ozone layer will begin to be restored in the early 21st century and the Antarctic ozone hole will disappear around the year 2045.

Further Reading (2)

Urbanization

Urbanization (also spelled urbanisation) is the physical growth of urban areas from rural areas as a result of population immigration to an existing urban area. Effects include change in density and administration services. While the exact definition and population size of urbanized areas varies among different countries, urbanization is attributed to growth of cities. Urbanization is also defined by the United Nations as movement of people from rural to urban areas with population growth equating to urban migration.

1. Urbanization Movement

As more and more people leave villages and farms to live in cities, urban growth results. The rapid growth of cities like Chicago in the late 19th century and Shanghai a century later can be attributed largely to people from rural communities migrating there. This kind of growth is especially commonplace in developing countries.

The rapid urbanization of the world's population over the twentieth century is described in the 2005 Revision of the UN World Urbanization Prospects report. The global proportion of urban population rose dramatically from 13% (220 million) in 1900, to 29% (732 million) in 1950, to 49% (3.2 billion) in 2005. The same report projected that the figure is likely to rise to 60% (4.9 billion) by 2030. However, French economist Philippe Bocquier, writing in THE FUTURIST magazine, has calculated that "the proportion of the world population living in cities and towns in the year 2030 would be roughly 50%, substantially less than the 60% forecast by the United Nations (UN), because the messiness of rapid urbanization is unsustainable. Both Bocquier and the UN see more people flocking to cities, but Bocquier sees many of them likely to leave upon discovering that there's no work for them and no place to live."

According to the UN State of the World Population 2007 report, sometime in the middle of 2007, the majority of people worldwide will be living in towns or cities, for the first time in history; this is referred to as the arrival of the "Urban Millennium". In regard to future trends, it is estimated 93% of urban growth will occur in developing nations, with 80% of urban growth occurring in Asia and Africa.

Urbanization rates vary between countries. The United States and United Kingdom have a far higher urbanization level than China, India, Swaziland or Niger, but a far slower annual urbanization rate, since much less of the population is living in a rural area.

Urbanization in the United States never reached the Rocky Mountains in locations such as Jackson Hole, Wyoming; Telluride, Colorado; Taos, New Mexico; Douglas County, Colorado and Aspen, Colorado. The state of Vermont has also been affected, as has the coast of Florida, the Birmingham-Jefferson County, AL area, the Pacific Northwest and the barrier islands of North Carolina. In the United Kingdom, two major examples of new urbanization can be seen in Swindon, Wiltshire and Milton Keynes, Buckinghamshire.

2. Urbanization Causes

Urbanization occurs naturally from individual and corporate efforts to reduce time and expense in commuting and transportation while improving opportunities for jobs, education, housing, and transportation. Living in cities permits individuals and families to take advantage of the opportunities of proximity, diversity, and marketplace competition.

People move into cities to seek economic opportunities. In rural areas, often on small family farms, it is difficult to improve one's standard of living beyond basic sustenance. Farm living is dependent on unpredictable environmental conditions, and in times of drought, flood or pestilence, survival becomes extremely problematic.

Cities, in contrast, are known to be places where money, services and wealth are centralised. Cities are where fortunes are made and where social mobility is possible. Businesses, which generate jobs and capital, are usually located in urban areas. Whether the source is trade or tourism, it is also through the cities that foreign money flows into a country. It is easy to see why someone living on a farm might wish to take their chance moving to the city and trying to make enough money to send back home to their struggling family.

There are better basic services as well as other specialist services that aren't found in rural areas. There are more job opportunities and a greater variety of jobs. Health is another major factor. People, especially the elderly are often forced to move to cities where there are doctors and hospitals that can cater for their health needs. Other factors include a greater variety of entertainment (restaurants, movie theaters, theme parks, etc) and a better quality of education, namely universities. Due to their high populations, urban areas can also have much more diverse social communities allowing others to find people like them when they might not be able to in rural areas.

These conditions are heightened during times of change from a pre-industrial society to an industrial one. It is at this time that many new commercial enterprises are made possible, thus creating new jobs in cities. It is also a result of industrialisation that farms become more mechanised, putting many labourers out of work. This is currently occurring fastest in India.

3. Economic Effects

Over the last few years urbanization of rural areas has increased. As agriculture, more traditional local services, and small-scale industry give way to modern industry the urban and related commerce with the city drawing on the resources of an ever-widening area for its own sustenance and goods to be traded or processed into manufactures.

Research in urban ecology finds that larger cities provide more specialized goods and services to the local market and surrounding areas, function as a transportation and wholesale hub for smaller places, and accumulate more capital, financial service provision, and an educated labor force, as well as often concentrating administrative functions for the area in which they lie. This relation among places of different sizes is called the urban hierarchy.

As cities develop, effects can include a dramatic increase in costs, often pricing the local working class out of the market, including such functionaries as employees of the local municipalities. For example, Eric Hobsbawm's book The age of the revolution: 1789—1848 (published 1962 and 2005) chapter 11, stated "Urban development in our period (1789—1848) was a gigantic process of class segregation, which pushed the new labouring poor into great morasses of misery outside the centres of government and business and the newly specialised residential areas of the bourgeoisie. The almost universal European division into a 'good' west end and a 'poor' east end of large cities developed in this period. "This is likely due the prevailing south-west wind which carries coal smoke and other airborne pollutants downwind, making the western edges of towns preferable to the eastern ones.

Urbanization is often viewed as a negative trend, but in fact, it occurs naturally from individual and corporate efforts to reduce expense in commuting and transportation while improving opportunities for jobs, education, housing, and transportation. Living in cities permits individuals and families to take advantage of the opportunities of proximity, diversity, and marketplace competition.

4. Environmental Effects

The urban heat island has become a growing concern. Urban sprawl creates a number of negative environmental and public health outcomes. For more than 100 years, it has been known that two adjacent cities are generally warmer than the surrounding areas. This region of city warmth, known as an urban heat island, can influence the concentration of air pollution. The urban heat island is formed when industrial and urban areas are developed and heat becomes more abundant. In rural areas, a large part of the incoming solar energy is used to evaporate water from vegetation and soil. In cities, where less vegetation and exposed soil exists, the majority of the sun's energy is absorbed by urban structures and asphalt. Hence, during warm daylight hours, less evaporative cooling in cities allows surface temperatures to rise higher than in rural areas. Additional city heat is given off by vehicles and factories, as well as by industrial and domestic heating and cooling units. This effect causes the city to become 2~10℉ (1~6℃) warmer than surrounding landscapes. Impacts also include reducing soil moisture and intensification of carbon dioxide emissions.

5. Changing Form of Urbanization

Different forms of urbanization can be classified depending on the style of architecture and planning methods as well as historic growth of areas.

In cities of the developed world urbanization traditionally exhibited a concentration of human activities and settlements around the downtown area, the so-called in-migration. In-migration refers to migration from former colonies and similar places. The fact that many immigrants settle in impoverished city centres led to the notion of the "peripheralization of the core", which simply describes that people who used to be at the periphery of the former empires now live right in the centre.

Recent developments, such as inner-city redevelopment schemes, mean that new arrivals in cities no longer necessarily settle in the centre. In some developed regions, the reverse effect, originally called counter urbanisation has occurred, with cities losing population to rural areas, and is particularly common for richer families. This has been possible because of improved communications, and has been caused by factors such as the fear of crime and poor urban environments. Later termed "white flight", the effect is not restricted to cities with a high ethnic minority population.

When the residential area shifts outward, this is called suburbanization. A number of researchers and writers suggest that suburbanization has gone so far to form new points of concentration outside the downtown. This networked, poly-centric form of concentration is considered by some an emerging pattern of urbanization. It is called variously exurbia, edge city, network city, or postmodern city. Los Angeles is the best-known example of this type of urbanization.

Rural migrants are attracted by the possibilities that cities can offer, but often settle in shanty towns and experience extreme poverty. In the 1980s, this was attempted to be tackled with the urban bias theory which was promoted by Michael Lipton who wrote: "... the most important class conflict in the poor countries of the world today is not between labour and capital. Nor is it between foreign and national interests. It is between rural classes and urban classes. The rural sector contains most of the poverty and most of the low-cost sources of potential advance; but the urban sector contains most of the articulateness, organization and power. So the urban classes have been able to win most of the rounds of the struggle with the countryside...". Most of the urban poor in developing countries able to find work can spend their lives in insecure, poorly paid jobs. According to research by the Overseas Development Institute pro-poor urbanisation will require labour intensive growth, supported by labour protection, flexible land use regulation and investments in basic services.

6. Planning for Urbanization

Urbanization can be planned urbanization or organic. Planned urbanization, ie: new town or the garden city movement, is based on an advance plan, which can be prepared for military, aesthetic, economic or urban design reasons. Examples can be seen in many ancient cities; although with exploration came the collision of nations, which meant that many invaded cities took on the desired planned characteristics of their occupiers. Many ancient organic cities experienced redevelopment for military and economic purposes, new roads carved through the cities, and new parcels of land were cordoned off serving various planned purposes giving cities distinctive geometric designs. UN agencies prefer to see urban infrastructure installed before urbanization occurs. landscape planners are responsible for landscape infrastructure (public parks, sustainable urban drainage systems, greenways etc) which can be planned before urbanization takes place, or afterward to revitalized an area and create greater livability within a region.

7. New Urbanism

New Urbanism was a movement which started in the 1990s. New Urbanism believes in shifting design focus from the car-centric development of suburbia and the business park, to concentrated pedestrian and transit-centric, walk able, mixed-use communities. New Urbanism is an amalgamation of old-world design patterns, merged with present day demands. It is a backlash to the age of suburban sprawl, which splintered communities, and isolated people from each other, as well as had severe environmental impacts. Concepts for New Urbanism include people and destinations into dense, vibrant communities, and decreasing dependency on vehicular transportation as the primary mode of transit.

Part B Urban Water Environment

Unit 5 Water Use and Availability

1. Water Use and Availability

Everyone knows that water is essential for sustaining life. It also plays a central role in the growth and environmental health of cities and towns. People depend on water for more than just drinking, cooking, and personal hygiene. Vast quantities are often required for industrial and commercial uses. In some parts of the country, large quantities of water for irrigation are necessary to support agriculture. Water resources are also essential for power generation, recreation, fish and wildlife conservation, and navigation.

Water use refers to the withdrawal of water from its source, which may be a river, lake or well, and the transport of that water to a specific location. For example, water used for cooling purposes in a power plant may be diverted from a nearby river, passed through the power plant, and then discharged back into the river without significant loss in quantity. The water would have to be cooled down before discharge to prevent thermal pollution. Navigation and recreation are other examples of non-withdrawal use. However, it is necessary to make a distinction between water use and water consumption. Water that is used for drinking or combined with a product and is not directly available for use again is consumed water.

More than 100 million cubic meters (m^3) of water per day is withdrawn for public water supplies in the United States. More than 500 million m^3 is withdrawn each day for irrigation. Industrial use accounts for the largest share of water demand, almost 1 billion m^3 per day. Most of this is used as cooling water at electric power utilities. These approximate figures are presented to give an appreciation of the tremendous quantities of water needed.

Water is present in abundant quantities on and under Earth's surface, but only less than 1 percent of Earth's water is actually available for use in economically satisfying the needs mentioned. Most of Earth's water is salt water or is frozen in the polar ice caps.

Many freshwater lakes and rivers have been deteriorating in quality because of land development and pollution, limiting the availability of water for use, particularly for public water supplies. Even groundwater is affected by pollution in some areas, although much of it is just too deep to pump out of wells economically.

In 1998, a UN conference of 84 nations was held to discuss management of the world's limited supply of fresh water. It was estimated at the conference that about one quarter of the world's 5.9 billion people have no access to clean drinking water. The conference delegates agreed that water should be paid for as a commodity rather than considered an essential staple to be supplied virtually free of cost. Water shortages are so important that governments may need to rely on private funds for the large investments needed for water networks and treatment systems.

2. The Distribution of Water

In addition to the limited availability of usable water, another basic problem in managing water resources is that it is not evenly distributed geographically. In some regions there is ample precipitation, including rain, snow, hail, sleet and dew, and water is readily available for use. On the average, about one third of this precipitation becomes available in lakes and rivers, and some makes it's way into the groundwater. But where there is little precipitation, water is scarce. The fact that there is a close relationship between the amount of rain or snow and the amount of water available for use should be self-evident.

There are different annual precipitation amounts across the United States. Except for the extreme northwestern corner of the country, where the total annual amount of rainfall may exceed 2500mm (100in.), it can be seen that the eastern half of the country gets significantly more rainfall than the western half. In some areas of the Southwest, less than 100mm (4in.) of rain may fall in any one year. In the Northeast, an annual rainfall of about 1000mm (40in.) is moderate compared to the two previously mentioned extremes.

The amount of rainfall and the availability of water can vary considerably even within a relatively small area. California is an example of a state with a very uneven distribution of water. Although southern California is very dry, the growing population there generates a large demand for water. Most of the needed water must be transported to the south from the northern part of the state, where water is more readily available. A huge system of reservoirs, open channels, pumping stations, and tunnels is used to accomplish this transfer of water.

Part of the system, called the California Aqueduct, can convey about 2800m^3 (100000ft^3) of water per second. The aqueduct is an open channel, about 40m (130ft) wide at the surface and about 9m (30ft) deep. At one point in the system, the water is pumped up about 600m (2000ft) to get over a mountain, quite an engineering undertaking.

The uneven distribution of water from one graphic location to another is only part of the problem in hydrology and water resources management. The occurrence and availability of water also vary with time.

In any given location there may be occasional periods of little rainfall or drought, and severe water shortages may result as water in storage reservoirs is used up during these dry periods.

On the other hand, the same area may sometimes experience periods of above-average rainfall. Serious flooding problems may result, with accompanying loss of lives and property, as well as environmental pollution problems. In any given area, then, there can be too little water or too much water, depending on natural climatic conditions.

3. The Hydrologic Cycle

Water is in constant motion on, under, and above Earth's surface. Even in what appears to be a stagnant pond, the water is evaporating, changing into a vapor and moving into the atmosphere. Powered by energy from the sun and from gravity, there is a constant circulation of water and water vapor. This natural process is called the hydrologic cycle. It is illustrated in schematic form in Figure 5-1.

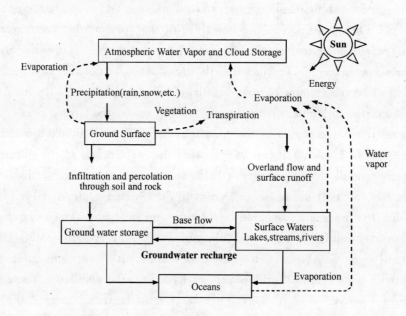

Figure 5-1   Schematic diagram of the natural hudrologic cycle. The constant circulation of water is powered by energy from the sun and by gravity.

Surprisingly, there was a time when people did not have an understanding of the cyclical motion of water through the environment and had misconceptions about the origin of water in streams or lakes. Even today, some people still have misconceptions, particularly with respect to groundwater.

Although the hydrologic cycle looks simple when sketched in schematic form as in Figure 5-1, there is more to it than initially meets the eye. The science of hydrology gets quite complicated, applying a good deal of statistics and higher mathematics. The basic objective is to measure and analyze the relationships controlling the form, quantity, and distribution of water.

When these relationships are understood, reliable predictions may be made concerning the occurrence of future floods or droughts. It is important that technicians involved in environmental control have all appreciation of the basic structure of the hydrologic cycle.

Precipitation begins when atmospheric moisture (water vapor) is cooled and condensed into water droplets. The precipitation can follow three different paths after it reaches the ground. First, some of it may be intercepted by vegetation or small surface depressions. In other words, it is temporarily stuck on the surfaces of leaves or grass or it is retained in puddles. Second, a portion of the water can infiltrate through the earth's surface and seep (or percolate, as it is called) downward into the ground. Third, a portion of the water can flow over the ground's surface. Measuring and predicting the relative amounts of water that follow each of these paths is of importance in hydrology.

Some of the intercepted water soon evaporates, and some of it is absorbed by the vegetation. A process called transpiration takes place as water is used by the vegetation and passes through the leaves of grass, plants, and trees, returning to the atmosphere as vapor.

The combined process of evaporation and transpiration is called evapotranspiration. Overall, more than half of the precipitation that reaches the ground is returned to the atmosphere by this process before reaching the oceans.

Overland flow and surface runoff occur when the rate of precipitation exceeds the combined rates of infiltration and evapotranspiration. Eventually, the overland flow finds its way into stream channels, rivers, and lakes, and finally the oceans. The ocean can be thought of as the final "sink" to which the water flows. As previously mentioned, about one third of the average annual rainfall in the United States becomes surface runoff in streams and rivers. This, of course, varies from region to region. In some areas of the Southwest, for example, there is no runoff for years at a time, since the rate of precipitation does not often exceed the rate of infiltration and evapotranspiration in that area.

The water that infiltrates the ground surface will percolate into saturated soil and porous rock layers, forming vast groundwater reservoirs. A "groundwater reservoir" should not be visualized as an underground lake-the water actually fills the tiny voids or spaces between the soil particles and fractures in the rock in what maybe called an aquifer. The groundwater may later seep out onto the ground surface in springs or into streams. (Groundwater flowing into streams is referred to as the base flow, which may be the sole source of streamflow during dry weather periods.) Eventually, the groundwater makes its way to the ocean, either directly or via surface streams. Evaporation from the ocean surface substantially replenishes the water vapor in the atmosphere, winds carry the moist air over land, and the hydrologic cycle continues.

4. Urban Hydrologic Cycle

This description of the hydrologic cycle is only a brief summary of a complex natural phenomenon. Some of the details of this natural cycle are discussed in the following sections of this chapter. But one water cycle should be mentioned here-the urban water cycle, illustrated in Figure 5-2.

In human communities there is a constant circulation of water. Water is withdrawn from its source in the natural hydrologic cycle-surface waters or groundwater - and is pumped through treatment and distribution systems. After use, the wastewater is collected in sewer systems, treated to reduce the effect of pollution, and finally disposed of back into surface water or groundwater. A most significant aspect of environmental technology is the maintenance of this urban water cycle while protecting public health and environmental quality. Much of this textbook focuses on this topic.

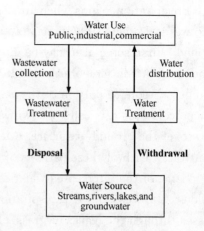

Figure 5-2 The urban hydrologic cycle

Vocabulary and Phrases

hygiene   n. 卫生［学］
irrigation   n. 灌溉
power generation   n. 电力生产，发电
conservation   n. (1)保存，保护；(2)守恒，不灭
navigation   n. (1)航运，航行，导航；(2)导航学，航海术，航空术
withdrawal   (1)收回，撤销；(2)提取，抽取；(3)放水，放油
thermal   adj. 热的，温(热)的
deteriorate   v. (1)变坏(质)，降低(品质)；(2)损坏(耗，伤)，消耗，磨损
staple   n. (1)主要产物，主要商品，主要成分；(2)U 形钉，骑马钉，环钩 adj. 主要的；常产的
ample   adj. 广大的，丰富的，充分的，有富裕的
precipitation   n. (1)降水；(2)沉淀，析出；(3)降水量
hail   n. 雹，冰雹   v. 下雹
sleet   n. 冻雨；雨夹雪
dew   n. 露
self-evident   adj. 自明的，不言而喻的
reservoir   n. (1)水库，蓄水池
channel   n. (1)河槽，河床，河道，航道，渠道；(2)通道，信道，频道
pump   n. 泵，抽水机，打气筒
     v. (用泵)抽水(油)
pumping station   (水)泵站
tunnel   n. 隧道，隧洞
aqueduct   n. 输水管，渡槽
convey   vt. (1)输(传，运)送，转运，递交；(2)传达，通知；(3)让予，转让
hydrology   n. 水文学
hydrologic cycle   水文循环，水循环
drought   n. 干旱
stagnant   adj. 停滞的
evaporate   v. 蒸发
vapor   n. 蒸气，蒸汽，汽，水蒸气，蒸发，汽化
schematic   adj. 图解的，示意的，概略的，纲要的，按照图式的
statistics   n. 统计，统计学
prediction   n. 预测，预报
appreciation   n. 估价，评价，鉴定，了解，判断，涨价，增值
condense   v. 冷凝，压缩，凝结，凝缩，缩合
droplet   n. 液滴
depression   n. 沉降，洼地，降低，减低，不景气

puddle   *n.* 稠黏土浆，胶泥，黏土涂料，夯实，水坑，冰上融水坑，搅炼，捣拌
infiltrate   *v.* 渗入，渗透
           n. 渗入水(液)，渗透水(液)
seep   *v.* 渗出，渗漏，渗滤
percolate   *v.* 渗滤，渗透，渗漏
           n. 滤出液，渗出液
transpiration   *n.* 蒸腾，散发
evaporation   *n.* 蒸发作用，蒸发
evapotranspiration   *n.* 蒸发蒸腾，蒸发，蒸散(作用)
overland flow   直接径流，坡面流
surface runoff   地面径流
saturated   *adj.* 饱和的
porous   *adj.* 多孔的，有空的，疏松的
aquifer   *n.* 蓄水层，含水层
spring   *n.* 泉，弹簧，春天，大潮，春潮，跳，跃，弹跳
replenish   *vt.* 补给，灌注
distribution system   配水系统，配电系统，分布式系统
sewer   *n.* 污水管，污水道
sewer system   排水工程，污(下)水道系统
base flow   基(本)流(量)

Notes

Base flow: Dry-weather flow in a stream, fed by groundwater seeping out of the ground and into the stream channel.

Runoff: Water from rain or snowmelt that flows overland to lakes, streams, and rivers.

Question and Exercises

1. List two uses of water other than for public supplies. Which use requires the greatest amount of water?
2. Briefly outline the basic features of the hydrologic cycle.
3. What is the origin of subsurface water (ground water)?
4. What is the problem of distribution of water resources in time and space?

Further Reading (1)

Aquifer Storage and Recovery: Need for Critical Analysis of the Technical, Economic, and Regulatory Issues

Competition for water will increase in the twenty-first century as we strive to meet the demands resulting from continued population and economic growth, and from efforts to protect and enhance aquatic ecosystems. Anticipated changes in the hydrologic cycle and hydrologic variability caused by global climate changes, the public's growing disaffection with dams, and the fact that virtually all surface waters are fully allocated, strongly suggests that ground water will be an increasingly important component of water supplies in the future. The fact that ground water accounts for two-thirds of the freshwater resources of the world, and that its long residence time serves to protect it from short-term contamination problems, should instill confidence that it will be available to meet future agricultural, domestic, and industrial needs. However, large-scale ground water development throughout the nation has resulted in many ill effects, including lowering of water tables, saltwater intrusion, subsidence, and lowered baseflow of streams. The burgeoning examples of local and regional ground water overdrafts, and point source and nonpoint source contamination from planned and inadvertent actions, call into question the sustainability of this resource.

In the face of the concern about the depletion of ground water reserves, thousands of aquifer recharge wells and aquifer storage and recovery (ASR) wells, and innumerable recharge basins, have been constructed to replenish water in aquifers. Such efforts are generally intended to prevent saltwater intrusion and land subsidence, and maintain baseflow in streams. The ASR wells are specifically intended to augment drinking water supplies. Most ASR wells being used today recharge drinking water. Changes in water quality during storage have proven to be minimal so that disinfection is the only treatment required on recovery. Thus, it appears that the use of ASR wells to inexpensively store drinking water underground in order to provide a secure community water supply for several months is an option that many water managers would probably support. The principal need with regard to the recharge of drinking water is to develop guidance for ASR legislation and regulations, possibly a model ASR code, so that issues and regulatory experiences in states with operating ASR systems are more readily available to those states that may wish to develop their own ASR regulatory framework.

Expanding ASR to store and recover treated surface water, untreated ground water or treated waste water would serve to conserve waters that would ultimately go unused. However, such ground water augmentation efforts are currently being resisted. Legislation authorizing the use of non-drinking water for ASR recharge has been blocked in several states at least in part due to concerns about aquifer contamination and human health. Current federal regulations requiring that recharge waters meet all primary drinking water standards at the wellhead prior to recharge also may make it prohibitively expensive to recharge anything but potable water. Acceptance of this potentially important source of drinking water requires answers to many technical, economic, and regulatory questions.

For instance:

What type and degree of treatment is necessary to ensure that no pathogens will survive in ground water? Will disinfection lead to the formation of carcinogenic compounds that will move to broader ground water areas?

What information is needed to ensure that the water being recharged is geochemically and microbiologically "compatible" with native ground water? Unanticipated reactions may lead to poor-quality water, biomass formation, pathogen growth, and well clogging.

What are the energy costs associated with ASR?

What monitoring will be required to ensure that unforeseen water-quality problems do not affect broader ground water resources?

How will communities be assured that the recharged water will not adversely affect other aquifers or surface water bodies?

What are the economic benefits that are to be expected from ASR? Maintaining high baseflow that supports ecosystem function, for instance, may be viewed as an important indirect outcome of ASR. Likewise, higher baseflow from ASR may increase dilution of surface-water contamination and bring into compliance stream reaches that may otherwise exceed Total Maximum Daily Load requirements.

What scale of ASR is technically and economically feasible?

What are the impacts of ASR on property values? On land use patterns?

What are the regulatory barriers to ASR?

What alternatives exist to augment drinking supplies? How do they compare with ASR?

The need exists for leadership by those in the water management community to answer these, and other, technical, economic, and regulatory questions and disseminate the information to the public, water managers and regulators. Federal laws and regulations must be reviewed to ensure that they reflect the state of knowledge about the risks and benefits of ASR.

The answer to such questions requires an analysis of the direct and indirect benefits and risks of ASR. The analysis would transcend traditional approaches by recognizing that water is to be valued not only for its extractive worth but also for its "in-situ" worth, a worth based on the essential role water plays in supporting life and ecosystem function. The analysis would take into consideration the role ASR plays in water management during times of floods and drought and as a strategic source of supply during regional or national emergencies.

The purpose of this evaluation is to provide policymakers, regulators, and water managers an objective summary of the technical, ecological, and economic factors that affect the efficacy of ASR and the options for its implementation. The initiative would use case studies to determine actual examples of benefits and risk.

Further Reading (2)

Rainfall

Water in streams, rivers, and lakes, as well as water in the ground, is the residue of precipitation. It is possible, and often necessary, to refer to records of rainfall in order to estimate the quantity of water that will be found on and under the ground. Other factors, such as topography and land use, play a role in the relationship between rainfall and water availability. These will be considered later. In this section, basic concepts related to rainfall intensity and volume are discussed.

1. Depth, Volume, and Intensity

The collection of rainfall data is the responsibility of the U. S. National Weather Service, a government agency that maintains rain-gage stations throughout the country. Rainfall amounts are expressed in terms of the depth of water accumulated in the rain gage during a storm. The units can be expressed in millimeters or inches. It is usually necessary to compute weighted averages of rainfall amounts over a region, using the data from several rain gages. The data may be weighted in proportion to the area covered by each gage.

Sometimes it is necessary to compute the total volume of water that falls on an area during a storm. The volume is computed by multiplying the land area by the rainfall depth, as follows:

$$\text{volume} = \text{depth} \times \text{area} \tag{5-1}$$

In SI metric units, the volume is usually expressed in terms of cubic meters, but rainfall depth is expressed in terms of millimeters. To keep the units consistent when applying Equation (5-1), area should be expressed in square meters and rainfall depth should be converted to meters. Relatively large areas that are expressed in units of hectares (ha) should first be converted to m^2 (1ha=10000m^2).

More important than total volume of rain is the rate at which the rain falls. This is called rainfall intensity. As discussed later, the rainfall intensity that occurs during a storm is of particular in the design of urban drainage facilities.

Rainfall intensity is expressed in terms of depth per unit time, as in. /h, mm/h. The National Weather Service gathers this kind of data using automatic rain gages that record of rainfall duration as well as depth; a continuous record of rainfall amount and intensity is plotted on a revolving drum. It is generally observed that short-duration storms have higher average rainfall intensities than longer duration storms. This will be of significance when we consider problems in stormwater control.

When using U. S. Customary units, cubic feet is a common unit for volume. But in hydrologic applications, large volumes of water are usually expressed in terms of acre-feet (ac-ft). This may seem to be a strange term at first, but, as illustrated in Figure 5-3, 1ac-ft can easily be visualized as the volume required to cover 1ac of land to a depth of 1ft. Since 1 ac is equivalent to 43560ft^2, 1ac-ft is equal to 43560ft^2 × 1ft, or 43560ft^3 (32590gal).

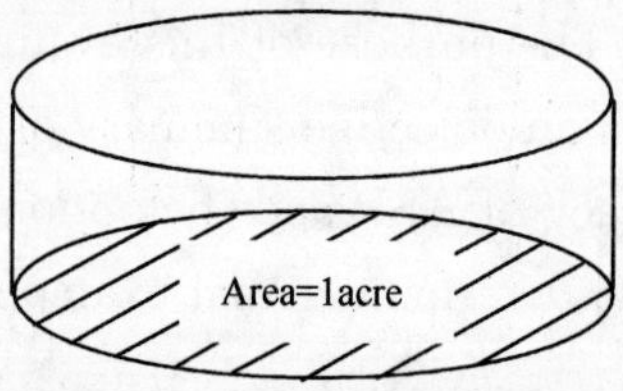

Figure 5-3 One acre-foot of water is equivalent to the volume that would cover 1 ac of land at a depth of 1ft, or 43560 cubic feet (ft^3) of water. (Not to scale.)

It is important to consider that the rainfall intensity is not constant over the duration of a storm, although the average intensity is a very useful number in a wide variety of hydrology problems and applications. In some hydrologic analyses, though, it is necessary to have more detailed information regarding the rainfall intensity. These data can be depicted in a hyetograph, which is a graph of rainfall intensity (or volume) versus time. An example hyetograph is depicted in Figure 5-4. Note that the average rainfall intensity over the 60-min duration of the storm is about 2.2in./h, whereas the peak intensity is about 8in./h.

2. Recurrence Interval

Common experience shows that hydrologic events, such as rain storms, do not occur with any definite regularity. The time span or period between storms is not constant. The occurrence of rainfall, its intensity, and its duration are random natural events. Despite the random nature of precipitation events, it is possible to determine average frequencies of occurrence of storms having specific intensities and durations. It would be convenient if the exact dates on which identical storms would occur in the future could be predicted, but obviously

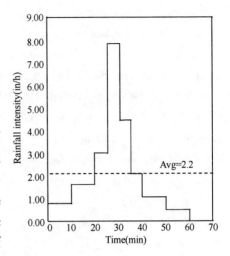

Figure 5-4　Example of a hyetograph

that is not possible. For instance, even though the data of next 20min, 25mm storm cannot be determined in advance, it is possible to predict how many times a similar storm can be expected to occur over the next year or several years. In addition, a prediction of the likelihood or probability of observing that storm again in any given period can be made.

By examining many years of rainfall records and applying statistical analyses, the average number of years between storms of specific intensities and durations can be determined. This time span between identical storms is called the recurrence interval or return period of the storm. These return periods are determined and reported by the National Weather Service, and designers of environmental facilities must know how to interpret and use the data.

When applying these data, the expression *N-year storm* is used., where N stands for the recurrence interval in years. For example, a storm with a return period of 5 years is called a 5-year storm. This means that over a long period of time, the average time span between storms of that particular intensity and duration is 5 years. It does not mean that a similar storm will occur once exactly every 5 years. In fact, it is possible that more than one of these 5-year storms could occur within a shorter time span, even within a single year, but the chances for this are slim. Note, too, that the probability of the 5-year storm occurring in any given 5-year period is not quite 100 percent. In other words, none can say for sure that what is called a 5-year storm will actually take place within, say, the next 5 years. But over a long time span, 500 years, for example, it is a good bet that there will be about 100 of these 5-year storms.

3. Probability of Occurrence

Data on storm intensity, duration, and return period are important in the design of urban drainage structures and for predicting peak flows in rivers. On the other end of the hydrologic spectrum, knowing the severity of droughts and their frequency of occurrence is of importance in designing water supply reservoirs.

Because of the uncertain and irregular nature of hydrologic events, there is always some risk of failure when designing a structure or facility involving water resources. For example, a river used for water supply may not provide enough water for a growing community during dry periods. Even if a small reservoir were built to overcome this deficiency, there would remain the risk that a more severe (though less frequent) drought would cause the reservoir to run dry. This risk can be reduced by building a larger reservoir, but this would be more expensive. Designers must be able to balance the economics and the risks, using probability concepts.

The probability or chance that a given event will occur can be expressed as a fraction, a decimal, or a percent. For example, the probability of a tossed coin coming up heads is one chance out of two, or $1/2=0.5=50$ percent. In the long run, 50 of 100 tosses can be expected to come up heads. A probability of 1 or 100 percent represents a certainty, and a probability of 0 represents an impossibility.

There is a simple relationship between the return period of a hydrologic event and the probability of occurrence of the event. If N is the recurrence interval of the event (in years), then the probability P of that event being equaled or exceeded in any given year is the reciprocal of N. Expressed as a formula, this is

$$P=\frac{1}{N} \tag{5-2}$$

For example, the probability of a 5-year storm occurring in any single year is $P=1/5=0.2$ or 20 percent. In effect, this also means that there is less than a 20 percent chance that a worse or more intense storm will occur in any given year.

Relying on common experience again, it can be seen that the really intense storms are few and far between. In other words, the more extreme the hydrologic event, the larger is its recurrence interval. And the larger the recurrence interval N, the lower is the probability of occurrence P, because of the inverse relationship between the two. For example, there is only a 1 percent chance that a 100-year storm will occur in a given year. It is much less likely to observe a severe 100-year storm than a 5-year storm. (Although in many regions of the country rainfall records do not go as far back as 100 years, statistics and probability theory can be used to extrapolate or extend the existing data beyond the actual period of record.)

To summarize, the larger the recurrence interval N, the less likely it is for a hydrologic event to be equaled or exceeded in a given year. This is an important concept. Generally, the more critical a project is in terms of potential loss of life, economic damage, or adverse environmental effects, the larger is the value of N used in design computations.

A dam, for instance, may be designed to accommodate a 100-year flood, whereas a local storm drain may be designed to handle only the flow from a 2-year storm. In the former case, designing the dam for the big flow will reduce the chance of failure or breach of the dam and ensure the protection of human lives and property down stream. In the latter case, a trade-of is made between saving money for construction and taking more of a chance on the storm drain backing up or overflowing of a chance on the storm drain backing up or overflowing once every 2 years or so.

4. Intensity-Duration-Frequency Relationships

In these discussions, terms such as *storm intensity*, *storm duration*, and *recurrence interval* have been examined as if they were independent quantify. But these three factors are related to each other and must be considered together. The term *frequency* is often used instead of return period. The frequency of a storm or other hydrologic event varies inversely with its return period. A 10-year storm, for example, will occur less frequently than a 5-year storm.

The rainfall data collected by the National Weather Service are compiled, analyzed, and published in various forms. The relationships among rainfall intensity, duration, and frequency may be shown graphically in curves or maps, or they may be expressed as formulas. As shown in Chapter 9 on stormwater management, these data are used by designers to estimate storm runoff and peak streamflow or discharge.

Further Reading (3)

Surface Water

Water that flows over the ground is often called runoff. Runoff that has not yet reached a definite stream channel is called overland flow or sheet flow (on a smooth surface, such as pavement). This type of surface water is important in the discussion of stormwater drainage systems. For the most part, the term surface water refers to water flowing in streams and rivers as well as water stored in natural or artificial lakes.

1. Watersheds

Runoff occurs when the rate of precipitation exceeds the rate of interception and evapotranspiration. The total land area that contributes runoff to a stream or river is called a watershed. It may also be called a drainage basin or catchment area, particularly if the water flows toward or in an urban drainage system. Generally, engineers are interested in determining the amount of runoff at a specific point in the natural stream or engineered drainage system. This point is called the basin outlet or point of concentration.

The natural boundary or perimeter of the watershed may be determined from a topographic map, using the ground elevation contour lines. Viewed on a topographic map, water flowing freely over the ground's surface would move in a direction perpendicular to the contour lines, which is the direction of the steepest slope at any given point. By examining the contour map and visualizing the pattern of overland flow, it is possible to locate the boundary of the watershed. This boundary is called the drainage divide line or ridge line; it separates adjacent watersheds.

A simplified picture of a watershed is that of a funnel (Figure 5-5). The wide rim at the top of the funnel represents the ridge line and the circular area encompassed by the rim represents the catchment area. As water falls within the rim, it flows downward toward the narrow outlet at the bottom, which represents the point of concentration. In practical applications, the ridge line must be located and drawn on a topographic (topo) map by the engineer or technician. Invariably, the ridge line forms an irregular shape rather than a circle like the rim of a funnel, and the point of concentration lies on the line rather than in the center of the area because the plan view of a watershed is depicted.

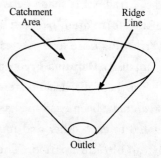

Figure 5-5 Simplified view of a watershed or drainage basin.

To draw a drainage divide line on a topo map, the following procedure may be followed:

(1) Start at the point of concentration. This might be at the intersection of two streams, at a point where a stream flows through a highway culvert, or at the location of a dam. The divide line will begin and end at this point.

(2) Examine the contours to determine flow patterns. Imagine a drop of water on the ground at any given point and visualize which way it will flow. Start to sketch sections of the divide line that clearly separate the watershed from an adjacent watershed. These sections of the line will follow ridges and pass through topographic saddles. Remember, the natural drainage divide line is always perpendicular to the contour lines.

(3) Fill in any gaps that may be left in the line being sketched. Occasionally, the divide line will turn sharply on the top of a ridge to pass through one of the saddles on the line.

A perspective sketch depicting flow patterns and a drainage divide line is shown in Figure 5-6 (a), and a plan view of the same area is shown in Figure 5-6 (b).

The sharp turns that may characterize a divide line as it passes through adjacent ridges and saddles are illustrated in Figure 5-7.

The point at which two streams converge or intersect is called a point of confluence. As small streams converge, larger streams and eventually rivers are formed. The catchment area for a particular stream may be only a part of a larger watershed; the smaller area is called a subbasin of the watershed. A typical drainage network is shown in Figure 5-8. The streams may be classified by their position in the overall network. Typical classifications are first-order streams, second-order streams, and so on. A first-order stream does not have any tributaries or smaller streams flowing into it.

A Watershed for a large river may encompass thousands of square miles and include many smaller tributaries. These large Watersheds are also called river basins. The Raritan River Basin in New Jersey, for example, encompasses about $2850 km^2$ ($1100 mi^2$). The USGS has divided the United States into 2149 basic watershed units, the smallest of which encompasses about $1800 km^2$ ($700 mi^2$).

Part B Urban Water Environment   63

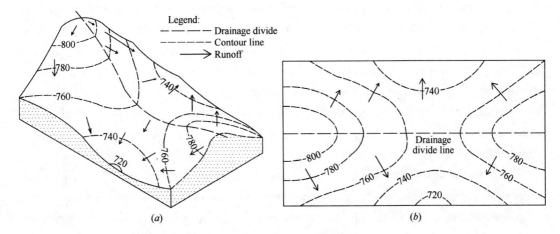

Figure 5-6 The direction of sheet flow and drainage divide line

(a) *A perspective view of runoff patterns; arrows show the direction of sheet flow. The drainage divide line passes through the ridges and the saddle, separating two adjacent watershededs. The direction of sheet flow is perpendicular to the contour lines.*

(b) *A plan view topographic map that shows the same drainage divide line and contours as depicted in (a).*

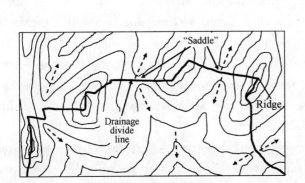

Figure 5-7 A drainage divide line will sometimes turn sharply on a ridge or saddle, as shown here. The dashed arrows show the direction of overland flow.

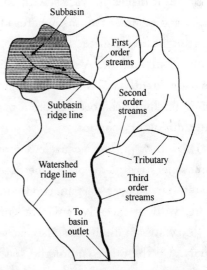

Figure 5-8 A large watershed usually comprises several smaller catchment areas or subbasins.

The size of a drainage basin refers to its total horizontal surface area. Relatively small basins maybe expressed in terms of acres or hectares. A mechanical device called a Planimeter is often used to measure the area by simply tracing the boundary of the watershed. Modern electronic planimeters can be calibrated to display the area digitally, based on the scale of the map being used.

The volume and rate of runoff in a watershed are functions of many variables. The basin area and the intensity and duration of rainfall have a direct effect on the amount and rate of runoff. Other factors include the slope of the ground, the type of soil and vegetative cover, and the type of land use. For example, a flat area with sandy soil would produce less runoff than a sloping area with clay soil. More of the water would infiltrate the ground surface through the porous sand in the former case, leaving a smaller fraction of the rain to become surface flow. Also, densely populated urban areas generate more runoff than suburban or rural areas.

2. Streamflow

The amount or volume of water that flows in a stream is called the flow rate or discharge of the stream. The discharge is expressed in terms of volume per unit time passing any given point in the stream. The SI units for discharge are usually cubic meters per second (m^3/s), cubic meters per hour (m^3/h), or megaliters per day (ML/d). In U.S. Customary units, discharge may be expressed as cubic feet per second (ft^3/s or cfs), gallons per minute (gpm), million gallons per day (mgd), or acre-feet per day (ac-ft/d). A section of a stream that has a relatively constant slope, cross section, and discharge may be called a *reach* of the stream.

Stream discharge varies with time. Generally, higher flow rates are observed in the spring and summer months, whereas lower discharges occur in the fall and winter. This is particularly the case for the northeastern United States. Snowmelt can contribute significantly to streamflow. Variations in discharge that occur on a weekly, daily, or even hourly basis are directly associated with rainfall events. In some streams, an extremely wide variation in discharge can occur, from a raging torrent during wet weather to hardly a trickle of water during dry periods.

Low flow rates can cause environmental problems in streams receiving discharges from wastewater treatment plants because there is less water in the stream to dilute the wastewater. Under certain conditions, a stream can assimilate or absorb biodegradable wastes without excessive environmental damage.

Low stream discharges also cause problems if the stream is used as a source for water supply. On the other end of the spectrum, excessively high discharges usually necessitate the construction of flood control facilities.

Unit 6 Parameters of Water Quality

The topic of water quality focuses on the presence of foreign substances in water and their effects on people or the aquatic environment. Water of good quality for one purpose may be considered to be of poor quality for some other use. For example, water suitable for swimming may not be of good enough quality for drinking. But even drinking water may not be suitable for certain industrial or manufacturing purposes that require pure water.

What exactly is pure water? Just how pure does it have to be for drinking or for other uses? Obviously, it is not enough to simply describe water quality as being "good" or "poor". Some quantitative measures for determining and describing the condition of the water are needed. It is necessary to determine what substances are in the water and in what concentrations they are present. Some knowledge of the effects of those substances on public and environmental health is also needed. Finally, some yardsticks or standards against which to compare the results of our analysis and thereby judge the suitability of the water for a particular use are needed.

Water has a remarkable tendency to dissolve other substances. Because of this, it is rarely found in nature in a pure condition. Even water in a mountain stream, far from civilization, contains some natural impurities in solution and in suspension.

Changes in water quality begin with precipitation. As rain falls through the atmosphere, it picks up dust particles and such gases as oxygen and carbon dioxide. In some industrialized regions, the quality of rainwater is altered significantly before it ever touches the ground.

Surface runoff picks up silt particles, bacteria, organic material, and dissolved minerals. Groundwater usually contains more dissolved minerals than surface water because of its longer contact with soil and rock. Finally, water quality is very much affected by human activities, including land use (such as agriculture) and the direct discharge of municipal or industrial wastewaters to the environment.

Protecting water quality and modifying it for a particular purpose are major objectives in the field of environmental technology. It is therefore necessary to make use of technical terms in discussing the various aspects of water quality and pollution. In particular, reference to the different parameters of physical, chemical, and biological quality needs to be made.

In the discussion of specific water quality parameters, only brief reference is made to actual laboratory analysis procedures. These are thoroughly described in Standard Methods for the Examination of Water and Wastewater (published by the American Water Works Association) and other professional publications.

Portable field test kits are particularly useful for conducting preliminary water quality surveys. But for most water quality analyses to be official and able to stand up to legal scrutiny if challenged, they must be done by qualified personnel in certified laboratories, following Standard Methods.

Physical Parameters of Water Quality

The parameters that are commonly used to describe the physical quality of water include turbidity, solids, temperature, color, taste, and odor.

Turbidity

When small particles are suspended in water, they tend to scatter and absorb light rays. This gives the water a murky or turbid appearance, and this effect is called turbidity. Clay, silt, tiny fragments of organic matter, and microscopic organisms are some of the substances that cause turbidity. They occur in water naturally or because of human activities and pollution.

Turbidity is a particularly important parameter of drinking water quality. Suspended particles can provide hiding places for harmful microorganisms and thereby shield them from the disinfection process in a water treatment plant. Because of this shielding effect, the microbes can be consumed by people who drink the water, and the spread of disease may result.

Turbidity in drinking water is also unacceptable for esthetic reasons-it makes the water look very unappetizing. Most people find even a slight degree of turbidity in their water objectionable. Even when told that the water is safe to drink in spite of its turbidity, people tend to seek alternative water supplies (which could possibly be of poorer quality).

Turbidity is measured in units that relate the clarity of the water sample to that of a standardized suspension of silica. The interference in the passage of light caused by a suspension of 1mg/L of silica is equivalent to one turbidity unit (TU). For example, a water sample that has the same degree of cloudiness as a 10mg/L suspension of silica has a turbidity of 10 TU.

The standard for calibrating turbidimeters is defined as

$$1\text{mg/L of } SiO_2 = 1 \text{ normalized turbidity unit} \qquad (6\text{-}1)$$

To interpret turbidity data, it is useful to be familiar with the typical ranges that occur. Turbidity in excess of 5 TU is just noticeable to the average person; most people do not complain about the clarity of the water at TU values less than 5. Turbidity in what most people would consider to be a relatively clear lake may be as high as 25 TU. In muddy water, turbidity generally exceeds 100 TU. Modern water treatment plants can routinely produce crystal clear water with turbidities of less than 1 TU.

Groundwater normally has very low turbidity because of the natural filtration that occurs as the water percolates through the soil. Most streams and rivers, though, have relatively high turbidities. This is particularly true during and just after rainstorms, which cause soil erosion. The treatment of turbid stream water for drinking supplies can be an expensive process; the greater the turbidity, the greater is the amount of chemicals needed and the more frequently must the filters be cleaned.

For drinking water, instruments called nephelometers are used to measure the turbidity after purification. These devices measure the amount of scattered light electronically and do not depend on human vision or judgment in making comparisons to standard suspensions. Measurements made with nephelometric turbidimeters may be expressed in terms of NTU instead of just TU, to indicate how the measurement was made. Turbidity in filtered drinking water in the United States must be equal to or less than 0.5 NTU in at least 95% of the samples tested each month.

A conventional Jackson candle turbidimeter is illustrated in Figure 6-1. It may be used to measure raw (untreated) water turbidities. The water is added to a vertical glass tube until the candle flame is just obscured from view. The glass tube is graduated with turbidity units; the higher the water column required to obscure flame, the less is the turbidity. Turbidity values obtained using the candle turbidimeter may be expressed as JTU.

Excessive turbidity in a lake reduces the depth to which sunlight penetrates the water. This has an effect on the photosynthesis of microscopic plants, or algae, and on the overall environmental balance of the lake. In field surveys, small, white Secchi disks may be lowered into the water on a line marked off in meters until the disk disappears from view. The depth of the disk at that point can be correlated with the turbidity of the lake water.

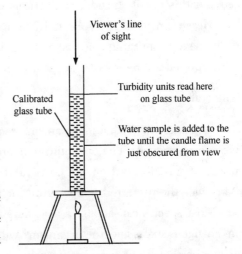

Figure 6-1 The candle turbidimeter.

Solids

Wastewater treatment is complicated by the dissolved and suspended inorganic material the water contains. In discussion of water treatment, both dissolved and suspended materials are called solids. The separation of these solids from the water is one of the primary objectives of treatment.

Strictly speaking, in wastewater anything other than water is classified as solid. The usual definition of solids, however, is the residue after evaporation at 103°C (slightly higher than the boiling point of water). The solids thus measured are known as total solids. Total solids may be divided into two fractions: the total dissolved solids (TDS) and the total suspended solids (TSS). The difference is illustrated in the following example:

A teaspoonful of table salt dissolves in a glass of water, forming a water-clear solution. However, the salt remains behind if the water evaporates. Sand, however, does not dissolve and remains as sand grains in the water and forms a turbid mixture. The sand also remains behind if the water evaporates. The salt is an example of a dissolved solid, whereas the sand is a suspended solid.

Suspended solids are separated from dissolved solids by filtering the water through a filter paper. The suspended material is retained on the filter paper, while the dissolved fraction passes through. If the initial dry weight of the filter paper is known, the subtraction of this from the total weight of the filter and the dried solids caught in the filter paper yields the weight of suspended solids, expressed in milligrams per liter.

Solids may be classified in another way: those that are volatilized at a high temperature and those that are not. The former are known as volatile solids, the latter as fixed solids. Volatile solids are usually organic compounds. Obviously, at 600℃, the temperature at which the combustion takes place, some of the inorganics are decomposed and volatilized, but this is not considered a serious drawback.

Measurement of the volatile fraction of suspended material, the volatile suspended solids, is made by burning the suspended solids and weighing them again. The loss in weight is interpreted as the volatile suspended solids.

Temperature

Fish and other aquatic organisms require certain condition of temperature in order to live and reproduce. The optimum temperature for trout, for example, is 15℃. A temperature of about 24℃ is best for perch, and carp do very well at a cozy 32℃, which is more than twice the preferred temperature for trout (a coldwater fish).

Most species can adapt to a moderate change from their optimum temperature, but if the change is excessive, the organisms will perish or migrate to a new location. Generally, a change of about 5℃ can significantly alter the balance and health of an aquatic environment. Sudden drops in temperature can be harmful, but usually an increase in temperature will cause more damage than a decrease (thus, rivers must be protected from warm water discharges from power plants).

A basic reason for this, as discussed previously, is that the solubility of oxygen in water decreases markedly as the temperature of the water goes up. Fish and other organisms need the oxygen to survive, and higher temperatures increase their rate of metabolism. In other words, the rate at which the organisms use oxygen to burn food for energy increases at higher temperatures. The combined effect of there being less available oxygen and the organisms having faster metabolism rates can eventually be very damaging.

Other than the fact that most people prefer cold drinking water, temperature is of little direct significance in public water supplies. Temperature plays a more important role, however, in wastewater treatment and water pollution control. Biological wastewater treatment systems are more efficient at higher temperatures. In colder regions of the country, treatment plants may be sheltered in heated enclosures to maintain optimum temperature ranges.

Color, Taste, and Odor

Color, taste, and odor are physical characteristics of drinking water that are important for esthetic reasons. They do not cause any direct harmful effect on health, but no matter how safe the water may be to drink, most people object strongly to water that offends their sense of sight, taste, or smell.

Color may be caused by dissolved or suspended colloidal particles, primarily from decaying leaves or microscopic plants. This tends to give the water a brownish-yellow hue. Streams or rivers with tributaries in swampy areas may have this problem. Color is measured by comparing the water sample with standard color solutions or colored glass disks. One color unit is equivalent to the color produced by a 1-mg/L solution of platinum. It is not practical to isolate and identify specific chemicals that cause the color.

Hydrogen sulfide gas, H_2S, is a common cause of odor in water supplies. The rotten-egg smell of this gas may be encountered in water that has been in contact with naturally occurring deposits of decaying organic matter. Groundwater supplies sometimes have this problem; the wells are called sulfur wells.

Odor is measured and expressed in terms of a threshold odor number. The threshold odor number is the ratio by which the sample has to be diluted for the odor to become virtually unnoticeable. For example, if a 50-mL volume of water sample has to be diluted to a volume of 200mL for the odor to be just barely detectable, the threshold number would be 200/50=4. A similar technique may be applied in measuring the taste of the water. Taste and odor measurements are very subjective and depend on the sensitivity of the person conducting the test.

Vocabulary and Phrases

aquatic *adj*. 水的；水生的；水中的；*n*. 水生生物
substance *n*. (1)物质；材料；(2)本质；实物
concentration *n*. (1)浓度；(2)浓缩
yardstick *n*. 尺度；码尺
impurity *n*. 杂质
suspension *n*. (1)悬置；悬浮；悬移；(2)悬浮体；悬浮液；(3)停止
dioxide *n*. 二氧化物
silt *n*. (1)淤泥；(2)粉砂；粉粒；(3)泥沙
bacteria *n*. 细菌
organic *adj*. 有机的；*n*. 有机物(质)
dissolved *adj*. 溶解的

discharge *v. & n*. (1)卸货；(2)排出；出流；(3)放电 *n*. 流量
municipal *adj*. 城市的；市政的
parameter *n*. (1)参数；参变量；(2)特性；(3)(复)极限；边界
portable *adj*. 轻便的；手提式的；可移动的；可携带的
kit *n*. 成套工具；工具箱；软件包；成套零件(部件；配件；设备)
survey *v. & n*. 测量；勘测；勘察；调查
scrutiny *n*. 详细审查
turbidity *n*. (浑)浊度；浑浊性
taste *n*. 味道；味觉，*v*. 品尝；辨味 (of)有…味道；领略

odor n. 臭(xiu)；气味(指香和臭味)；臭(chou)
scatter v. 分散；散开；散射；撒开；驱散
absorb vt. 吸收；吸引
ray n. 射线；光线；鳐或魟
murky adj. 黑暗的
clay n. 黏土；泥土
fragment n. 碎片，断片，片段
microscopic organism n. 微生物
microorganism n. 微生物
microbe n. 微生物
disinfection n. 消毒
esthetic adj. 感觉的；美学的；审美的
appetizing adj. 引起食欲(欲望)的
silica n. 二氧化硅；硅石
interference n. 干扰；干涉
calibrating turbidimeter 校准浊度计
normalize v. 正常化；正则化；标准化
rainstorm n. 暴雨
erosion n. 腐蚀，侵蚀
nephelometer n. (散射)浊度计
purification n. 净化
Jackson candle turbidimeter 杰克逊烛光浊度仪
obscure adj. 暗的，朦胧的，模糊的，晦涩的，vt. 使暗，使不明显
graduate v. 刻度；标度；校准；(蒸发)浓缩；(准予)毕业；逐步消失(away)，n. 毕业生；量杯；分度器；adj. 分等级的
penetrate v. 穿透；渗透；看穿；洞察
photosynthesis n. 光合作用
algae n. 藻类，海藻
secchi disk 西奇盘；透明度板

residue n. 残余；剩余物；渣滓；滤渣；残数；余数；余式；余项
total solids 总固体
total dissolved solids 总溶解固体
total suspended solids 总悬浮固体
filter paper 滤纸
subtraction n. 减少
volatile solids 挥发性固体
fixed solids 不挥发固体
drawback n. 缺点；障碍；退还的关税；退税
trout n. [鱼] 鲑，鲑鱼
perch n. 栖木，人所居的高位，有利的地位，杆，河鲈；v. (使)栖息，就位，位于
carp n. 鲤鱼；vi. 吹毛求疵
cozy adj. 舒适的，安逸的，惬意的
perish vi. 毁灭，死亡，腐烂，枯萎；vt. 毁坏，使麻木
migrate vi. 移动，移往，移植，随季节而移居，(鸟类的)迁徙；vt. 使移居，使移植
metabolism n. 新陈代谢，变形
colloidal adj. 胶体的，胶态的
decay v. & n. 衰变；衰减；腐烂；腐败
brownish adj. 呈褐色的
hue n. 色调；样子；颜色；色彩
tributary n. 支流
swampy adj. 沼泽的，沼泽多的，湿地的，松软的，沼泽中发现的
platinum n. 白金，铂
hydrogen sulfide 氢化硫
sulfur n. 硫磺；硫黄；vt. 磺化
threshold odor number 气味阈数

Notes

The term *constituent* is used to refer to an individual compound or element, such as suspended solids or ammonia nitrogen. However, the terms of *contaminants*, *impurities*, and *pollutants* are often used interchangeably and refer to constituents added to the water supply through use.

Question and Exercises

1. The following data were obtained for a sample:
Total solids=3500mg/L
Suspended solids=4500mg/L
Volatile suspended solids=1500mg/L
Fixed suspended solids=1000mg/L
Which of these numbers is questionable and why?
2. What is the difference between TDS and TSS? How are they measured?
3. How does the solubility of solids in water change with increasing temperature? Is the situation the same for gases?
4. Briefly describe the significance of temperature in water quality.
5. Please describe how odor and color is measured in water.
6. What do the terms JTU and NTU mean?

Further Reading (1)

Chemical Parameters of Water Quality

Many organic and inorganic chemicals affect water quality. In drinking water, these effects may be related to public health or to esthetics and economics. In surface waters, chemical quality can affect the aquatic environment. Several chemical parameters are also of concern in wastewater. In this section, the most common chemical parameters of water quality are discussed.

1. Dissolved Oxygen

Dissolved oxygen is generally considered to be one of the most important parameters of water quality in streams, rivers, and lakes. It is usually abbreviated simply as DO. Just as people need oxygen in the air they breathe, fish and other aquatic organisms need DO in the water to survive. With most other substances, the less there is in the water, the better is the quality. But the situation is reversed for DO. The higher the concentration of dissolved oxygen, the better is the water quality.

Oxygen is only slightly soluble in water. For example, the saturation concentration at 20℃ is about 9mg/L or 9 ppm. (Remember that this is equivalent to the relationship between 9 in. and 16 mi.) Because of this very slight solubility, there is usually quite a bit of competition among aquatic organisms, including bacteria, for the available dissolved oxygen. As discussed in some detail later, bacteria will use up the DO very rapidly if there is much organic material in the water. Trout and other fish soon perish when the DO level drops. Another factor to remember is that oxygen solubility is very sensitive to temperature. Changes in water temperature have a significant effect on DO concentrations.

Dissolved oxygen has no direct effect on public health, but drinking water with very little or no oxygen tastes flat and may be objectionable to some people. Dissolved oxygen does play a part in the corrosion or rusting of metal pipes; it is an important factor in the operation and maintenance of water distribution networks.

Dissolved oxygen is used extensively in biological wastewater treatment facilities. Air, or sometimes pure oxygen, is mixed with sewage to promote the aerobic decomposition of the Organic wastes. The role of dissolved oxygen in water pollution and wastewater treatment is discussed in subsequent chapters.

The DO concentration can be determined by using standard wet chemistry methods of analysis or membrane electrode meters in the lab or in the field. Field instruments are available that have probes that can be lowered directly into a stream or treatment tank. The electrode probe senses small electric currents that are proportional to the dissolved oxygen level in the water.

2. Biochemical Oxygen Demand

Bacteria and other microorganisms use organic substances for food. As they metabolize organic material, they consume oxygen. The organics are broken down into simpler compounds, such as CO_2 and H_2O, and the microbes use the energy released for growth and reproduction.

When this process occurs in water, the oxygen consumed is the DO. If oxygen is not continually replaced in the water by artificial or natural means, then the DO level will decrease as the organics are decomposed by the microbes. This need for oxygen is called the *biochemical oxygen demand*. In effect, the microbes "demand" the oxygen for use in the biochemical reactions that sustain them. The abbreviation for biochemical oxygen demand is BOD; this is one of the most commonly used terms in water quality and pollution control technology.

Organic waste in sewage is one of the major types of water pollutants. It is impractical to isolate and identify each specific organic chemical in these wastes and to determine its concentration. Instead, the BOD is used as an indirect measure of the total amount of biodegradable organics in the water. The more organic material there is in the water, the higher the BOD exerted by the microbes will be.

In addition to being used as a measure of organic pollution in streams or lakes, the BOD is used as a measure of the strength of sewage. As seen in chapter 10, this is one of the most important parameters for the design and operation of a water pollution control plant. A strong sewage has a high concentration of organic material and a correspondingly high BOD. A *weak* sewage, with a low BOD, may not require as much treatment.

The complete decomposition of organic material by microorganisms takes time, usually 20d or more under ordinary circumstances. The amount of oxygen used to completely decompose or stabilize all the biodegradable organics in a given volume of water is called the *ultimate* BOD, or BOD_L. For example, if a 1-L volume of municipal sewage requires

300mg of oxygen for complete decomposition of the organics, the BOD_L would be expressed as 300mg/L. One liter of wastewater from an industrial or food processing plant may require as much as 1500mg of oxygen for complete stabilization of the waste. In this case, the BOD_L would be 1500mg/L, indicating a much stronger waste than ordinary municipal or domestic sewage. In general, then, the BOD is expressed in terms of mg/L of oxygen.

The BOD is a function of time. At the very beginning of a BOD test, or time=0, no oxygen will have been consumed and the BOD=0. As each day goes by, oxygen is used by the microbes and the BOD increases. Ultimately, the BOD_L is reached and the organics are completely decomposed. A graph of the BOD versus time has the characteristic shape. This is called the BOD curve.

The BOD curve can be expressed mathematically by the following equation:

$$BOD_t = BOD_L \times (1 - 10^{-kt}) \qquad (6\text{-}2)$$

where BOD_t = BOD at any time t, mg/L;

BOD_L = ultimate BOD, mg/L;

k = a constant representing the rate of the BOD reaction;

t = time, d.

3. Chemical Oxygen Demand

The BOD test provides a measure of the biodegradable organic material in water, that is, of the substances that microbes can readily use for food. There also might be nonbiodegradable or slowly biodegradable substances that would not be detected by the conventional BOD test.

The *chemical oxygen demand*, or *COD*, is another parameter of water quality, which measures all organics, including the nonbiodegradable substances. It is a chemical test using a strong oxidizing agent (potassium dichromate), sulfuric acid, and heat. The results of the *COD* test can be available in just 2h, a definite advantage over the 5d required for the standard BOD test.

COD values are always higher than BOD values for the same sample, but there is generally no consistent correlation between the two tests for different wastewaters. In other words, it is not feasible to simply measure the *COD* and then predict the BOD. Because most wastewater treatment plants are biological in their mode of operation, the BOD is more representative of the treatment process and remains a more commonly used parameter than the *COD*.

Among many drawbacks of the BOD test, the most important is that it takes five days to run. If the organic compounds are oxidized chemically instead of biologically, the test can be shortened considerably. Such oxidation is accomplished with the chemical oxygen demand (*COD*) test. Because nearly all organic compounds are oxidized in the *COD* test and only some are decomposed during the BOD test, *COD* values are always higher than BOD values. One example of this is wood pulping waste, in which compounds such as cellulose are easily oxidized chemically (high *COD*) but are very slow to decompose biologically (low BOD).

4. pH

The pH of a solution is a measure of hydrogen ion concentration, which in turn is a measure of its acidity. Pure water dissociates slightly into equal concentrations of hydrogen and hydroxyl (OH^-) ions.

$$H_2O \rightleftharpoons H^+ + OH^- \tag{6-3}$$

An excess of hydrogen ions makes a solution acidic, whereas a dearth of H^+ ions, or an excess of hydroxyl ions, makes it basic. The equilibrium constant for this reaction, K_w, is the product of H^+ and OH^- concentrations and is equal to 10^{-14}. This relationship may be expressed as

$$[H^+][OH^-] = K_w = 10^{-14} \tag{6-4}$$

where $[H^+]$ and $[OH^-]$ are the concentrations of hydrogen and hydroxyl ions, respectively, in moles per liter. Considering Equation (6-2) and solving Equation (6-3), in pure water,

$$[H^+][OH^-] = 10^{-7} \text{moles/L} \tag{6-5}$$

The hydrogen ion concentration is so important in aqueous solutions that an easier method of expressing it has been devised. Instead of as moles as per liter, we define a quantity pH as the negative logarithm of $[H^+]$ so that

$$pH = -\log_{10}[H^+] = \log_{10}\frac{1}{[H^+]} \tag{6-6}$$

or

$$[H^+] = 10^{-pH} \tag{6-7}$$

For a neutral solution, $[H^+]$ is 10^{-7}, or pH=7. For larger hydrogen ion concentrations, then, the pH of the solution is <7. For example, if the hydrogen ion concentration is 10^{-4}, the pH=4 and the solution is acidic. In this solution, we see that the hydroxyl ion concentration is $10^{-14}/10^{-4} = 10^{-10}$. Since $10^{-4} \gg 10^{-10}$, the solution contains a large excess of H^+ ions, confirming that it is indeed acidic. A solution containing a dearth of ions would have $[H^+] < 10^{-7}$, or pH>7, and would be basic. The pH range of dilute solutions is from 0 (very acidic; 1 mole of H^+ ions per liter) to 14 (very alkaline). Solutions containing more than 1 mole of H^+ ions per liter have negative pH.

The measurement of pH is now almost universally by electronic means. Electrodes that are sensitive to hydrogen ion concentration (strictly speaking, the hydrogen ion activity) convert the signal to electric current. pH is important in almost all phases of water and wastewater treatment. Aquatic organisms are sensitive to pH changes, and biological treatment requires either pH control or monitoring. In water treatment as well as in disinfection and corrosion control, pH is important in ensuring proper chemical treatment. Mine drainage often involves the formation of sulfuric acid (high H^+ concentration), which is extremely detrimental to aquatic life. Continuous acid deposition from the atmosphere may substantially lower the pH of a lake.

5. Alkalinity

A parameter related to pH is alkalinity, or the buffering capacity of the water against acids. Water that has a high alkalinity can accept large doses of an acid without lowering the pH significantly. Waters with low alkalinity, such as rainwater, can experience a drop in the pH with only a minor addition of hydrogen ion.

In natural waters much of the alkalinity is provided by the carbonate/bicarbonate buffering system. Carbon dioxide (CO_2) dissolves in water and is in equilibrium with the bicarbonate and carbonate ions.

$$CO_2(gas) \rightleftharpoons CO_2(dissolved)$$
$$CO_2(dissolved) + H_2O \rightleftharpoons H_2CO_3$$
$$H_2CO_3 \rightleftharpoons H^+ + HCO_3^-$$
$$HCO_3^- \rightleftharpoons H^+ + CO_3^{2-}$$

where H_2CO_3 = carbonic acid;
HCO_3^- = bicarbonate ion;
CO_3^{2-} = carbonate ion.

Any change that occurs in the components of this equation influences the solubility of CO_2. If acid is added to the water, the hydrogen ion concentration is increased, and this combines with both the carbonate and bicarbonate ions, driving the carbonate and bicarbonate equilibria to the left, releasing carbon dioxide into the atmosphere. The added hydrogen ion is absorbed by readjustment of all the equilibria, and the pH does not change markedly. Only when all of the carbonate and bicarbonate ions are depleted will the additional acid added to the water cause a drop in pH.

6. Nitrogen

Nitrogen, N_2, is an important element in biological reactions.

The four forms of nitrogen that are of particular significance in environmental technology are organic nitrogen, ammonia nitrogen, nitrite nitrogen, and nitrate nitrogen.

Organic nitrogen may be bound in high-energy compounds such as amino acids and amines. Ammonia is one of the intermediate compounds formed during biological metabolism and, together with organic nitrogen, is considered an indicator of recent pollution. Therefore, these two forms of nitrogen are often combined in one measure, called *Kjeldahl nitrogen*, after the scientist who first suggested the analytical procedure. Aerobic decomposition (oxidation) eventually produces nitrite (NO_2^-) and finally nitrate (NO_3^-) from organically bound nitrogen and ammonia. A high-nitrate and low-ammonia nitrogen therefore suggests that pollution occurred, but some time before.

These forms of nitrogen can all be measured analytically by colorimetric techniques. In colorimetry, the ion in question combines with a reagent to form a colored compound; the color intensity is proportional to the original concentration of the ion. For example, ammonia can be measured by adding an excess amount of a compound called *Nessler re-*

agent (potassium mercuric iodine, K_2HgI_4) to the unknown sample. Nessler reagent combines with ammonia in solution to form a yellow-brown colloid. The amount of colloid formed is proportional to the concentration of ammonium ions in the sample.

The color is measured photometrically. A photometer, illustrated in Figure 6-2, consists of a light source, a filter, the sample, and a photocell. The filter allows only those wavelengths of light to pass through that the compounds being measured will absorb. Light from the light source passes through the sample to the photocell, which converts light energy into electric current. An intensely colored sample will absorb a considerable amount of light and allow only a limited amount of light to pass through and thus create little current. On the other hand, a sample containing very little of the chemical in question will be lighter in color and allow almost all of the light to pass through, and will set up a substantial current.

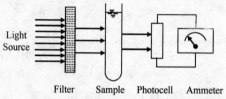

Figure 6-2 Elements of a filter photometer

The intensity of light transmitted by the colored solution obeys the Beer-Lambert Law:

$$I/I_0 = e^{-acx} \quad (6\text{-}8)$$

or

$$\ln(I_0/I) = acx \quad (6\text{-}9)$$

where I = intensity of light after it has passed through the sample;

I_0 = intensity of light incident on the sample;

a = "absorption coefficient" of the colored compound;

x = path length of light through the sample;

c = concentration

A photometer, as shown in Figure 6-2, measures the difference between the intensity of light passing through the sample [I in Equation (6-7)] and the intensity of light passing through clear distilled water (I_0). This difference may be read out as I_0/I on a logarithmic scale; $\ln(I_0/I)$ is called the absorbance. The absorbance of samples containing known ammonia concentrations is plotted against the known concentrations, and the absorbance of an unknown sample is then compared with that of these standards.

7. Phosphates

Like nitrogen, phosphorus, P, is an essential nutrient that contributes to the growth of algae and the eutrophication of lakes, although its presence in drinking water has little effect on health. Phosphorus can enter water from sewage or from agricultural runoff containing fertilizers and animal wastes.

Phosphorus in wastewater may be either inorganic or organic. Although the greatest single source of inorganic phosphorus is synthetic detergents, organic phosphorus is found in food and human waste as well. All phosphates in nature will, by biological action, eventually revert to inorganic forms to be used again by the plants in making high-energy material. Total phosphates may be measured by first boiling the sample in acid solution, which converts all the phosphates to their inorganic forms. From that point the test is colorimetric, like the tests for nitrogen, using a chemical that when combined with phosphates produces a color directly proportional to the phosphate concentration.

8. Heavy Metals and Trace Toxic Organic Compounds

Some *heavy metals* that are toxic are cadmium, Cd, chromium, Cr, lead, Pb, mercury, Hg, and silver, Ag. Arsenic, As, barium, Bar, and selenium, Se, are also poisonous inorganic elements that must be monitored in drinking water.

Heavy metals such as arsenic and mercury can harm fish even at low concentrations. Consequently, the method of measuring these ions in water must be very sensitive. The method of choice is *atomic absorption spectrophotometry*, in which a solution of lanthanum chloride is added to the sample, and the treated sample is sprayed into a flame using an atomizer. Each metallic element in the sample imparts a characteristic color to the flame, whose intensity is then measured spectrophotometrically.

A wide variety of toxic inorganic and organic substances may be found in water in very small or trace amount. Even in trace amounts, they can be a danger to public health. Some toxic substances are from natural sources, but many come from industrial activities and improper management of hazardous waste.

A toxic chemical may be a poison, causing death, or it may cause disease that is not noticeable until many tears after exposure. A carcinogenic substance is one that causes cancer; substances that are also *mutagenic* cause harmful effects in the offspring of exposed people.

Many toxic organic chemicals have been identified and are currently monitored in public water supplies. Among these are the trihalomethanes formed after chlorination, as previously discussed in the section on chlorine residual. Pesticides such as endrin and toxaphene are toxic chlorinated hydrocarbons that are monitored; DDT and chlordane are not routinely checked for in drinking water because they have been banned from use.

Very low concentrations of chlorinated hydrocarbons and other agrichemical residues in water can be assayed by gas *chromatography*. Oil residues in water are generally measured by extracting the water sample with Freon and then evaporating the Freon and weighing the residue from this evaporation.

Further Reading (2)

Bacteriological Parameters of Water Quality

1. Bacteriological Quality

From the public health standpoint, the bacteriological quality of water is as important as the chemical quality. A large number of infectious diseases may be transmitted by water, among them typhoid and cholera. However, it is one thing to declare that water must not be contaminated by pathogens (disease-causing organisms) and another to determine the existence of these organisms. First, there are many pathogens. Each has a specific detection procedure and must be screened individually. Second, the concentration of these organisms, although large enough to spread disease, may be so small as to make their detection impossible, like the proverbial needle in a haystack.

How then can we measure for bacteriological quality? The answer lies in the concept of indicator organisms that, while not particularly harmful, indicate the possible presence of bacteria that are pathogenic. The indicator most often used is a group of microbes of the family *Escherichia coli* (*E. coli*), often called coliform bacteria, which are organisms normal to the digestive tracts of warm-blooded animals. In addition, *E. coli are*

- Plentiful and hence not difficult to find,
- Easily detected with a simple test,
- Generally harmless except in unusual circumstances,
- Hardy, surviving longer than most known pathogens.

Coliforms have thus become universal indicator organisms. The presence of collorms does not prove the presence of pathogens. If a large number of coliforms are present, there is a good chance of recent pollution by wastes from warm-blooded animals, and therefore the water may contain pathogenic organisms.

This last point should be emphasized. The presence of coliforms does not prove that there are pathogenic organisms in the water, but indicates that such organisms might be present. A high coliform count is thus suspicious, and the water should not be consumed, even though it may be safe.

Coliforms are measured by first filtering the sample through a sterile micropore filter by suction, thereby capturing any coliforms on the filter. The filter is then placed in a petri dish containing a sterile agar that soaks into the filter and promotes the growth of the coliforms while inhibiting other organisms. After 24 or 48 hours of incubation the number of shiny black dots, indicating coliform colonies, is counted. If we know how many milliliters of sample poured through the filter, the concentration of coilforms may be expressed as coliforms/mL.

2. Viruses

Viruses can cause a variety of illnesses in humans, including chicken pox, rabies, yellow fever, polio, influenza, gastroenteritis, and the common cold. They can be transmitted among people in a variety of ways, including by ingestion of water contaminated with sewage. Viruses that can infect cells of the intestinal tract of humans are called *enteric viruses* or *enteroviruses*. Many types of viral infection can be controlled by use of vaccines or by eradication of insect vectors that transmit the viruses to humans. Most waterborne viruses can be inactivated by water treatment methods, which include coagulation, filtration, and disinfection; the inactivation occurs during the disinfection process, after coagulation and filtration remove substances that can interfere with disinfection.

Viruses are extremely small pathogens, so small that they can pass through filters that do not permit the passage of bacteria; most viruses can be seen only with the aid of a powerful electron microscope. Since they are incapable of independent metabolism and reproduction, there is debate as to whether viruses should be called "living" organisms. To reproduce, viruses must invade a suitable host cell and take over the cell's metabolic processes for their own use.

Because of their minute size and extremely low concentration and the need to culture them on living tissues, pathogenic (or animal) viruses are fiendishly difficult to measure. Moreover, there are as yet no standards for viral quality of water supplies, as there are for pathogenic bacteria.

One possible method of overcoming this difficulty is to use an indicator organism, much like the coliform group is used as an indicator for bacterial contamination. This can be done by using a *bacteriophage*-a virus that attacks only a certain type of bacterium. For example, *coliphages* attack coliform organisms and, because of their association with wastes from warm-blooded animals, seem to be an ideal indicator. The test for coliphages is performed by inoculating a petri dish containing an ample supply of a specific coliform with the wastewater sample. Coliphages will attack the coliforms, leaving visible spots, or plaques, that can be counted, and an estimation can be made of the number of coliphages per unit volume.

Unit 7　Safe Drinking Water Act

Water withdrawn directly from rivers, lakes, or reservoirs is rarely clean enough for human consumption if it is not treated to purify it. Even water pumped from underground aquifers often requires some degree of treatment to render it potable, that is, suitable for drinking.

The nature and extent of treatment required to preparing water from surface or subsurface sources depend on the quality of the raw (untreated) water.

Better-quality water needs less treatment. Generally, a source of raw water with a coliform count of up to 5000/100mL and a turbidity of up to 10 units is considered good. Water with coliform counts that frequently exceed 20000/100mL and turbidities that exceed 250 units is considered a very poor source and requires expensive treatment to render it potable.

The primary objective of water purification is to remove harmful microorganisms or chemicals, thereby preventing the spread of disease and protecting public health. In addition to being safe to drink, the water must also be esthetically pleasing. It should be crystal clear, and it should not have any objectionable color, taste, or odor.

Generally, groundwater may require some degree of treatment. It is usually free of bacteria and suspended or colloidal particles because of the natural filtration that occurs as the water percolates through the soil. But because it is in direct contact with soil or rock, groundwater often contains dissolved minerals, such as calcium or iron.

As a minimum, most states require that public groundwater supplies be disinfected with chlorine to ensure the absence of pathogens. If dissolved minerals are present in excessive amounts, some combination of chemical treatment, aeration, filtration, and other processes may be needed to purify the water.

Some groundwater supplies have recently been found to be contaminated with very low or trace amounts of toxic organic chemicals, usually from improper land disposal of hazardous wastes. If purification using aeration, activated carbon, or other processes is not feasible, the contaminated wells may have to be abandoned.

Surface water supplies generally require more extensive treatment than groundwater supplies because most streams, rivers, and lakes are contaminated to some extent with domestic sewage and runoff. Even in areas far removed from human activity, surface water contains suspended soil particles (silt and clay) and organics and bacteria (from decaying vegetation and animal wastes).

The most common type of treatment for surface water includes clarification and disinfection. Clarification is usually accomplished by a combination of coagulation-flocculation, sedimentation, and filtration; the most common method for disinfection in the United States is chlorination. The individual treatment step is also called unit process.

The Safe Drinking Water Act (SDWA), enacted by Congress in 1974 and amended several times since then, establishes minimum drinking water standards in the United States. While stream standards serve to protect surface water quality, the SDWA standards are for water that people actually consume, and thereby serve directly to protect public health and welfare. SDWA standards ensure that drinking water supplied to the public is safe and wholesome by setting limits on the amounts of various substances sometimes found in the water supply.

The SDWA applies to public water systems, defined as having 15 or more service connections or serving 25 or more people each day, at least 60d per year. A system can be owned by a private company and still be classified as a public system if it meets this definition. Most states have been delegated the authority for making sure that the SDWA standards are met; some states in turn have delegated their authority to county health departments, which routinely keep track of water quality testing results, conduct inspections, and take enforcement actions when necessary. The EPA provides guidance and technical assistance to the states, conducts research, and periodically revises the standards.

A major revision of the SDWA was enacted in 1996. In addition to authorizing billions of dollars of expenditures for drinking water systems, it focuses water program spending on contaminants that pose the greatest risk to human health and that are most likely to be present in a public water system. The amended Act requires stricter controls on microbial contaminants as well as on the by-products of chlorination. Health risk reduction analyses must now include cost/benefit considerations. Additional revisions include the requirement that water utilities notify the public of water safety violations within 24h, that all water system operators be certified to meet the EPA's minimum certification standards, and that the EPA establish a database to monitor the presence of unregulated contaminants in water.

Based on the results of public health research and scientific judgment, the EPA has established two types of drinking Water standards: primary and secondary. Primary standards are designed to protect public health by setting maximum permissible levels of potentially harmful substances in the water. Secondary standards are guidelines that apply to the esthetic aspects of drinking water, which do not pose a health risk (for example, color and odor). Primary standards are enforceable by law; secondary standards are not.

Most primary standards are specified as maximum contaminant levels, or MCLs; these are the enforceable limits. Primary standards may also be specified as treatment technique (TT) requirements, which are set for those contaminants that difficult or costly to measure; specific treatment processes (for example, filtration or corrosion control) may be required in lieu of an MCL to remove those contaminants. The EPA has also issued maximum contaminant level goals (MCLGs). A MCLG is a level of a contaminant not expected to cause any adverse health effects; it is a goal, not an enforceable standard. MCLs, which are revised periodically, are set as close to MCLGs as current technology and economics allow. MCLGs are set at zero for carcinogenic chemicals because there are no known safe levels for them.

1. Primary MCLs (1)

MCLs for potentially toxic or harmful substances reflect levels that can be safely consumed in water, taking into account exposure to substances from other sources. They are based on consumption of 2L (roughly 2 quarts) of water-based fluids every day for a lifetime. The states can establish MCLs that are more stringent than those set by the EPA.

Categories of primary contaminants include organic chemicals, inorganic chemicals, microorganisms, turbidity, and radionuclides. Except for some microorganisms and nitrate, water that exceeds the listed MCLs pose no immediate threat to public health. However, all these substances must be controlled because drinking water that exceeds the standards over long periods of time maybe harmful.

Organic Chemicals

Many synthetic organic chemicals (SOCs) are included in the primary regulations. Some of them (like benzene and carbon tetrachloride) readily become airborne and are known as volatile organic compounds (VOCs). Table 7-1 shows a partial allowable levels for several selected organic contaminants. As more is learned from research about the health effects of various contaminants, the number of regulated organics is likely to grow. Public drinking water supplies must be sampled and analyzed for organic chemicals at least every 3 years.

It is seen from Table 7-1 that extremely small concentrations can have public health significance. Levels are expressed in terms of mg/L; 1mg/L is equivalent to one part per million. The MCL for the insecticide lindane, for example, is 0.0002mg/L; this value can also be expressed as 0.2μg/L (micrograms per liter) and is equivalent to 0.2 parts per billion. [One part per billion is roughly proportional to the first 0.4m (about 1.3ft) of a trip to the moon.]

Selected primary standard MCLs and MCLGs for organic chemicals　　Table 7-1

Contaminant	Health effect	MCL (mg/L)	Typical source	MCLG
Aldicarb	Nervous system effects	0.003	Insecticide	0.001
Benzene	Possible cancer	0.005	Industrial chemical, pesticides, paints, plastics	0.000
Carbon tetrachloride	Possible cancer	0.005	Cleaning agents, industrial wastes	0.000
Chlordane	Possible cancer	0.002	Insecticide	0.000
Endrin	Nervous system, liver, kidney effects	0.002	Insecticide	0.002

Continued

Contaminant	Health effect	MCL (mg/L)	Typical source	MCLG
Heptachlor	Possible cancer	0.0004	Insecticide	0.00
Lindane	Nervous system, liver, kidney effects	0.0002	Insecticide	0.0002
Pentachlorophenol	Possible cancer, liver, kidney effects	0.001	Wood preservative	0.00
Styrene	Liver, Nervous system effects	0.1	Plastics, rubber, drug industry	0.1
Toluene	Kidney, Nervous system, liver, circulatory effects	1	Industrial solvent, gasoline additive, chemical manufacturing	1
Total trihalomethanes (TTHM)	Possible cancer risk	0.1	Chloroform, drinking water chlorination by-product	0.00
Trichloroethylene (TCE)	Possible cancer	0.005	Waste from disposal of dry cleaning materials and manufacture of pesticides, paints, waxes, metal degreaser	0.00
Vinyl chloride	Possible cancer	0.002	May leach from PVC pipe	0.00
Xylene	Liver, kidney, nervous system effects	10	Gasoline refining by-product, paint, ink, detergent	10

Inorganic Chemicals

Several inorganic substances (that is, containing no carbon), particularly heavy metals, are of public health importance. Some of these inorganics are listed in Table 7-2. Treated water is sampled and tested for inorganics at least once per year in public supplies. For most inorganics, MCLs are the same as the MCLGs, but the MCLG for lead is zero; the use of lead pipe and lead solder of flux for installation or repair of public water systems is no longer allowed in the United States.

Arsenic, a well-known poison, can contaminate drinking water supplies naturally if the raw water has been in contact with certain rocks and minerals; arsenic can also enter water sources from industrial and mining activities. It is found at higher levels in groundwater than in surface waters, such as lakes and rivers. The 0.05mg/L (50ppb) MCL was set by the EPA in 1975, based on a U.S. Public Health Service standard originally set in 1942. In 1999, the National Academy of Sciences completed a review of updated scientific data on arsenic and recommended that the EPA lower the MCL as soon as possible. Early in 2001, the EPA established a new standard of 0.01mg/L (10ppb), by 2006, all water utilities will have to comply with the new standard (10ppb of arsenic is equivalent to one teaspoon per 5ML, or 1.3mil gal, of water).

Nitrate levels above 10mg/L pose an immediate threat to children under 1 year of age. Excessive levels of nitrate can react with hemoglobin in blood to produce an anemic condition known as blue baby. Nitrates can enter water supplies naturally from soil and mineral deposits as well as from fertilizers and sewage pollution. The sources and health effects of other inorganic drinking water contaminants are summarized in Table 7-2.

Selected primary standard MCLs for inorganic chemicals Table 7-2

Contaminant	Health effect	MCL (mg/L)	Typical source
Arsenic	Nervous system effects	0.05	Geological, pesticide residues, industrial waste, smelter operations
Asbestos	Possible cancer	7 MFL①	Natural mineral deposits, air conditioning pipe
Barium	Circulatory system effects	2	Natural mineral deposits, paint
Cadmium	Kidney effects	0.005	Natural mineral deposits, metal finishing
Chromium	Liver, kidney, digestive system effects	0.1	Natural mineral deposits, metal finishing, textile and leather industries
Copper	Digestive system effects	TT②	Corrosion of household plumbing, natural deposits, wood preservatives
Cyanide	Nervous system effects	0.2	Electroplating, steel, plastics, fertilizer
Fluoride	Dental fluorosis, skeletal effects	4	Geological deposits, drinking water additive, aluminum industries
Lead	Nervous system and kidney effects, toxic to infants	TT	Corrosion of lead service lines and fixtures
Mercury	Kidney, nervous system effects	0.002	Industrial manufacturing, fungicide, natural mineral deposits
Nickel	Heart, liver effects	0.1	Electroplating, batteries, metal alloys
Nitrate	Blue-baby effect	10	Fertilizers, sewage, soil and mineral deposits
Selenium	Liver effects	0.05	Natural deposits, mining, smelting

Source: Environmental Protection Agency
①Million fibers per liter; ②Treatment technique.

Lead and Copper Rule

Treatment techniques have been set for lead and copper because the occurrence of these chemicals in drinking water usually results from corrosion of plumbing materials. All systems that do not meet the action level at the tap are required to improve corrosion control treatment to reduce the levels. The action level for lead is 0.015mg/L and for copper is 1.3mg/L.

2. Primary MCLs (2)

Microorganisms

This group of contaminants includes bacteria, viruses, and protozoa. The total coliform group of bacteria is used indicate the possible presence of pathogenic organisms. In testing for total coliforms, the number of monthly samples required is based on the population served and the size of the distribution system. The SDWA standards now require that coliforms not be found in more than 5 percent of the samples examined during a 1-month period. This is now known as the presence/absence concept; it replaces previous MCLs based on the number of coliforms detected in the sample. All coliform-positive samples have to be further tested for fecal coliforms (or E. coli), the presence of which is strong evidence of recent sewage contamination and indicates an urgent public health risk.

Legionella (which causes an upper respiratory disease), intestinal viruses, and Giardia lamblia (a protozoan cyst that causes intestinal illness) are also regulated under the SDWA, using treatment technique requirements. The Surface Water Treatment Rule (SWTR) requires that all public systems using surface water properly filter the water unless they can meet certain strict criteria. These systems must also disinfect the water, without exception, to kill disease-causing microorganisms.

The SWTR requires source water reductions of 99.9 percent for Giardia and virus concentrations, in lieu of setting MCLs. The MCLGs, though, are zero because ingestion of even very low numbers of those organisms can cause illness. Microbial standards will be improved and strengthened to control Cryptosporidium and other pathogens after the Enhanced Surface Treatment Rule is fully implemented, as required by the 1996 amendments to the SDWA.

Turbidity. The presence of suspended particles in the water is measured in nephelometric turbidity units (NTUs); NTUs measure the amount of fight scattered or reflected from the water. Turbidity testing is not required for groundwater sources.

Turbidity affects more than just the appearance of water; it can be a health hazard in drinking water and is therefore controlled as a primary contaminant. Turbidity interferes with disinfection by shielding microorganisms. MCLs for turbidity depend on the type of treatment used to clarify the water. Conventional and direct filtration systems must be monitored at least every 4h and have turbidity levels less than 0.5NTU. (Beginning in 2002, the EPA will require continuous monitoring of turbidity for surface water systems serving 10000 people or more.)

Radionuclides

Water can be contaminated with substances from nuclear facilities and radioactive wastes or from natural radioactive materials. The radioactive gas radon-222, for example, occurs in certain types of rock and can get into groundwater. People can be exposed to radon in water by drinking it, showering in it, or using it to wash dishes (The primary source of exposure to radon in the home is radon seeping out of the soil and into the basement air). Tests for radioactivity, which can cause a cancer risk, are required at least every 4 years. The MCL for radon in water is 300pCi/L (picocuries per liter) and the MCL for radium is 20pCi/L. Limits are also set for emitters of beta particles, photons, and alpha particles.

3. Secondary MCLs

Under the Secondary Drinking Water Standards, a range of concentrations is established for substances that affect only the esthetic qualities of drinking water (for example, taste, odor, and color), but have no direct effect on public health. Secondary standards are presented in Table 7-3. These standards are guidelines or suggestions related to the general acceptability of the water to consumers. States may adopt their own enforceable regulations governing these substances.

National secondary drinking water standards   Table 7-3

Contaminant or adverse effect	Suggested level	Contaminant effect
Aluminum	0.05~0.2mg/L	Discoloration of water
Chloride	205mg/L	Salty taste; corrosion of pipes
Color	15 color units	Visible tint
Copper	1.0mg/L	Metallic taste; blue-green staining of porcelain
Corrosivity	Noncorrosive	Metallic taste; fixture staining, corroded pipes (corrosive water can leach pipe materials, such as lead, into drinking water)
Fluoride	2.0mg/L	Dental fluorosis (a brownish discoloration of the teeth)
Foaming agents	0.5mg/L	Esthetic: frothy, cloudy, bitter taste, odor
Iron	0.3mg/L	Bitter metallic taste; staining of laundry, rusty color, sediment
Manganese	0.05mg/L	Taste; staining of laundry, black-to-brown color, black staining
Odor	3 threshold odor number	Rotten-egg, musty, or chemical smell
pH	6.5~8.5	Low pH: bitter metallic taste, corrosion High pH: slippery feel, soda taste, deposits
Silver	0.1mg/L	Argyria (discoloration of skin), graying of eyes
Sulfate	250mg/L	Salty taste; laxative effects
Total dissolved solids	500mg/L	Taste and possible relation between low hardness and cardiovascular disease; also an indicator of corrosivity (related to lead levels in water); can damage plumbing and limit effectiveness of soaps and detergents
Zinc	5mg/L	Metallic taste

Source: Environmental Protection Agency

4. Sampling Procedures

Sampling frequency requirements vary for each contaminant group as well as for individual contaminants within each group. They also depend on the population served and whether surface water or groundwater is used. The sampling frequencies can range from once every 4h (for turbidity) to once every 9 years (for asbestos). Detection of a contaminant above a certain level sometimes triggers increased sampling requirements, even if the MCL is not exceeded.

Most samples must be collected at points representative of water quality throughout the distribution system. Generally, drinking water samples are *fully flushed*. This means that the water has run for a long enough time to household plumbing. The exception to this is monitoring for copper and lead, for which a *first draw* sample is required at the consumer's taps, where contamination is more likely to occur.

Some samples must be collected in glass containers; others must be collected in plastic. Sample volumes vary for each contaminant, ranging from 100mL for a coliform sample to 1 L for some radionuclide sample. Certain samples must be kept cold for preservation; others can be delivered to the lab at ambient temperature. Sample bottles for VOCs must be filled to the top with no air space. The maximum allowable time between sample collection and analysis in the laboratory can range from 1d for coliforms to 1 year for a radionuclide sample. Details of all these requirements can be obtained from the laboratory doing the testing.

5. Record Keeping and Reporting

Good record-keeping procedures are important for proper operation of a public water system. They also provide data for future planning, public information, and legal protection. Under the requirements of the SDWA, each public system must maintain records of water quality test results, reports, and actions taken to correct deficiencies; depending on the type of record, it may have to be kept on file for up to 12 years.

Records of MCL analysis data for bacteria and chemicals are particularly important. Records should include the name, data, and place of sampling; the name of the technician who took the sample; the type of sample; and the place, data, method, and results of analysis.

Public Notification

The SDWA requires that public water systems submit reports to consumers as well as to an appropriate local regulatory agency. Results of routine sampling and testing must be sent to the agency every month. Also, the state must be notified within 48h of a violation of any of the primary regulatory MCLs. The requirement for routine sample and violation reports helps to ensure that water system deficiencies and potential heath hazards will be identified and corrected.

Public notification is required to advise consumers of the potential health hazards and to educate them about the importance of adequate financing and support for drinking water systems. The public is notified only of confirmed MCL violations by mail, newspaper, and radio and television broadcasts (for acute health risks). Timely public notice must describe the nature of the problem and include any steps people should take to protect their health. An illustration of a notice for violation of the nitrate MCL is shown in Figure 7-1.

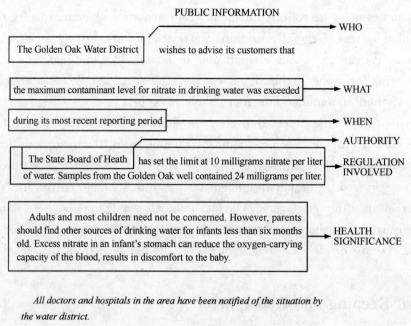

Figure 7-1 A typical MCL violation notice that would appear in local newspapers.

Vocabulary and Phrases

consumption　*n.* 消费；消耗量
purify　*v.* 使纯净；净化
potable　*adj.* 适于饮用的
　　　　n. 饮料
subsurface　*adj.* 地下的；液面下的；表面下的
coliform　*n.* 大肠菌
　　　　adj. 大肠菌的
calcium　*n.* 钙（元素符号 Ca）
iron　*n.* 铁；熨斗；(pl.)铁粉
　　　v. 烫平；熨；用铁包
pathogen　*n.* [微生物]病菌；病原体
aeration　*n.* 曝气；充气
toxic organic chemical　有毒有机化学物
activated carbon　活性炭

contaminate　*v.* 污染；玷污；沾染（放射性）
domestic sewage　生活污水；家庭污水
coagulation　*n.* 凝聚；混凝
flocculation　*n.* 絮凝(作用)
sedimentation　*n.* 沉淀；沉积；沉降；沉积学
filtration　*n.* 过滤(作用)
chlorination　*n.* 氯化(作用)；加氯处理
enact　*vt.* 制定法律；颁布；扮演
wholesome　*adj.* 卫生的；有益的；健康的
by-product　*n.* 副产品；副产物
primary standards　一级标准；基本标准
secondary standard　二级标准
maximum contaminant level　最大污染物

浓度；最大玷污物浓度
maximum contaminant level goal　最大玷污物浓度目标值
carcinogenic　adj. 致癌的
radionuclides　n. 放射性核素
nitrate　n. 硝酸盐；硝酸（根）；硝酸酯
synthetic organic chemicals（SOC）合成有机化学品
benzene　n. 苯
carbon tetrachloride　n. 四氯化碳
airborne　adj. 空气传播的；空运的；空降的
volatile organic compounds（VOCs）挥发性有机化合物
insecticide　n. 杀虫剂
lindane　n. 林丹；高丙体六六六（农药）
solder　n. 焊料；焊剂；焊锡
　　　　v. 焊接；锡焊
arsenic　n. 砷；砒霜
hemoglobin　n. 血色素；血红蛋白
anemic　adj. 贫血的；患贫血症的
blue baby　蓝婴
fertilizer　n. 化肥
aldicarb　n. 涕灭威（农药）
chlordane　n. 六氯
endrin　n. 异狄氏剂
heptachlor　n. 七氯（农药）
pentachlorophenol　n. 五氯苯酚
styrene　n. 苯乙烯
toluene　n. 甲苯
total trihalomethanes（TTHM）　n. 总三卤甲烷
trichloroethylene（TCE）　n. 三氯乙烯

vinyl chloride　n. 氯乙烯
xylene　n. 二甲苯
liver　n. 肝脏；居住者；生活优裕的人
kidney　n. 肾；（动物可食用的）腰子；个性；性格
pesticide　n. 杀虫剂；农药
wood preservative　木材防腐（剂的）
chloroform　n. 氯仿；三氯甲烷
chlorination by-product　氯化副产物
degreaser　n. 去（油）污剂；脱脂剂
gasoline refining　汽油精制
detergent　n. 清洁剂；去垢剂
asbestos　n. 石棉
barium　n. 钡
cadmium　n. 镉
chromium　n. 铬
copper　n. 铜；警察（美俚）
cyanide　n. 氰化物
fluoride　n. 氟化物
lead　n. 铅；导线
mercury　n. 水银；汞
nickel　n. 镍；镍币
selenium　n. 硒
digestive system　n. 消化系统
dental fluorosis　n. 牙氟中毒
smelter　n. 熔炉；冶金厂；冶炼工人
metal finishing　金属表面精整
electroplating　n. 电镀；电镀术
fixture　n. 配件；附件；固定器；夹紧装置；卫生器具
fungicide　n. 杀真菌剂
metal alloy　金属合金

Notes

　　Unit processes: Methods of treatment in which the removal of contaminants is brought about by chemical or biological reactions are known as *unit processes*.

　　Unit operations: Methods of treatment in which the application of physical forces predominate are known as *unit operations*.

Treatment technique: A specific treatment process (for example, filtration) required to remove certain contaminants from drinking water, used in lieu of an MCL for substances that are very difficult or costly to measure; TT.

Question and Exercises

1. Which water usually requires more extensive treatment for purification-groundwater or surface water? Why?
2. What is a *public water system*?
3. What do SDWA and MCL stand for?
4. What are the five general groups or types of contaminants that are controlled and limited under the SDWA? Briefly discuss the requirements for organic chemicals and inorganic chemicals.
5. What is the difference between the SDWA primary regulations and the secondary regulations?

Further Reading

Understanding the Safe Drinking Water Act

Overview

The Safe Drinking Water Act (SDWA) was originally passed by Congress in 1974 to protect public health by regulating the nation's public drinking water supply. The law was amended in 1986 and 1996 and requires many actions to protect drinking water and its sources, rivers, lakes, reservoirs, springs, and ground water wells. (SDWA does not regulate private wells that serve fewer than 25 individuals.) SDWA authorizes the United States Environmental Protection Agency (US EPA) to set national health-based standards for drinking water to protect against both naturally-occurring and man-made contaminants that may be found in drinking water. US EPA, states, and water systems then work together to make sure that these standards are met.

Millions of Americans receive high quality drinking water every day from their public water systems, (which may be publicly or privately owned). Nonetheless, drinking water safety cannot be taken for granted. There are a number of threats to drinking water: improperly disposed of chemicals; animal wastes; pesticides; human wastes; wastes injected deep underground; and naturally-occurring substances can all contaminate drinking water. Likewise, drinking water that is not properly treated or disinfected, or which travels through an improperly maintained distribution system, may also pose a health risk.

Originally, SDWA focused primarily on treatment as the means of providing safe drinking water at the tap. The 1996 amendments greatly enhanced the existing law by recognizing source water protection, operator training, funding for water system improvements, and public information as important components of safe drinking water. This approach ensures the quality of drinking water by protecting it from source to tap.

Roles and Responsibilities

SDWA applies to every public water system in the United States. There are currently more than 160000 public water systems providing water to almost all Americans at some time in their lives. The responsibility for making sure these public water systems provide safe drinking water is divided among US EPA, states, tribes, water systems, and the public. SDWA provides a framework in which these parties work together to protect this valuable resource.

US EPA sets national standards for drinking water based on sound science to protect against health risks, considering available technology and costs. These National Primary Drinking Water Regulations set enforceable maximum contaminant levels for particular contaminants in drinking water or required ways to treat water to remove contaminants. Each standard also includes requirements for water systems to test for contaminants in the water to make sure standards are achieved. In addition to setting these standards, US EPA provides guidance, assistance, and public information about drinking water, collects drinking water data, and oversees state drinking water programs.

The most direct oversight of water systems is conducted by state drinking water programs. States can apply to US EPA for "primacy", the authority to implement SDWA within their jurisdictions, if they can show that they will adopt standards at least as stringent as US EPA's and make sure water systems meet these standards. All states and territories, except Wyoming and the District of Columbia, have received primacy. While no Indian tribe has yet applied for and received primacy, four tribes currently receive "treatment as a state" status, and are eligible for primacy. States, or US EPA acting as a primacy agent, make sure water systems test for contaminants, review plans for water system improvements, conduct on-site inspections and sanitary surveys, provide training and technical assistance, and take action against water systems not meeting standards.

To ensure that drinking water is safe, SDWA sets up multiple barriers against pollution. These barriers include: source water protection, treatment, distribution system integrity, and public information. Public water systems are responsible for ensuring that contaminants in tap water do not exceed the standards. Water systems treat the water, and must test their water frequently for specified contaminants and report the results to states. If a water system is not meeting these standards, it is the water supplier's responsibility to notify its customers. Many water suppliers now are also required to prepare annual reports for their customers. The public is responsible for helping local water suppliers to set priorities, make decisions on funding and system improvements, and establish programs to protect drinking water sources. Water systems across the nation rely on citizen advisory committees, rate boards, volunteers, and civic leaders to actively protect this resource in every community in America.

Protection and Prevention

Essential components of safe drinking water include protection and prevention. States and water suppliers must conduct assessments of water sources to see where they may be vulnerable to contamination. Water systems may also voluntarily adopt programs to protect their watershed or wellhead and states can use legal authorities from other laws to prevent pollution. SDWA mandates that states have programs to certify water system operators and make sure that new water systems have the technical, financial, and managerial capacity to provide safe drinking water. SDWA also sets a framework for the Underground Injection Control (UIC) program to control the injection of wastes into ground water. U. S. EPA and states implement the UIC program, which sets standards for safe waste injection practices and bans certain types of injection altogether. All of these programs help prevent the contamination of drinking water.

Setting National Drinking Water Standards

U. S. EPA sets national standards for tap water that help ensure consistent quality in our nation's water supply. U. S. EPA prioritizes contaminants for potential regulation based on risk and how often they occur in water supplies. (To aid in this effort, certain water systems monitor for the presence of contaminants for which no national standards currently exist and collect information on their occurrence). U. S. EPA sets a health goal based on risk (including risks to the most sensitive people, e. g. , infants, children, pregnant women, the elderly, and the immuno-compromised). U. S. EPA then sets a legal limit for the contaminant in drinking water or a required treatment technique. This limit or treatment technique is set to be as close to the health goal as feasible. U. S. EPA also performs a cost-benefit analysis and obtains input from interested parties when setting standards. U. S. EPA is currently evaluating the risks from several specific health concerns, including: microbial contaminants (e. g. , *Cryptosporidium*); the byproducts of drinking water disinfection; radon; arsenic; and water systems that don't currently disinfect their water but get it from a potentially vulnerable ground water source.

Funding and Assistance

U. S. EPA provides grants to implement state drinking water programs, and to help each state set up a special fund to assist public water systems in financing the costs of improvements (called the drinking water state revolving fund). Small water systems are given special consideration, since small systems may have a more difficult time paying for system improvements due to their smaller customer base. Accordingly, U. S. EPA and states provide them with extra assistance (including training and funding) as well as allowing, on a case-by-case basis, alternate water treatments that are less expensive, but still protective of public health.

Compliance and Enforcement

National drinking water standards are legally enforceable, which means that both U. S. EPA and states can take enforcement actions against water systems not meeting safety standards. U. S. EPA and states may issue administrative orders, take legal actions, or fine utilities. U. S. EPA and states also work to increase water systems, understanding of, and compliance with, standards.

Public Information

SDWA recognizes that since everyone drinks water, everyone has the right to know what's in it and where it comes from. All water suppliers must notify consumers quickly when there is a serious problem with water quality. Water systems serving the same people year-round must provide annual consumer confidence reports on the source and quality of their tap water. States and U. S. EPA must prepare annual summary reports of water system compliance with drinking water safety standards and make these reports available to the public. The public must have a chance to be involved in developing source water assessment programs, state plans to use drinking water state revolving loan funds, state capacity development plans, and state operator certification programs.

1996 SDWA Amendment Highlights

Consumer Confidence Reports

All community water systems must prepare and distribute annual reports about the water they provide, including information on detected contaminants, possible health effects, and the water's source.

Cost-Benefit Analysis

U. S. EPA must conduct a thorough cost-benefit analysis for every new standard to determine whether the benefits of a drinking water standard justify the costs.

Drinking Water State Revolving Fund

States can use this fund to help water systems make infrastructure or management improvements or to help systems assess and protect their source water.

Microbial Contaminants and Disinfection Byproducts

U. S. EPA is required to strengthen protection for microbial contaminants, including Cryptosporidium, while strengthening control over the byproducts of chemical disinfection. The Stage 1 Disinfectants and Disinfection Byproducts Rule and the Interim Enhanced Surface Water Treatment Rule together address these risks.

Operator Certification

Water system operators must be certified to ensure that systems are operated safely. U. S. EPA issued guidelines in February 1999 specifying minimum standards for the certification and recertification of the operators of community and non-transient, noncommunity water systems. These guidelines apply to state Operator Certification Programs. All States are currently implementing EPA-approved operator certification programs.

Public Information & Consultation

SDWA emphasizes that consumers have a right to know what is in their drinking water, where it comes from, how it is treated, and how to help protect it. U. S. EPA distributes public information materials (through its Safe Drinking Water Hotline, Safewater web site, and Water Resource Center) and holds public meetings, working with states, tribes, water systems, and environmental and civic groups, to encourage public involvement.

Small Water Systems

Small water systems are given special consideration and resources under SDWA, to make sure they have the managerial, financial, and technical ability to comply with drinking water standards.

Source Water Assessment Programs

Every state must conduct an assessment of its sources of drinking water (rivers, lakes, reservoirs, springs, and ground water wells) to identify significant potential sources of contamination and to determine how susceptible the sources are to these threats.

Unit 8 Drinking Water Purification

Many aquifers and isolated surface water are high in water quality and may be pumped from the supply and transmission network directly to any number of end uses, including human consumption, irrigation, industrial processes, and fire control. However, clean water sources are the exception in many parts of the world, particularly regions where the population is dense or where there is heavy agricultural use. In these places, the water supply must receive varying degrees of treatment before distribution.

Impurities enter water as it moves through the atmosphere, across the earth's surface, and between soil particles in the ground. These background levels of impurities are often supplemented by human activities. Chemicals from industrial discharges and pathogenic organisms of human origin, if allowed to enter the water distribution system, may cause health problems. Excessive silt and other solids may make water aesthetically unpleasant and unsightly. Heavy metal pollution, including lead, zinc, and copper, may be caused by corrosion of the very pipes that carry water from its source to the consumer.

The method and degree of water treatment are site specific. Although water from public water systems is used for other uses, such as industrial consumption and firefighting, the cleanest water that is needed is for human consumption and therefore this requirement defines the degree of treatment. Thus, we focus on treatment techniques that produce *potable water*, or water that is both safe and pleasing.

A typical water treatment plant is diagrammed in Figure 8-1. It is designed to remove odors, color, and turbidity as well as bacteria and other contaminants. Raw water entering a treatment plant usually has significant turbidity caused by colloidal clay and silt particles. These particles carry an electrostatic charge that keeps them in continual motion and prevents them from colliding and sticking together. Chemicals like alum (aluminum sulfate) are added to the water both to neutralize the particles electrically and to aid in making them

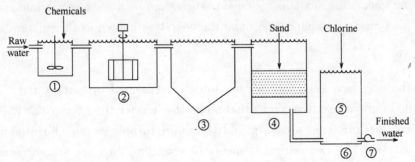

Figure 8-1 Movement of water through a water treatment facility
(1—Rapid mixing; 2—Flocculation; 3—Settling; 4—Sand filtration;
5—Chlorination; 6—Clear well storage; 7—Pumping to distribution system)

"sticky" so that they can coalesce and form large particles called flocs. This process is called coagulation and flocculation and is represented in stages 1 and 2 in Figure 8-1.

1. Coagulation and Flocculation

Naturally occurring silt particles suspended in water are difficult to remove because they are very small, often colloidal in size, and possess negative charges; thus they are prevented from coming together to form large particles that can more readily be settled out. However, the charged layers surrounding the particles form an energy barrier between the particles. The removal of these particles by settling requires reduction of this energy barrier by neutralizing the electric charges and by encouraging the particles to collide with each other. The charge neutralization is called coagulation, and the building of larger flocs from smaller particles is called flocculation.

One means of accomplishing this end is to add trivalent cations to the water. These ions would snuggle up to the negatively charged particle and, because they possess a stronger charge, displace the monovalent cations. The effect of this would be to reduce the net negative charge and thus lower the repulsive force seen in Figure 8-2. In this condition, the particles will not repel each other and, upon colliding, will stick together. A stable colloidal suspension has thus been made into an unstable colloidal suspension.

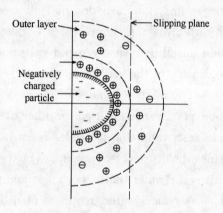

Figure 8-2 A colloidal particle is negatively charged and attracts positive counter ions to its surface.

The usual source of trivalent cations in water treatment is alum. Alum has an additional advantage in that some fraction of the aluminum may form aluminum oxides/hydroxides, represented simply as

$$Al^{3+} + 3OH^- \rightarrow Al(OH)_3 \downarrow$$

These complexes are sticky and heavy and will greatly assist in the clarification of the water in the settling tank if the unstable colloidal particles can be made to come into contact with the floc. This process is enhanced through the operation known as flocculation.

2. Settling

When the flocs have been formed they must be separated from the water. This is invariably done in gravity-settling tanks that allow the heavier-than-water particles to settle to the bottom. Settling tanks are designed to minimize turbulence and allow the particles to fall to the bottom. The two critical elements of a settling tank are the entrance and exit configurations because this is where turbulence is created and where settling can be disturbed. Figure 8-3 shows one type of entrance and exit configuration used for distributing the flow entering and leaving the water treatment settling tank.

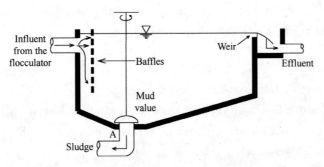

Figure 8-3 Settling tank used in water treatment.

The particles settling to the bottom become what is known as *alum sludge*. Alum sludge is not very biodegradable and will not decompose at the bottom of the tank. After some time, usually several weeks, the accumulation of alum sludge at the bottom of the tank is such that it has to be removed. Typically, the sludge exits through a mud valve at the bottom and is wasted either into a sewer or to a sludge holding and drying pond. In contrast to alum sludge from water treatment, sludges collected in wastewater treatment plants can remain in the bottom of the settling tanks only a matter of hours before starting to produce odoriferous gases and floating some of the solids.

Settling tanks can be analyzed by assuming an ideal settling tank. In this tank an imaginary column of water enters at one end, moves through the tank, and exits at the other end. Solid particles within this column settle to the bottom, and all those that reach the bottom before the column reaches the far end of the tank are assumed to be removed (settled out). If a solid particle enters the tank at the top of the column and settles at a velocity of v_o, it should have settled to the bottom as the imaginary column of water exits the tank, having moved through the tank at velocity v.

Consider now a particle entering the settling tank at the water surface. This particle has a settling velocity of v_o and a horizontal velocity v. In other words, the particle is just barely removed (it hits the bottom at the last instant). Note that if the same particle enters the settling tank at any other height, such as height h, its trajectory always carries it to the bottom. Particles having this velocity are termed critical particles in that particles with lower settling velocities are not all removed. For example, the particle having velocity v_s, entering the settling tank at the surface, will not hit the bottom and escape the tank. However, if this same particle enters at some height h, it should just barely hit the bottom and be removed. Any of these particles having a velocity v_s that happen to enter the settling tank at height h or lower are thus removed, and those entering above h are not. Since the particles entering the settling tank are assumed to be equally distributed, the proportion of those particles with a velocity of v_s removed is equal to h/H, where H is the height of the settling tank.

The retention time of a settling tank is the amount of time necessary to fill the tank at some given flow rate, or the amount of time an average water particle spends in the tank. Mathematically, the residence time is defined as:

$$\bar{t} = \frac{V}{Q}$$

or the volume divided by the flow rate.

Although the water leaving a settling tank is essentially clear, it still may contain some turbidity and may carry pathogenic organisms. Thus, additional polishing is performed with a rapid sand filter.

3. Filtration

Even with the help of chemical coagulation, sedimentation by gravity is not sufficient to remove all the suspended impurities from water. About 5 percent of the suspended solids may still remain as nonsettleable floc particles. These remaining flocs can cause noticeable turbidity and may shied microorganisms from the subsequent disinfection process. To produce crystal clear potable water that satisfies the SDWA requirement of $0.5NTU$ (the MCL for turbidity), an additional treatment step following coagulation and sedimentation is typically needed.

As we know, groundwater is usually of excellent quality. This is primarily because of the natural filtration that occurs in the layers of soil through which the water slowly moves. Picture the extremely clear water that bubbles up from "underground streams" as spring water. Soil particles help filter the groundwater, and this principle is applied to water treatment. In almost all cases, filtration is performed by a rapid sand filter.

As the sand filter removes the impurities, the sand grains get dirty and must be cleaned. The process of rapid sand filtration therefore involves two operations: filtration and backwashing. Figure 8-4 shows a cutaway of a slightly simplified version of the rapid sand filter. Water from the settling basins enters the filter and seeps through the sand and gravel bed, through a false floor, and out into a clear well that stores the finished water. Valves A and C are open during filtration.

The cleaning process is done by reversing the flow of water through the filter. The operator first shuts off the flow of water to the filter, closing valves A and C, then opens valves D and B, which allow wash water (clean water stored in an elevated tank or pumped from the clear well) to enter below the filter bed. This rush of water forces the sand and gravel bed to expand and jolts individual sand particles into motion, rubbing against their neighbors. The light colloidal material trapped within the filter is released and escapes with the wash water. After 5 to 10 minutes of washing, the wash water is shut off and filtration is resumed.

Filter beds might contain filtration media other than sand. Crushed coal, for example, is often used in combination with sand to produce a dual media filter which can achieve greater removal efficiencies.

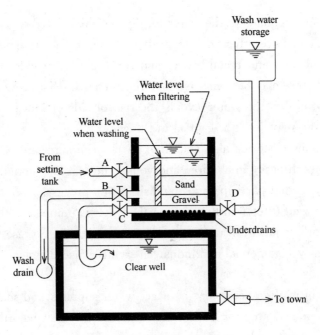

Figure 8-4　Rapid sand filter

4. Disinfection

After filtration, the finished water is disinfected, often with chlorine (step 5 in Figure 8-1), called chlorination. Disinfection kills the remaining microorganisms in the water, some of which may be pathogenic. Chlorine gas from bottles or drums is fed in correct proportions to the water to obtain a desired level of chlorine in the finished water. When chlorine comes in contact with organic matter, including microorganisms, it oxidizes this material and is in turn itself reduced. Chlorine gas is rapidly hydrolyzed in water to form hypochlorous acid, by the reaction

$$Cl_2 + H_2O \rightleftharpoons HOCl + H^+ + Cl^- \qquad (8-1)$$

The hypochlorous acid itself ionizes further to the hypochlorous ion

$$HOCl \rightleftharpoons H^+ + OCl^- \qquad (8-2)$$

At the temperatures usually found in water supply systems, the hydrolysis of chlorine is usually complete in a matter of seconds, while the ionization of HOCl is instantaneous. Both HOCl and OCl$^-$ are effective disinfectants and are called free available chlorine in water. Free available chlorine kills pathogenic bacteria and thus disinfects the water. Many water plant operators prefer to maintain a residual of chlorine in the water; that is, have some available chlorine left over once the chlorine has reacted with the currently available organics. Then, if organic matter like bacteria enters the distribution system, there is sufficient chlorine present to eliminate this potential health hazard. Tasting chlorine in drinking water indicates that the water has maintained its chlorine residual.

Chlorine may have adverse secondary effects. It is thought to combine with trace amounts of organic compounds in the water to produce chlorinated organic compounds that may be carcinogenic or have other adverse health effects. Some studies have shown an association between bladder and rectal cancer and consumption of chlorinated drinking water, indicating that there may be some risk of carcinogenesis. Disinfection by ozonation, the bubbling of ozone through the water, avoids the risk of side effects from chlorination, but ozone disinfection does not leave a residual in the water.

A number of municipalities also add fluorine to drinking water because it has been shown to prevent tooth decay in children and young adults. The amount of fluorine added is so small that it does not participate in the disinfection process.

From the clear well (step 6 in Figure 8-1) the water is pumped to the distribution system, a closed network of pipes, all under pressure. Users tap into these pipes to obtain potable water. Similarly, commercial and industrial facilities use the clean water for a variety of applications.

Water treatment is often necessary if surface water supplies, and sometimes groundwater supplies, are to be available for human use. Because the vast majority of cities use one water distribution system for households, industries, and fire control, large quantities of water often must be made available to satisfy the highest use, which is usually drinking water.

But does it make sense to produce drinkable water and then use it for other purposes, such as lawn irrigation? Growing demands for water have prompted serious consideration of dual water supplies: one high-quality supply for drinking and other personal use and one of lower quality, perhaps reclaimed from wastewater, for urban irrigation, firefighting, and similar applications. The growing use of bottled water for drinking is an example of a dual supply. In many parts of the world the population concentrations have stretched the supply of potable water to the limit, and either dual systems will be necessary or people will not be allowed to move to areas that have limited water supply. The availability of potable water often dictates land use and the migration of populations.

Vocabulary and Phrases

background level （自然）背景值；（自然）本底值
electrostatic adj. 静电的；静电学的
collide vi. 碰撞；抵触
aluminum sulfate 硫酸铝
neutralize vt. 使中和；平衡；使失效；抑制
coalesce vi. 聚结；凝聚；结合
floc n. 絮凝体；絮状体；絮体
energy barrier 能障；能垒
trivalent adj. 三价的
repulsive force 排斥力

aluminum oxide 氧化铝
aluminum hydroxide 氢氧化铝
turbulence n. 湍流；紊流；湍动（性）；紊动（性）
alum sludge 铝钒污泥；硫酸铝污泥
mud valve 泥阀
sludge n. 泥渣；污泥；淤渣
odoriferous adj. 散发气味的；臭的
floating adj. 漂浮的；浮动的；浮游的
ideal settling tank 理想沉淀池
trajectory n. （射线的）轨道；弹道；轨线

critical *adj.* 临界的
retention time 滞水时间；滞洪时间；停留时间；保留时间
residence time 停留时间；滞留时间
polishing *n.* （水的）最终处理（净化）；精（处）理
rapid sand filter 快滤池
grain *n.* 颗粒；晶粒；谷粒；（木材）纹理
backwashing *n.* 反洗；反冲洗
  adj. 反洗的
cutaway *n. & adj.* 剖面的
gravel *n.* 砾石；卵石
false floor 假底
finished water 成品水
jolt *v. & n.* 振动；摇动
rub *v.* (rubbed; rubbing) 摩擦；擦（净，亮，光）；磨损；研磨
  n. （磨）擦；磨损处

crushed coal 碎煤
dual media filter 双层滤料滤池
pathogenic *adj.* 致病的，病原的，发病的
hydrolyze *v.* （进行）水解
hydrolysis *n.* 水解；水解作用
hypochlorous acid 次氯酸
free available chlorine 游离性有效氯
residual *n.* 剩余；残差
  adj. 剩余的；残留的
chlorinated organic compound 氯化有机物
bladder *n.* 气囊；膀胱
rectal *adj.* 直肠的
carcinogenesis *n.* 致癌（作用）
bubbling *n.* 冒泡；沸腾
fluorine *n.* 氟（9号元素，符号 F）
dual water supply 双质供水（系统）

Notes

Rapid filter：A water purification system that removes suspended solids as the water flows through a granular bed of sand or other material and that is cleaned by backwashing the filter bed.
Floc：A particle large enough to settle out of water or wastewater, formed during the coagulation-flocculation process; also, settleable particles of activated sludge.

Question and Exercises

1. For the house or dormitory where you live, suggest which water uses require potable water and which require a lower water quality. What minimum requirements must be met for the lower water quality supply?

2. What is the difference between coagulation and flocculation?

3. Propose some use and disposal options for water treatment sludges collected in the settling tanks following flocculation basins. Remember that these sludges consist mostly of aluminum oxides and clay.

4. A settling tank has dimensions of 3m deep, 12m long, and 5m wide. The flow entering the tank is $10m^3/min$.
  (a) What is the residence time in this tank?
  (b) What is the velocity of a *critical particle* in this settling tank?

5. A settling tank is to settle out a slurry that has particles with a settling velocity of 0.05m/min. An engineer decides that a tank of 5m with, 16m long, and 3m deep is ade-

quate. What is the maximum allowable water flow rate into the tank if 100% removal of particles is to be achieved?

6. Briefly describe the configuration of a typical rapid filter.

7. Briefly describe the operation of a typical rapid filter.

8. What is considered to be the most important water treatment process with respect to preventing the spread of waterborne disease? Is there any potential harmful side effect from chlorination?

9. What are the effective disinfectants that kill bacteria when chlorine is added to water? What is the purpose of chlorine residual in chlorine disinfection?

Further Reading (1)

Other Treatment Processes

1. Boiling

In an emergency, boiling is the best way to purify water that is unsafe because of the presence of protozoan parasites or bacteria.

If the water is cloudy, it should be filtered before boiling. Filters designed for use when camping, coffee filters, towels (paper or cotton), cheesecloth, or a cotton plug in a funnel are effective ways to filter cloudy water.

Place the water in a clean container and bring it to a full boil and continue boiling for at least 3 minutes (covering the container will help reduce evaporation). If you are more than 5000 feet above sea level, you must increase the boiling time to at least 5 minutes (plus about a minute for every additional 1000 feet). Boiled water should be kept covered while cooling.

The advantages of Boiling Water include:
- Pathogens that might be lurking in your water will be killed if the water is boiled long enough.
- Boiling will also drive out some of the Volatile Organic Compounds (VOCs) that might also be in the water. This method works well to make water that is contaminated with living organisms safe to drink, but because of the inconvenience, boiling is not routinely used to purify drinking water except in emergencies.

The disadvantages of Boiling Water include:
- Boiling should not be used when toxic metals, chemicals (lead, mercury, asbestos, pesticides, solvents, etc.), or nitrates have contaminated the water.
- Boiling may concentrate any harmful contaminants that do not vaporize as the relatively pure water vapor boils off.
- Energy is needed to boil the water.

2. Distillation

In many ways, distillation is the reverse of boiling. To remove impurities from water by distillation, the water is usually boiled in a chamber causing water to vaporize, and the pure (or mostly pure) steam leaves the non volatile contaminants behind. The steam moves to a different part of the unit and is cooled until it condenses back into liquid water. The resulting distillate drips into a storage container.

Salts, sediment, metals-anything that won't boil or evaporate-remain in the distiller and must be removed. Volatile organic compounds (VOCs) are a good example of a contaminant that will evaporate and condense with the water vapor. A vapor trap, carbon filter, or other device must be used along with a distiller to ensure the more complete removal of contaminants.

The advantages of distillation include (Figure 8-5):

- A good distillation unit produces very pure water. This is one of the few practical ways to remove nitrates, chloride, and other salts that carbon filtration can not remove.
- Distillation also removes pathogens in the water, mostly by killing and leaving them behind when the water vapor evaporates. If the water is boiled, or heated just short of boiling, pathogens would also be killed.

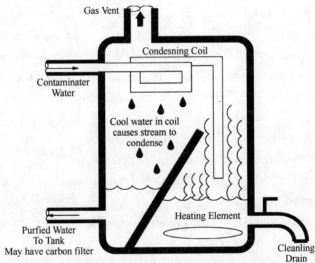

Figure 8-5 Distillation process

- As long as the distiller is kept clean and is working properly the high quality of treated water will be very consistent regardless of the incoming water-no drop in quality over time.
- No filter cartridges to replace, unless a carbon filter is used to remove volatile organic compounds.

The disadvantages of distillation include:

- Distillation takes time to purify the water, It can take two to five hours to make a gallon of distilled water.
- Distillers uses electricity all the time the unit is operating.
- Distillers requires periodic cleaning of the boiler, condensation compartment, and stor-

age tank.
- Countertop Distillation is one of the more expensive home water treatment methods, using $0.25 to $0.35 of electrical energy per gallon of distilled water produced-depending on local electricity costs. The cost of ownership is high because you not only have the initial cost of the distillation unit to consider, but you also must pay for the electrical energy for each gallon of water produced.
- Most home distillation units require electricity, and will not function in an emergency situation when electrical power is not available.

3. Water Softeners and Deionizer

Water softeners operate on the ion exchange process. In this process, water passes through a media bed, usually sulfonated polystyrene beads. The beads are supersaturated with sodium (a positive ion). The ion exchange process takes place as hard water passes through the softening material. The hardness minerals (positively charged Calcium and Magnesium ions) attach themselves to the resin beads while sodium on the resin beads is released simultaneously into the water. When the resin becomes saturated with calcium and magnesium, it must be recharged. The recharging is done by passing a concentrated salt (brine) solution through the resin. The concentrated sodium replaces the trapped calcium and magnesium ions which are discharged in the waste water. Softened water is not recommended for watering plants, lawns, and gardens due to its elevated sodium content.

Several factors govern the efficiency of a cationic softener:
- Type & quality of resin used;
- Amount of salt per cubic foot of resin for regeneration;
- Brine concentration in the resin bed during regeneration;
- Brine flow rate through the resin bed (contact time) during regeneration;
- Raw water hardness;
- Raw water temperature-softeners perform better at higher temperatures; and
- Optimal flow rate of hard water through the resin bed.

Although not commonly used, potassium chloride can be used to create the salt brine for softeners designed to use KCl. In that case potassium rather than sodium is exchanged with calcium and magnesium. Before selecting an ion exchange water softener, test water for hardness and iron content. When selecting a water softener, the regeneration control system, the hardness removal capacity, and the iron limitations are three important elements to consider.

The advantages of water softeners include:
- The nuisance factor of hard water is reduced.
- some other cations like barium, radium and iron may be reduced depending on the manufacturer's specifications.

The disadvantages of water softeners include:
- The process of regenerating the ion exchange bed dumps salt water into the environment.

- The elevated sodium concentration of most softened water can affect the taste and may not be good for people on low sodium diets, although sodium concentrations are typically quite low relative to sodium levels in most food.
- Cation exchange does not reduce the level of anions (like nitrates), or biological contaminants (bacteria, viruses, cysts); nor does the process reduce the levels of most organic compounds.
- Typically, approximately 50 gallons of rinse water per cubic foot of resin is required to totally remove hardness and excess salt from the resin after each regeneration.

Water deionizers use both Cation and Anion Exchange to exchange both positive and negative ions with H^+ or OH^- ions respectively, leading to completely demineralized water. Deionizers do not remove uncharged compounds from water, and are often used in the final purification stages of producing completely pure water for medical, research, and industrial needs.

A potential problem with deionizers is that colonies of microorganisms can become established and proliferate on the nutrient-rich surfaces of the resin. When not regularly sanitized or regenerated, ion-exchange resins can contaminate drinking water with bacteria.

4. Activated Carbon Filters:

Activated carbon (AC) is particles of carbon that have been treated to increase their surface area and increase their ability to adsorb a wide range of contaminants, activated carbon is particularly good at adsorbing organic compounds. You will find two basic kinds of carbon filters Granular Activated Carbon (GAC) and Solid Block Activated Carbon (SBAC).

Contaminant reduction in AC filters takes place by two processes, physical removal of contaminant particles, blocking any that are too large to pass through the pores (obviously, filters with smaller pores are more effective), and a process called adsorption by which a variety of dissolved contaminants are attracted to and held (adsorbed) on the surface of the carbon particles. The characteristics of the carbon material (particle and pore size, surface area, surface chemistry, density, and hardness) influence the efficiency of adsorption.

AC is a highly porous material; therefore, it has an extremely high surface area for contaminant adsorption. One reference mentions "The equivalent surface area of 1 pound of AC ranges from 60 to 150 acres (over 3 football fields)". Another article states, "Under a scanning electron microscope the activated carbon looks like a porous bath sponge. This high concentration of pores within a relatively small volume produces a material with a phenomenal surface area: one tea spoon of activated carbon would exhibit a surface area equivalent to that of a football field."

AC is made of tiny clusters of carbon atoms stacked upon one another. The carbon source is a variety of materials, such as peanut shells, coconut husks, or coal. The raw carbon source is slowly heated in the absence of air to produce a high carbon material. The carbon is activated by passing oxidizing gases through the material at extremely high temperatures. The activation process produces the pores that result in such high adsorptive properties.

The adsorption process depends on the following factors: (1) physical properties of the AC, such as pore size distribution and surface area; (2) the chemical nature of the carbon source, or the amount of oxygen and hydrogen associated with it; (3) chemical composition and concentration of the contaminant; (4) the temperature and pH of the water; and (5) the flow rate or time exposure of water to AC.

- The effectiveness of carbon filters to reduce contaminants is affected by the factors affecting adsorption listed above and three additional characteristics of the filter, contact time between the water and the carbon material, the amount of carbon in the filter, and pore size.

The length of contact time between the water and the carbon material, governed by the rate of water flow and the amount/volume of activated carbon, has a significant effect on adsorption of contaminants. More contact time results in greater adsorption.

The amount of carbon present in a cartridge or filter affects the amount and type of contaminant removed. Less carbon is required to remove taste-and odor-producing chemicals than to remove trihalomethanes.

Pore size characteristics will be discussed in greater detail on the GAC and SBAC, but GAC filters contain loose granules of activated carbon while in SBAC filters, the activated carbon is in the form of very small particles bound into a solid, matrix with very small pores.

- Because of the filter characteristics discussed above, the most effective point of Use activated carbon filters are large SBAC filtration systems, and the least effective are the small, pour-through pitcher filters.

- Activated carbon filter cartridges will, over time, become less effective at reducing contaminants as the pores clog with particles (slowing water flow) and the adsorptive surfaces in the pores become filled with contaminants (typically not affecting flow rate). There is often no noticeable indication that a carbon filter is no longer removing contaminants, so it is important to replace the cartridge according to the manufacturer's instructions. The overall water quality (turbidity or presence of other contaminants) also affects the capacity of activated carbon to adsorb a specific contaminant.

- It is important to note, particularly when using counter-top and faucet-mount carbon filtration systems, that hot water should never be run through a carbon filter. Perhaps more importantly, hot water will tend to release trapped contaminants into the water flow potentially making the water coming out of the filter more contaminated than the water going in.

Granular Activated Carbon (GAC)

In this type of filter, water flows through a bed of loose activated carbon granules which trap some particulate matter and remove some chlorine, organic contaminants, and undesirable tastes and odors. The three main problems associated with GAC filters are: channeling, dumping, and an inherently large pore size. Most of the disadvantages discussed below are not the fault of the activated carbon filtration media, rather, the problem is the design of the filters and the use of loose granules of activated carbon.

The advantages of GAC filters include:
- Simple GAC filters are primarily used for aesthetic water treatment, since they can reduce chlorine and particulate matter as well as improve the taste and odor of the water.
- Loose granules of carbon do not restrict the water flow to the extent of Solid Block Activated Carbon (SBAC) filters. This enables them to be used in situations, like whole house filters, where maintaining a good water flow rate and pressure is important.
- Simple, economical maintenance. Typically an inexpensive filter cartridge needs to be changed every few months to a year, depending on water use and the manufacturer's recommendation.
- GAC filters do not require electricity, nor do they waste water.
- Many dissolved minerals are not removed by activated carbon. In the case of calcium, magnesium, potassium, and other beneficial minerals, the taste of the water can be improved and some (usually small) nutrient value can be gained from the water.

The bottom line is that GAC filters are effective and valuable water treatment devices, but their limitations always need to be considered. A uniform flow rate, not to exceed the manufacture's specifications, needs to be maintained for optimal performance, and the filter cartridge must be changed after treating the number of gallons the filter is rated for.

The disadvantages of GAC filters include:
- Water flowing through the filter is able to "channel" around the carbon granules and avoid filtration. Water seeks the path of least resistance. When it flows through a bed of loose carbon granules, it can carve a channel where it can flow freely with little resistance. Water flowing through the channel does not come in contact with the filtration medium. The water continues to flow, however, so you do not realize that your filter has failed, you get water, but it is not completely filtered.
- Pockets of contaminated water can form in a loose bed of carbon granules. With changes in water pressure and flow rates, these pockets can collapse, "dumping" the contaminated water through the filter into the "filtered" flow.
- Since the carbon granules are fairly large (0.1mm to 1mm in one popular pitcher filter), the effective pore size of the filter is relatively large (20~30 microns or larger). GAC filters, by themselves, can not bacteria.
- As described above, hot water should never be run through a carbon filter.
- Also, if you think of a bed of charcoal that traps an occasional bacterium, picks up a bit of organic material, and removes the chlorine from the water, you can see how these fil-

ters might become breeding grounds for the bacteria they trap. You will see warnings about GAC filters suggesting you run water through them for a few minutes each morning to flush out any bacteria.
- Unless the filter plugs up or you notice an odor in the "filtered water", it may be difficult to know when the filter has become saturated with contaminants and ineffective. That is why it is necessary to change filter cartridges according to the manufacturer's recommendation.

Solid Block Activated Carbon (SBAC): Activated carbon is the primary raw material in solid carbon block filters; but instead of carbon granules comprising the filtration medium, the carbon has been specially treated, compressed, and bonded to form a uniform matrix. The effective pore size can be very small (0.5~1 micron). SBAC, like all filter cartridges, eventually become plugged or saturated by contaminants and must be changed according to manufacturer's specifications. Depending on the manufacturer, the filters can be designed to better reduce specific contaminants like arsenic, MTBE, etc.

The advantages of SBAC filters include:
- Provide a larger surface area for adsorption to take place than Granular Activated Carbon (GAC) filters for better contaminant reduction.
- Provide a longer contact time with the activated carbon for more complete contaminant reduction.
- Provide a small pore size to physically trap particulates. If the pore size is small enough, around 0.5 micron or smaller, bacteria that become trapped in the pores do not have enough room to multiply, eliminating a problem common to GAC filters.
- Completely eliminate the channeling and dumping problems associated with GAC filters.
- SBAC filters are useful in emergency situations where water pressure and electricity might be lost. They do not require electricity to be completely effective, and water can even be siphoned through them.
- SBAC filters do not waste water like reverse osmosis.
- Many dissolved minerals are not removed by activated carbon. In the case of calcium, magnesium, potassium, and other beneficial minerals, the taste of the water can be improved and some (usually small) nutrient value can be gained from the water.
- Simple, economical maintenance. Typically an inexpensive filter cartridge needs to be changed every few months to a year, depending on water use and the manufacturer's recommendation.
- This combination of features provides the potential for greater adsorption of many different chemicals (pesticides, herbicides, chlorine, chlorine byproducts, etc.) and greater particulate filtration of parasitic cysts, asbestos, etc. than many other purification process available. By using other specialized materials along with specially prepared ac-

tivated carbon, customized SBAC filters can be produced for specific applications or to achieve greater capacity ratings for certain contaminants like lead, mercury, arsenic, etc.

The disadvantages of SBAC filters include:

- SBAC filters, like all activated carbon filters, do not naturally reduce the levels of soluble salts (including nitrates), fluoride, and some other potentially harmful minerals like arsenic (unless specially designed) and cadmium. If these contaminants are present in your water, reverse osmosis would usually be the most economical alternative followed by distillation.
- As described above, hot water should never be run through a carbon filter.
- As SBAC filters remove contaminants from the water they gradually lose effectiveness until they are no longer able to adsorb the contaminants. There is no easy way to determine when a filter is nearing the end of its effective life except that the "filtered" water eventually begins to taste and smell like the unfiltered water. The manufacturer's guidelines for changing filter cartridges should always be followed.

5. Ultra Violet Light

Water passes through a clear chamber where it is exposed to Ultra Violet (UV) Light. UV light effectively destroys bacteria and viruses. However, how well the UV system works depends on the energy dose that the organism absorbs.

The advantages of using UV include:

- No known toxic or significant nontoxic byproducts introduced.
- Removes some organic contaminants.
- Leaves no smell or taste in the treated water.
- Requires very little contact time (seconds versus minutes for chemical disinfection).
- Improves the taste of water because some organic contaminants and nuisance microorganisms are destroyed.
- Many pathogenic microorganisms are killed or rendered inactive.
- Does not affect minerals in water.

The disadvantages of using UV include:

- UV radiation is not suitable for water with high levels of suspended solids, turbidity, color, or soluble organic matter. These materials can react with UV radiation, and reduce disinfection performance. Turbidity makes it difficult for radiation to penetrate water and pathogens can be "shadowed", protecting them from the light.
- UV light is not effective against any non-living contaminant, lead, asbestos, many organic chemicals, chlorine, etc.
- Tough cryptosporidia cysts are fairly resistant to UV light.
- Requires electricity to operate. In an emergency situation when the power is out, the purification will not work.

UV is typically used as a final purification stage on some filtration systems. If you are concerned about removing contaminants in addition to bacteria and viruses, you would still need to use a quality carbon filter or reverse osmosis system in addition to the UV system.

6. Ozonation

The formation of oxygen into ozone occurs with the use of energy. This process is carried out by an electric discharge field as in the CD-type ozone generators (corona discharge simulation of the lightning), or by ultraviolet radiation as in UV-type ozone generators (simulation of the ultra-violet rays from the sun). In addition to these commercial methods, ozone may also be made through electrolytic and chemical reactions.

Ozone is a naturally occurring component of fresh air. It can be produced by the ultraviolet rays of the sun reacting with the Earth's upper atmosphere (which creates a protective ozone layer), by lightning, or it can be created artificially with an ozone generator.

The ozone molecule contains three oxygen atoms whereas the normal oxygen molecule contains only two. Ozone is a very reactive and unstable gas with a short half-life before it reverts back to oxygen. Ozone is the most powerful and rapid acting oxidizer man can produce, and will oxidize all bacteria, mold and yeast spores, organic material and viruses given sufficient exposure.

The advantages of using Ozone include:
- Ozone is primarily a disinfectant that effectively kills biological contaminants.
- Ozone also oxidizes and precipitates iron, sulfur, and manganese so they can be filtered out of solution.
- Ozone will oxidize and break down many organic chemicals including many that cause odor and taste problems.
- Ozonation produces no taste or odor in the water.
- Since ozone is made of oxygen and reverts to pure oxygen, it vanishes without trace once it has been used. In the home, this does not matter much, but when water companies use ozone to disinfect the water there is no residual disinfectant, so chlorine or another disinfectant must be added to minimize microbial growth during storage and distribution.

The disadvantages of using Ozone include:
- Ozone treatment can create undesirable byproducts that can be harmful to health if they are not controlled (e. g. , formaldehyde and bromate).
- The process of creating ozone in the home requires electricity. In an emergency with loss of power, this treatment will not work.
- Ozone is not effective at removing dissolved minerals and salts.

Caution-The effectiveness of the process is dependent, on good mixing of ozone with the water, and ozone does not dissolve particularly well, so a well designed system that exposes all the water to the ozone is important.

In the home, ozone is often combined with activated carbon filtration to achieve a more complete water treatment.

Further Reading (2)

Membrane Filtration Processes

Filtration involves the separation (removal) of particulate and colloidal matter from a liquid. In membrane filtration the range of particle sizes is extended to include dissolved constituents (typically 0.0001 to 1.0 μm). The role of the membrane, as shown in Figure 8-6, is to serve as a selective barrier that will allow the passage of certain constituents and will retain other constituents found in the liquid. To introduce membrane technologies and their application, the following subjects are considered: (1) membrane process terminology, (2) membrane classification, (3) membrane configurations, (4) application of membrane technologies, (5) electrodialysis, (6) the need for pilot-plant studies, and (7) the disposal of concentrated waste streams.

Terminology used to describe membrane processes Table 8-1

Term	Description
Brine	Concentrate stream containing total dissolved solids greater than 36000mg/L
Concentration, retentate, retained phase, reject, residual stream	The portion of the feed stream that does not pass through the membrane that contains higher TDS than the feed stream
Feed stream, feedwater	Input stream to the membrane array
Flux	Mass or volume rate of transfer through the membrane surface
Fouling	Deposition of existing solid material in the element on the feed stream of the membrane. Fouling can be either reversible or irreversible
Lumen	The interior of a hollow fiber membrane
Mass transfer coefficient (MTC)	Mass or volume unit transfer through membrane based on driving force
Membrane element	A single membrane unit containing a bound group of spiral-wound or hollow fine-fiber membranes to provide a nominal surface area
Module	A complete unit comprised of the membranes, the pressure support structure for the membranes, the feed inlet and outlet permeate and retentate ports, and an overall support structure
Molecular weight cutoff (MWCO)	The molecular weight of the smallest material rejected by the membrane, usually expressed in Daltons (D)
Permeate, product, permeating stream	The portion of the feed stream that passes through the membrane that contains lower TDS than the feed stream
Reject ion	Percent solute concentration reduction of permeate stream relative to feed stream
Pressure vessel	A single tube that contains several membrane elements in series
Scaling	Precipitation of solids in the element due to solute concentration on the feed stream of the membrane
Size exclusion	Removal of particles by sieving
Solvent	Liquid containing dissolved constituents (TDS), usually water
Solute	Dissolved constituents (TDS) in raw, feed, permeate, and concentrate streams
Stage or bank	Pressure vessels arranged in parallel
System arrays	Number of arrays needed to produce the required plant flow
Train or array	Multiple interconnected stages in series

1. Membrane Process Terminology

Terms commonly encountered when considering the application of membrane processes are summarized in Table 8-1. Referring to Figure 8-6 and Table 8-1, the influent to the membrane module is known as the *feed stream* (also known as feedwater). The liquid that passes through the semipermeable membrane is known as *permeate* (also known as the product stream or permeating stream) and the liquid containing the retained constituents is known as the *concentrate* (also known as the retentate, reject, retained phase, or waste stream). The rate at which the permeate flows through the membrane is known as the rate of *flux*, typically expressed as $kg/(m^2 \cdot d)[gal/(ft^2 \cdot d)]$.

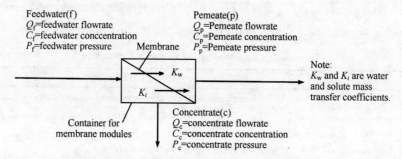

Figure 8-6 Definition sketch for a membrane process.

2. Membrane Process Classification

Membrane processes include microfiltration (MF), ultrafiltration (UF), nanofiltration (NF), reverse osmosis (RO), dialysis, and electrodialysis (ED). Membrane processes can be classified in a number of different ways including (1) the type of material from which the membrane is made, (2) the nature of the driving force, (3) the separation mechanism, and (4) the nominal size of the separation achieved. Each of these methods of classifying membrane processes is considered in the following discussion. The general characteristics of membrane processes including typical operating ranges are reported in Table 8-2. The focus of the following discussion is on pressure-driven membrane processes. Electrodialysis is considered separately following the discussion of the application of pressure-driven membranes.

Membrane Materials

Membranes used for the treatment of water and wastewater typically consist of a thin skin having a thickness of about 0.20 to 0.25 μm supported by a more porous structure of about 100 in thickness. Most commercial membranes are produced as flat sheets, fine hollow fibers, or in tubular form. The flat sheets are of two types, asymmetric and composite. Asymmetric membranes are cast in one process and consist of a very thin (less than 1 μm)

General characteristics membrane processes Table 8-2

Membrane process	Membrane driving force	Typical separation mechanism	Operating structure (pore size)	Typical operating range	Permeate description	Typical constituents removed
Microfiltration	Hydrostatic pressure difference	Sieve	Macropores (>50nm)	0.08~2.0	Water+ dissolved solutes	TSS, turbidity, protozoan oocysts and cysts, some bacteria and viruses
Ultrafiltration	Hydrostatic pressure difference	Sieve	Mescopores (2-50nm)	0.005~0.2	Water+ small molecules	Macromolecules, colloids, most bacteria, some viruses, proteins
Nanofiltration	Hydrostatic pressure difference	Sieve+ solution/ diffusion +exclusion	Micropores (<2nm)	0.001~ 0.01	Water+ very small molecules, ionic solutes	Small molecules, some harness, viruses
Reverse osmosis	Hydrostatic pressure difference	Solution/ diffusion +exclusion	Dense (<2nm)	0.0001~ 0.001	Water+ very small molecules, ionic solutes	Very small molecules, color, harness, sulfates, nitrate, sodium, other ions
Dialysis	Concentration difference	Diffusion	Mesopores (2-50nm)	—	Water+ small molecules	Macromolecules, colloids, most bacteria, some viruses, proteins
Electrodialysis	Electromotive force	Ion exchange with selective membranes	Micropores (<2nm)	—	Water+ ionic solutes	Ionized salt ions

layer and a thicker (up to 100μm) porous layer that adds support and is capable of high water flux. Thin-film composite (TFC) membranes are made by bonding a thin cellulose acetate, polyamide, or other active layer (typically 0.15 to 0.25μm thick) to a thicker porous substrate, which provides stability. Membranes can be made from a number of different organic and inorganic materials. The membranes used for water treatment are typically organic. The principal types of membranes used include polypropylene, cellulose acetate, ar-omatic polyamides, and TFC. The choice of membrane and system configuration is based on minimizing membrane clogging and deterioration, typically based on pilot-plant studies.

Driving Force

The distinguishing characteristic of the first four membrane processes considered in Table 8-2 (MF, UF, NF, and RO) is the application of hydraulic pressure to bring about the desired separation. Dialysis involves the transport of constituents through a semipermeable membrane on the basis of concentration differences. Electrodialysis involves the use of an electromotive force and ion-selective membranes to accomplish the separation of charged ionic species.

Removal Mechanisms

The separation of particles in MF and UF is accomplished primarily by straining (sieving), as shown on Figure 8-7(a). In NF and RO, small particles are rejected by the water layer adsorbed on the surface of the membrane which is known as a *dense* membrane (see Figure 8-7b). Ionic species are transported across the membrane by diffusion through the pores of the macromolecule comprising the membrane. Typically NF can be used to reject constituents as small as 0.001 whereas RO can reject particles as small as $0.0001 \mu m$. Straining is also important in NF membranes, especially at the larger pore size openings.

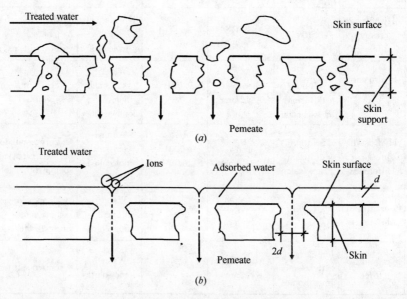

Figure 8-7 Definition sketch for the removal of wastewater constituents:
(a) removal of large molecules and particles by sieving (size exclusion) mechanism and
(b) rejection of ions by adsorbed water layer

Size of Separation

The pore sizes in membranes are identified as macropores ($>$50nm), mesopores (2 to 50nm), and micropores ($<$2nm). Because the pore sizes in RO membranes are so small, the membranes are defined as dense. The classification of membrane processes on the basis of the size of separation is shown in Table 8-2. There is considerable overlap in the sizes of particles removed, especially between NF and RO. Nanofiltration is used most commonly in water-softening operations in place of chemical precipitation.

Membrane Configurations and Membrane Operation

In the membrane field, the term *module* is used to describe a complete unit comprised of the membranes, the pressure support structure for the membranes, the feed inlet and outlet permeate and retentate ports, and an overall support structure. The principal types of membrane modules used for water treatment are (1) tubular, (2) hollow fiber, and (3) spiral wound. Plate and frame and pleated cartridge filters are also available but are used more commonly in industrial applications.

The operation of membrane processes is quite simple. A pump is used to pressurize the feed solution and to circulate it through the module. A valve is used to maintain the pressure of retentate. The permeate is withdrawn, typically at atmospheric pressure. As constituents in the feedwater accumulate on the membranes (often termed *membrane fouling*), the pressure builds up on the feed side, the membrane flux (i. e. , flow through membrane) starts to decrease, and the percent rejection also starts to decrease (see Figure 8-8). When the performance has deteriorated to a given level, the membrane modules are taken out of service and backwashed and/or cleaned chemically.

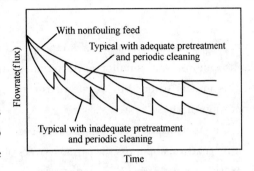

Figure 8-8 Definition sketch for the performance of a membrane filtration system

Unit 9 Water Distribution Systems

A water distribution system is an interconnected network of pipelines, storage tanks, pumps, and smaller appurtenances, including valves and flow meters. The purpose of this unit is to describe some of the practical aspects related to the design, analysis, and operation of these systems.

The unit begins with discussion of basic design factors, materials, and appurtenances. A section on centrifugal pumps-the prime movers of a distribution system is included. Conservation and distribution reservoirs are discussed, and the analysis of pipe network hydraulics is covered. Much of this material assumes some prior knowledge of hydraulics, so it would be best to study or review the knowledge of hydraulics before starting this unit.

In engineering practice, sophisticated computer modeling programs are used to analyze and design water distribution networks. Many of these programs are integrated with computer-aided drafting (CAD) software and provide graphical output as well as hydraulic analysis. But the best way to learn about water distribution system design and analysis is to do the computations "by hand", that is, with an electronic hand-held calculator. It is only after the student obtains a firm grasp of underlying computational methods and terminology that powerful software packages can be used effectively.

1. Design Factors

The design of a water distribution system begins after a study of community water requirements has been completed. A water distribution system must be able to deliver adequate quantities of water for various uses in a community. Also, sufficient pressures must be maintained throughout the system.

A survey of the service area is required so that maps of streets and topographical features can be prepared. On a relatively small scale map (about 1 : 24000, or 1 in. =2000 ft), the principal elements of the system can be planned, showing the general locations of water mains, pump stations, storage tanks, and so on. On larger scale maps (about 1 : 600, or 1 in. =50ft), the exact locations of the proposed facilities are shown in detail, as are existing utilities such as sewers or gas mains. These plan drawings are accompanied by written specifications describing the materials and methods of construction.

Required Flows and Pressures

It is convenient to classify water demands or water uses into four basic categories, as follows:
(1) *Domestic* water for drinking, cooking, personal hygiene, lawn sprinkling, and the like.
(2) *Public* water for fire protection and street cleaning and for use in schools or other public buildings.

(3) *Commercial and industrial water* for restaurants, laundries, manufacturing operation, and the like.

(4) *Loss* due to leaks in mains and house plumbing fixtures.

The total demand for water in a community varies, depending on the population, the industrial and commercial activity, the local climate, and the cost of the water. For example, in warm, dry climates, domestic use is generally a larger fraction of total consumption than it is in colder climate; lawn watering is much more common in dry climates. However, when the water bill is based on individual meter readings rather than on a flat rate, conservation is encouraged and water demand decreases.

Per Capita Demand

If the total annual water use of a community is divided by 365d, a value of average daily water consumption is obtained. If this value is further divided by the total population served, a *per capita* value is obtained. In SI units, this is expressed in terms of liters per day per person, in U. S. Customary units in gallons per capita per day (gpcd).

For example, if the average daily water demand is 5 megaliters per day (5ML/d) in a system serving 10000 people, the average per capita demand would be $(5000000L/d)/(10000 \text{ people}) = 500L/d$ per person. Keep in mind that a figure likes this includes each person's share of industrial, commercial, and public use and leakage; it is not just individual domestic use.

Since the exact water demands of a new service area may not be known, it is common to use average per capita values from similar communities in order to design the new distribution system. New systems are generally designed to accommodate populations and water demands that are anticipated 10 to 30 years in the future. Otherwise, the system would be too small soon after it was built. Table 9-1 presents overall average daily water demands in the United States.

Estimated average water requirements in the United State　　　Table 9-1

Type of use	L/d per person	gpcd	Percent of total
Domestic	300	80	44
Commercial/industrial	260	70	39
Public	60	16	9
Loss	50	14	8
Total	670	180	100

2. Variations in Water Demand

In any community, water demand will vary on a seasonal, daily, and hourly basis. For example, on a hot summer day it is not unusual for water consumption to be as much as 200 percent of the average daily demand. If the average demand is 670L/d, then we can estimate a peak daily demand to be $2 \times 670 = 1340L/d$ per person. Generally, the pipelines and pumps of a distubution systerm (as well as treatment plants and well) must be designed to accommodate peak daily flows rather than average flows. The minimum flow a system should be designed for is about 1000L/d per person (or about 250gpcd).

Water consumption also varies hourly throughout the day, according to a somewhat predictable pattern. Peak hourly demands in residential districts usually occur in the morning and evening hours, just before and after the normal workday. In commercial or industrial districts, water consumption may be uniformly high throughout the workday. Minimum flows typically occur around 4 AM, when almost no one is using water.

A graph illustrating typical hourly variations in water use is shown in Figure 9-1. On this graph, the peak hourly flow occurs at about 6 PM. In extreme cases, these maximum hourly flows could be as much as 10 times the average flow, but they are usually around 3.5 times the average flow rate. As discussed later, these peak hourly demands are generally accommodated by water from storage tanks instead of by the pumps in the system. Otherwise, the pumps and pipes would have to be excessively large just to handle flows that occur for a relatively short time.

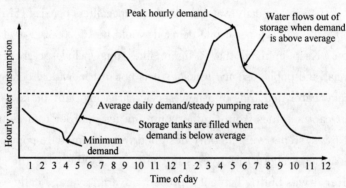

Figure 9-1 A graph that shows the typical variation in water demand or consumption throughout the day

There can be a wide variation in average, peak daily, and peak hourly flow rates among different communities. As far as is practical, the specific water demands should be determined or estimated for each service area. Generally, big cities have higher per capita water use than small communities, and small service areas are noted for their very high peak rates.

3. Fire Flows

Water for fire fighting is an important part of the total demand that must be provided for in a water distribution system. Fire flows are only required once in a while, and the total amount of water used to extinguish fires in any year is small compared to all other uses. But the rate and volume of water needed in the few hours of a fire emergency can be large in a local area. Sometimes, it can be the controlling factor affecting the size of the water mains.

Municipal insurance rates depend to a large extent on the fire protection provided by the distribution system. Factors involved in determining required fire flow capacity include type of building construction, occupancy, sprinkler protection, and so on. As a minimum 30L/s (475gpm) of fire flow is required for at least 2h. In more extreme cases, up to 760L/s (12000gpm) for a 10-h duration may be necessary. The required fire flow must be added to the peak daily demand in the system when sizing pipes and pumps.

4. Pressures

Water pressures in a distribution system should not drop below 350kPa (50psi) in order to provide for adequate operation of home plumbing fixtures and appliances as well as for fire fighting when pumper trucks are used at fire hydrants. Maximum pressures in water mains are generally kept below 760kPa, or 110psi, to reduce the chances for leaks or water main breaks. Pressures of about 550kPa (80psi) are considered optimum. Pressure-regulating valves must be installed in the distribution system to reduce pressures in low lying service areas; otherwise, pressure heads in the system would be too high.

5. Pipeline Layout

Water mains are generally not less than 150mm (6 in.) in diameter. They are usually located in the street right-of-way (ROW) so as to provide water to every potential customer. The *gridiron* arrangement of pipes is preferred to a layout that has many dead-end branches. In the gridiron system, water can circulate in interconnected loops, but in the dead-end system, the water may remain relatively stagnant in sections of the system causing taste and odor problems from bacterial growth. The two types of layouts are illustrated in Figure 9-2.

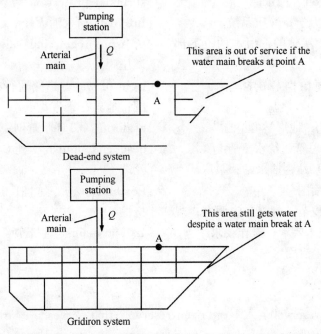

Figure 9-2 A gridiron pattern for water mains is preferable to a dead-end type of system; gridiron networks provide greater flexibility in operation and service.

In the dead-end layout, frequent flushing of the pipes at the fire hydrants is necessary to prevent consumer complaints about taste and odor. Another disadvantage of the dead-end system is that water service could be disrupted for long periods of time while repairs are made to a broken water main. But in a gridiron system, the broken section can be isolated by valves, and water can still reach consumers from the other side of the loop. Most distribution systems combine both layouts, depending on local conditions and economic factors.

Water mains may be referred to as primary feeders or secondary feeders. The primary feeders, also called arterial mains, carry large quantities of water from the treatment or pumping facility to areas of major water use. Secondary feeders are smaller pipes that provide a daily supply to local areas.

Vocabulary and Phrase

pipeline   n. 管线
storage tank   蓄水池；蓄水箱；贮槽
appurtenance   n. 附属物
appurtenances   n. 附属设备；附属建筑；附属机械
flow meter   流量计
centrifugal pump   离心泵
conservation reservoir   保护蓄水库
distribution reservoir   配水库
hydraulics   n. 水力学；液压系统
underlying   adj. 基础的；根本的；下部的；底层的
terminology   n. 术语；术语学
software package   软件包
topographical   adj. 地形学的
water main   给水总管；给水干管
pump station   水泵站；抽水站；泵房
plan drawing   平面图
specification   n. 规范；规格；说明书；规程；技术条件；技术要求；操作要求
domestic water   生活用水
sprinkling   n. 洒水；喷洒
public water   公用水
fire protection   防火
flat rate   统一收费率
per capita   每人
accommodate   v. 调节；使适应；适应；供应；容纳
fire fighting   消防
sprinkler protection   喷水保护
fire hydrant   消火栓
pressure-regulating valve   压力调节阀；调压阀
right-of-way   道路通行权；用地；用地范围；地界；管带
gridiron   n. 环状管网；格状管网
dead-end branch   死（水）端支管；支管死端
flexibility   n. 弹性；适应性；机动性；挠性
arterial mains   主干管

Notes

Conservation reservoir: A large open reservoir that serves primarily to store excess wet-weather stream flow for later use during periods of dry weather or drought.

Distribution reservoir: A water storage tank connected directly to a distribution system, providing about 1 d of capacity.

Question and Exercises

1. List four basic categories of water use. What is the effect of local climate and the use of individual water meters on water demand?

2. What does *gpcd* stand for?

3. Briefly discuss variations in water demand over time. Sketch a graph that would illustrate hourly variations over a 24-h period.

4. What is the range of working pressures in a typical water distribution main?

5. What is the minimum size of a water main in a public water system?

6. Why is a gridiron pipe layout preferable to a system with many dead ends?

7. What are the factors involved in determining required fire flow capacity?

Further Reading (1)

Water Transmission

Water can be transported from a ground or surface supply either directly to the water users in a community or initially to a water treatment facility. Water is transported by different types of conduits, including: (1) Pressure conduits: tunnel and pipelines; (2) Gravity-flow conduits: channels and canals.

The location of the river or well field as well as the location of the water treatment facility defines the length of these conduits. Long, gentle slopes allow canals and aqueducts to be used, but in most instances, pressurized systems are constructed for water transmission from the water supply watershed. The water then enters a water treatment facility where it is cleaned into potable water and subsequently distributed to the community of residential, commercial, and industrial users through a system of pressurized pipes. Because the demand for water is variable, we use more water during the daylight hours and for random fire control; for example, this distribution system must include storage facilities to even out the fluctuations.

Water distribution systems, particularly those serving densely populated cities, consist of a complex network of interconnected pipes and appurtenances.

1. Water Mains
Materials

The water mains in a distribution system must be strong and durable in order to resist applied forces and corrosion. The pipe is subjected to internal pressure from the water and to external pressure from the weight of the soil (backfill) and vehicles above it. Another force the pipe may have to withstand is called *water hammer*. This can occur when a valve is closed too fast, for example, causing waves of high pressure to surge through the pipe. Finally, damage due to corrosion or rusting may occur internally because of the water quality or externally because of the soil condition.

Ductile Iron Pipe: Ductile iron is one the most common materials used for the construction of water distribution pipelines. Because of its chemical composition, ductile iron is stronger and more elastic than gray cast iron, which was the predominant pipe material used until the mid-1900s. Cast iron (CI) is also strong and durable, with many older installations still in service after 100 years or more. Ductile iron, though, is less brittle than gray cast iron; it is less vulnerable to damage during construction and is considered to be more corrosion-resistant.

Asbestos Cement (AC) Pipe: A compacted mixture of sand, cement, and asbestos fibers provides a lightweight pipe material that is smooth and corrosion-resistant. Although it is not as strong as iron pipe, the absence of tuberculation and ease of installation make AC pipe desirable in many instances. AC pipe has long-lasting hydraulic properties with high carrying capacity ($C=140$).

Plastic Pipe: Poly vinyl chloride (PVC) plastic is sometimes used as a pipe material for construction of water distribution mains. These plastic pipes are strong and durable, yet they are very lightweight and are easily handled and installed. They are resistant to corrosion and they are very smooth, providing excellent hydraulic characteristics ($C=150$). Other plastic materials used for service connections and domestic plumbing include polyethylene (PE) and acrylonitrate-butadiene-styrene (ABS) plastic. These pipes may be joined using threaded screw couplings or chemical solvent welds.

Other Pipe Materials: Reinforced concrete pipes (RCPs) are made of welded steel cylinders wrapped with steel wire and embedded in concrete. They are used primarily in long water transmission lines of large diameter. They can be precast in sections up to 5m in length and up to about 6m in diameter. RCP pipes are very strong and durable, and have excellent hydraulic characteristics.

Steel pipe is sometimes used for water transmission lines, particularly for aboveground installation. It is very strong, yet it is lighter in weight than RCP. But it must be carefully protected against corrosion; this is usually done by lining in interior and painting and wrapping the exterior.

Appurtenances

Proper functioning of the water mains in a distribution system requires many different devices in addition to the sections of pipe. These devices called appurtenances include hydrants, shutoff valves, throttling valves, pressure-reducing valves, and other fittings.

Hydrants: The primary purpose of a hydrant is to provide convenient access to water for firefighting and other emergencies. A hydrant also serves for flushing out water mains, washing debris off public streets, and providing access to the underground pipe system for pressure testing. The spacing and location of hydrants depend primarily on fire protection and insurance needs. Hydrants are also placed at dead ends and at high and low points in the pipeline.

Service Connections: Water from the distribution main reaches the property line of individual consumers through a service pipe, usually made of copper or plastic, with a minimum diameter of 20mm.

Valves: Many different types of valves are used in water distribution systems to control the quantity and direction of flow. Many of these can be opened or closed manually by screw stems or gear train devices; large valves often are power-operated using electric or hydraulic systems.

The most common function of a valve is for complete shutoff of flow. *Gate valves* are usually used for this purpose. They are placed throughout the distribution network, allowing sections of pipeline to be shutoff and isolated during repairs of broken mains, pumps, or hydrants. Gate valves are usually either in the fully open or fully closed position; they are rarely used for throttling flow by blocking it only partially.

A type of valve commonly used for throttling and controlling flow rate is the *butterfly valve*. The fact that the disk is always in the flow is a disadvantage of the butterfly valve since it blocks the use of pipe-cleaning tools. Because the force of flowing water tends to close the valve, reducing gear drives are used for manual operation, and power operators are required for the large butterfly valves.

A device called a *check valve* is used to permit flow in only one direction in a pipe; it closes automatically when the flow stops or tends to flow in the opposite direction. Check valves are usually installed in the discharge piping of a pump to prevent backflow when the pump stops. They are called *foot valves* when installed at the end of a pump suction line in a well or tank. Foot valves prevent loss of prime in the centrifugal pumps. In plumbing systems, special double-check valves may be used to prevent backflow and possible contamination of a drinking water supply when a cross section with another system exists.

Other types of valves that find use in water distribution systems include pressure-reducing valves, air-release valves, and altitude valves. Pressure-reducing valves operate automatically to lower excessive hydrostatic pressure in water mains that at low elevation in the system. In effect, these valves form separate networks or pressure zones in a large distribution system.

Installation

Water mains must be installed at sufficient depths below the ground surface to provide protection against traffic loads and to prevent freezing. Generally, these depths are in the range of 1 to 2m. Since flow occurs under pressure instead of by gravity, the water mains can follow the general topographic shape of the ground, uphill as well as downhill.

Water mains should not be installed in the same trench with a sewer line. Generally, they should be at least 3m away from a sewer, horizontally, and they should be at least 0.5m higher than sewer lines when they cross.

Pressure Testing: No matter how well a pipeline is constructed, there may be some leakage at the joints. Within certain limits, some leakage is acceptable; the construction specifications should indicate the maximum allowable rate of leakage.

A pressure or leakage test is conducted on a newly installed water main by filling the pipe with water and maintaining a pressure of 1000kPa for 1 h. If excessive leakage occurs, an amount of water greater than the allowable leakage, Q_L, must be pumped into the line to maintain the pressure; repairs are necessary before the pipeline can be accepted for use.

Thrust Blocks: It is usually necessary to anchor the pipeline securely in the trench at dead ends and at bends or at changes in horizontal or vertical direction. This is because of the force, or thrust, caused by the internal pressure and kinetic energy of flow, which tends to move the pipe or fittings. Such movement can damage the joints and cause excessive leakage. One method for providing the necessary anchorage is to use concrete thrust blocks.

Disinfection: A newly installed water main must be flushed clean and disinfected before being put into service. Flushing velocities of about 1m/s are generally enough to remove dirt and debris that may accumulate in the pipe during construction. The pipe is disinfected to kill bacteria by filling it with a relatively concentrated chlorine solution for a certain period of time, as specified by local regulatory agencies. It is flushed again before being put into service.

Rehabilitation

The use of proper material and installation methods does not guarantee trouble-free operation of a water main for an unlimited period of time. Pipeline breaks and leaks occur periodically for several reasons, and emergency repairs must be made. Most water utilities have a plan of action for dealing with these emergencies and keep spare parts, tools, and equipment readily available.

Leaks that are not readily observable from wet or sunken spots in the street can be located by using sounding rods for electronic amplification of the sound of the escaping water. Relatively small leaks can be repaired without shutting off the water pressure. This not only avoids inconvenience to utility customers, it also prevents contamination of the distribution system by backflow (backflow of dirty water into the system can occur when the water pressure is turned off). All water is pumped out of the excavated trench. Repair clamps or sleeves are installed over the leakage section of pipe and tightened until the leakage stops. The pipe should be flushed, hydrostatically retested, and disinfected after the repair has been made.

Cleaning: Sudden water main breaks are not the only problems that can occur in a water distribution system. Loose deposits of sediment may accumulate in the pipeline, particularly in dead-end branches. These sediments, which cause taste, odor, and color problems, can be removed by periodic flushing through hydrants.

Many water mains suffer the effects of a gradual and persistent buildup of solid deposits on the inside wall of the pipe. The longer the pipeline is in service, the worse this problem becomes. These deposits reduce the hydraulic capacity of the pipeline and cause high pressure losses. Pumping costs increase and there is a greater chance for regrowth of bacteria in the distribution system.

The deposits may consist of tubercles or lumps of iron oxide if the pH of the water is low and the metal pipe is unlined. This problem is called *tuberculation*. When the pH of the water is high, the deposits consist of calcium carbonate scale. Maintaining a proper pH so that the water is neither corrosive nor scale forming may be accomplished by adding chemicals at the treatment plant. Although this can minimize the formation of additional deposits, it does not restore the lost capacity that has already occurred.

A common method for rehabilitating old water mains is to clean them using a mechanical or hydraulic scraper tool. During the cleaning operation, water supply service can be provided to customers by using small-diameter temporary bypass pipes or hoses. Thick deposits can be removed using power-driven mechanical cleaning devices. The cutting tool consists of rotating steel blades mounted on a series of body sections attached to center rod.

Another type of cleaning device is a bullet-shaped resilient foam object called a *pig*. It is wrapped with wearing strips in a spiral or crisscross arrangement. Propelled by water pressure through the pipeline, the pig serves to scrape the deposits off the wall of the pipe. It is inserted into the line either through a fire hydrant or through a specially installed *launcher*. Several passes of the pig may be needed to remove all the deposits. Afterward, the pipe is flushed out and disinfected before being placed into service again.

Lining: Newly installed metal pipes are supplied with a cement mortar lining to prevent tuberculation. But many iron water mains were installed in the past without these linings. When these mains are cleaned and the deposits removed, it is necessary to install a lining that will prevent the problem of tuberculation from recurring. One such rehabilitation method, called *sliplining*, involves placing a plastic pipe inside the cleaned pipe. The plastic pipe, of slightly smaller diameter than the original pipe, is pulled through straight sections of the transmission main.

Another method for protecting the cleaned pipe wall is to apply a cement-mortar lining to the pipe in place. For pipes less than 600mm in diameter, mortar can be pumped through a hose to a lining machine that is pulled through the pipeline. The mortar is sprayed centrifugally onto the pipe wall; the thickness of the lining can be controlled by the speed at which the machine is pulled through the pipe. A lining thickness of 6mm or less is preferred, so as not to reduce the inner diameter of the pipe excessively. For pipes over 600 mm in diameter, a worker can enter the water lining equipment.

Cleaning and lining can effectively rehabilitate a water main, increasing its carrying capacity and reducing pressure drops. But the pipe must not be structurally defective if this method of rehabilitation is to be used. A deteriorated pipeline that experiences frequent breaks and leaks may have to be replaced completely with a new pipeline. Large concrete water mains, however may be lined with pipe made of steel plate to restore structural integrity with minimum loss of capacity. It is always necessary to compare the relative economics of complete replacement versus cleaning and relining. If water demands in the area are increasing, new water mains may be needed; rehabilitation alone may not be sufficient to meet the higher demands.

2. Centrifugal Pumps

A pump is a mechanical device that adds energy to water or other liquids. In most water distribution systems, pumps are needed to raise the water in elevation and to move it through the network of water mains under pressure. One way of classifying pumps is by their application in the system. Pumps that lift the water from a river or lake and move it to a nearby treatment plant are called *low-lift pumps*. They move large quantities of water, but at relatively low discharge pressures. The pumps that discharge the treated drinking water into the transmission and distribution system are called *high-lift pumps*; they operate under relatively high heads or pressures.

Sometimes it is necessary to increase the pressure within the distribution system or to raise the water into an elevated storage tank; *booster pumps* can be used for this purpose. *Well pumps* lift water from an underground aquifer and often discharge directly into the distribution system.

Another way of classifying pumps is according to the mechanical principles on which they operate. The two basic types are positive-displacement pumps and centrifugal pumps. A positive-displacement pump will deliver a fixed quantity of water with each revolution of the pump rotor or piston. The water is physically pushed or displaced from the pump casing. The capacity of the pump is unaffected by changes in pressure in the system in which it operates.

Centrifugal pumps are the most common type used in water supply (as well as wastewater) systems. As discussed shortly, the capacity of the pump is very much a function of the pressure against which it operates in the system. A centrifugal pump adds energy to the water by accelerating it through the action of a rapidly rotating impeller. The water is thrown outward by the vanes of the impeller and passes through a shaped casing where its velocity is gradually slowed down. As its velocity drops in the expanding spiral volute, the kinetic energy is converted to pressure head, called the *discharge pressure*.

Centrifugal pumps can be further classified as *radial flow or axial flow*. In the radial-flow type, the water discharges at right angles to the direction of flow into the pump impeller; in the axial-flow type, the water discharges in the same direction as the axis of the impeller. Centrifugal pumps with more than one impeller are called *turbine pumps*.

Centrifugal type pumps have several advantages over positive-displacement pumps. They are simple, with only one moving part - the impeller; no internal valves are required, and there is no need for internal lubrication. Also, they operate very quietly. Disadvantages include the effect of pressure on the pump output and efficiency, and the necessity for priming the pump before it is operated. Priming involves filling the pump casing and suction line with water.

3. Distribution Storage

Elevated water storage tanks and towers are familiar sights in most communities. These relatively small water storage facilities serve two basic purposes: They provide *equalizing storage* and they provide *emergency storage*. They are called distribution storage tanks or reservoirs because they are part of the localized water distribution systems. Much large storage facilities, called *conservation reservoirs*, are generally located at a considerable distance form the distribution network and are meant to store water that can be used during long dry-weather periods.

Equalizing storage refers to the volume of water in the tank that is available to satisfy the peak hourly demands for water use in the community. The hourly variation in water demand is illustrated in Figure 9-1. During the late night and early morning hours, when water demand is very low, the high-lift pump move water into the distribution storage tanks. During the day, when water demand exceeds the average daily demand, water flows out of the tanks to help to met the peak hourly needs of the community.

Distribution storage tanks are often described as *floating on the line*. This means that the flow into or out of the tank varies directly with the demand for water in the system, and the hydrostatic pressure in the water main is equivalent to the pressure head or elevation of water in the tank. A distribution storage tank is illustrated in Figure 9-3. Automatically controlled altitude valves can maintain the water elevations, and therefore the water main pressure, within a desired range.

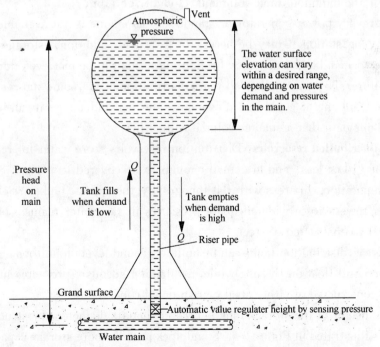

Figure 9-3 An elevated water storage tank will float on the line in a water distribution system.

The equalizing or averaging effect on flow rates provided by the stored water allows for a uniform or steady water treatment and pumping rate. When water demand is low, the extra water being pumped fills the storage tanks, and when demand exceeds the pumping rate, the tanks are emptied to make up the difference. This has the advantage of reducing the required sizes and capacities of the pipes, pumps, and treatment facilities, resulting in reduced construction and operating costs.

Generally, the volume of water needed to balance or equalize the peak hourly flows is about 20 percent of the average daily water demand in the service area. For example, if a community has an average water demand of 1ML/d, then at least 0.2ML, or 200000L, of storage volume should be provided for equalizing purposes. In communities where adequate records of water use and demand are available, the summation hydrograph method can be used for a more accurate determination of the required equalizing volume in a new tank.

Distribution storage tanks are not constructed for the sole purpose of providing equalizing storage. Emergency storage volume is also provided by these tanks. This furnishes additional water for firefighting needs, or overcoming problems due to power blackouts or pump station failure.

The amount of water required for fire control varies depending on the type of service area and the capacity of the water pumping station. For example, if storage for a fire flow of 30 L/s for a 2-h duration is needed, the required volume is 30L/s × 3600s/h × 2h = 216000L. This would be added to the volume needed for equalizing purposes to determine the total tank volume. If 200000L is needed for equalizing storage and 216000L is needed for fire control, the minimum tank volume is 416000L, or 416m^3.

If an emergency power generator or standby diesel engine is not available at the high-lift water pumping station, it may be necessary to provide additional storage so that both domestic water demand and emergency firefighting needs can be met even during a temporary power failure. To accommodate all these distribution needs, some states simply require that the storage volume be equal to 1-d average water demand, unless it can be demonstrated with available data that a smaller volume will suffice.

Types of distribution reservoirs: Distribution reservoirs store water for relatively short periods of time (1d or less) and are small enough to be covered to prevent contamination and reduce evaporation. In areas with flat topography, the storage tanks are elevated above the ground on towers to provide adequate pressures in the water mains. These elevated tanks are usually constructed of steel.

In hilly areas, distribution tanks can be built at ground level on hilltops higher than the service area and still float on the line while maintaining adequate pressures in the system. They may be constructed of either steel or reinforced concrete.

When the height of the storage tank is greater than its diameter, the structure is called a standpipe, as illustrated in Figure 9-4. Standpipes provide more storage capacity than elevated tanks. But the storage capacity that is useful for equalizing purposes is only that volume above the elevation required for minimum pressure in the water main. The water below that elevation can be used for fire protection with pumper trucks or during other emergency conditions. Standpipes are constructed of steel, with thicker walls at the bottom to withstand the hydrostatic pressure.

Location of storage tanks: Using several small storage tanks near the major centers of water withdrawal is preferable to using one large tank near the pumping station. Also, it is best to locate the tanks on the opposite side of the demand center from the pumping station. This allows for more uniform pressures throughout the distribution network, as il-

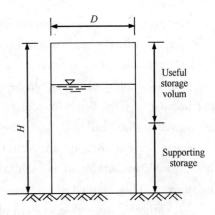

Figure 9-4 A water storage tank is called a standpipe when its height is greater than its diameter ($H > D$).

lustrated in Figure 9-5. It also allows the use of smaller-diameter mains and pumps than would otherwise be needed.

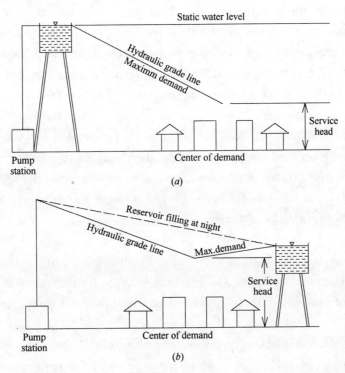

Figure 9-5 Location of distribution storage tanks opposite the source (b) is preferable to location at the source (a).

Maintenance: Distribution storage reservoirs should be inspected frequently. Cracks and leaks must be repaired, and air vents should be kept clear. Blocked air vents can cause excessive pressures or vacuums to develop in the tank, which can result in structural damage.

Storage tanks are occasionally drained, cleaned, and disinfected. Accumulated silt should be removed. Sometimes the inside of a steel tank will have to be painted with an approved bituminous or vinyl coating. All traces of rust must first be removed. Disinfection can be accomplished by spraying the tank walls with a 500-mg/L chlorine solution.

Corrosion of steel tanks can be a major problem. About 10kg of steel per year can be lost because of corrosion. In addition to protecting the steel with high-quality bituminous coatings, a method called *cathodic protection* is sometimes used. Corrosion involves a flow of electric current through the water in the form of positively charged metal ions. In cathodic protection, a voltage is maintained in the tank that tends to reverse the direction of the current, This applied voltage keeps the metal from ionizing and thus prevents corrosion. A cathodic protection system must be custom designed by specialists for each tank.

Further Reading (2)

An Introduction of the CWLP Water Transmission & Distribution

The process of transmitting and distributing drinking water to our customers begins as the clean water leaves the filtration gallery of the Water Purification Plant.

From the water purification plant, the water flows to two underground "clear well" storage tanks on the grounds of the lakeside plant complex. These tanks, which have a combined capacity of 6 million gallons, hold the water for approximately four hours before it is pumped out to the city and surrounding communities via four primary, 16-to 36-inch diameter transmission mains and four high-service pumps. This transmission system has a combined capacity of 65 million gallons per day (mg/d). On a typical day, however, only about a third of this capacity is called upon. On average, the utility treats more than 22 mg/d. The maximum amount ever treated in one day, 39.7mg/d, occurred on July 23, 1999. From the transmission mains, the water is further conveyed through six-to twelve-inch distribution mains to CWLP's Springfield customers.

Water Main

Treated water is carried from the Water Purification Plant to the ultimate users through a network of water mains. The current CWLP water transmission and distribution system consists of over 600 miles of water mains, which measure from 2-to 36-inches in diameter.

The history of the expansion of the water system closely parallels the growth of the city. There were just less than 100 miles of water mains in the system at the beginning of 1930. About 25 miles were added during that decade. Although very little main expansion took place during World War II, rapid development following the war resulted in the addition of over 45 miles during the late 1940s. In response to the continued growth of both Springfield's population and the needs of our state government over the next five decades, 60 miles of new main were added in the 1950s; 65 in the 1960s; just under 100 miles in the 1970s; 60 miles in the 1980s; and 130 miles in the 1990s. From 2000 through

2006, approximately 64 miles of new main were added to the system.

Most of Springfield's water main system is composed of durable cast iron pipe, which has a very long useful life cycle under most conditions. Newer mains are constructed of ductile iron, an even more durable product.

Storage Tanks

Approximately 85% of the water distributed to the city each day is used directly out of the distribution system. The remaining 15% is stored in one of three above-ground storage tanks located throughout the city. Most water storage takes place during the night, when customer demand is lowest, ensuring plenty of water will be available to our customers during the high-demand daylight hours. To keep stored water fresh, the tanks are cycled at least partially each day.

By providing storage for water that is not immediately needed, these tanks help to create an equalizing effect on the water system by allowing some pumpage to be shifted from peak-demand to low-demand hours. They also contribute to public safety, by helping to ensure adequate flows will be available for fire fighting purposes.

The oldest of the above-ground storage tanks is located on Old Jacksonville Road. Built in 1966, this elevated model has a capacity of one million gallons. The second tank, a ground-level model built in 1976, is located on Factory Street near Sangamon Avenue. It has a capacity of four million gallons. The newest tank, built in 1998, is a five million gallon "standpipe" located off of Cockrell Lane south of Wabash Avenue.

CWLP's three storage tanks fill and discharge their water in different ways. The elevated Old Jacksonville Road tank "floats on the system", meaning it discharges or refills based on reductions or increases in the system's water pressure. As pressure lowers, the tank discharges water; as pressure rises, the tank fills. No pumps are required by this tank. The Factory Street ground-level tank relies on system pressure for refilling, but requires pumps for discharging. Water can be discharged from this tank at a rate of about 6 million gallons per day. The Cockrell Lane standpipe can either float on the system or use pumpage, as desired, for both refilling and discharge.

System Maintenance

CWLP devotes substantial time and resources each year for the maintainence and repair of the water distribution system, as well as for the construction of new facilities required to accommodate system growth. However, because of the age of the water distribution system, we believe it is important to provide funding to rebuild deteriorated sections of the water system rather than continue to incur the expense of repair. The Water System Rebuilding Program, consists of the following:

1. Replacement of deteriorating water mains;
2. Valve replacement and insertion;
3. Dead end ties;
4. Replacement of galvanized risers;

5. Meter pit cut-ins;
6. Replacement of large water meter.

Water Main Replacements

Each year, CWLP identifies a number of sections of water distribution mains that are in poor condition and in need of repair. These water mains have deteriorated due to age, the use of substandard materials in the past, soil conditions which have caused the mains to corrode, or a combination of these factors.

Valve Replacement and Insertion

When water main breaks occur, valves are used to turn off the water so that repairs can be made. If the valves closest to the main break are inoperable, other valves must be closed, necessitating the interruption of water service to additional customers. To reduce the likelihood of this occuring, the CWLP distribution system has been surveyed to determine where older valves should be replaced and where the installation of additional valves would enhance system control.

Dead End Ties

In some areas of the distribution system, water mains dead end without being looped to other mains. Dead end ties loop dead end mains together to improve water pressure.

Galvanized Riser Replacement

Galvanized iron risers to hold water meters in meter pits were considered state of the art between 1945 and 1965. About 23000 were installed in Springfield. However, galvanized risers deteriorate with age and often break when meters are changed. As meters are changed, CWLP replaces galvanized risers with copper setters.

Meter Pit Cut-Ins

The existence of approximately 10000 water meter settings located inside structures within the CWLP distribution system presents a problem to our customers, meter readers and service personnel, alike. Not only does this complicate the meter reading process, but the "curb-stop" valve boxes controlling these meters are frequently difficult to locate. And even if they can be located, the valves are often inoperable. To rectify these problems, CWLP is actively involved in the replacement of inside meter settings with meter pits, which are equipped with their own control valves.

Meter Replacements

As a meter ages, it tends to record less than the full amount of water used by the customer. CWLP's ongoing meter replacement program pays for itself through accurate meter readings.

Service Connections

Customers have interactions most with the service line portion of the CWLP water system that brings water to their home or business. Currently, there are over 52000 metered water service connections to the system. These connections range from 5/8 inch to 10 inches in diameter, with 97% being one inch or smaller. A service is generally composed of: a connection to the main, a control valve and connective piping, a water meter and (for 75% of services) a meter pit. In 2006, a total of 369 new service connections were installed on the CWLP water system.

A variety of materials have been used for water services and are still present in the CWLP system. Galvanized iron, lead and copper have been used for connective piping.

Fire Hydrants

There are over 6360 fire hydrants on the CWLP system. Hydrants are used for fire suppression, pressure tests, and main flushing. Hydrant function can have a direct and immediate impact on the safety of human life and property. CWLP tests and maintains all fire hydrants in the city at least once each year and has an excellent record of keeping hydrants in service. Rarely is there more than a handful of hydrants inoperable at any time. Significant resources are expended in order to maintain that high level of service which contributes to achieving the City's Class 1 Insurance Service Office (ISO) rating. In 2006, CWLP added 104 new hydrants to the water transmission and distribution system.

Unit 10 Water Pollution

Although people intuitively relate filth to disease, the transmission of disease by pathogenic organisms in polluted water was not recognized until the middle of the nineteenth century. The Broad Street pump handle incident demonstrated dramatically that water could carry diseases. In 1854, a public health physician named John Snow, assigned to try to control the spread of cholera, noticed a curious concentration of cholera cases in one part of London. Almost all of the people affected drew their drinking water from a community pump in the middle of Broad Street. However, people who worked in an adjacent brewery were not affected. Snow recognized that the brewery workers' apparent immunity to cholera occurred because the brewery drew its water from a private well and not from the Broad Street pump (although the immunity might have been thought due to the health benefits of beer). Snow's evidence convinced the city council to ban the polluted water supply, which was done by removing the pump handle so that the pump was effectively unusable. The source of infection was cut off, the cholera epidemic subsided, and the public began to understand the importance of having clean drinking water supplies.

Until recently, water pollution was viewed primarily as a threat to human health because of the transmission of bacterial and viral waterborne diseases. In less developed countries, and in almost any country in time of war, waterborne diseases remain a major public health threat. In the United States and other developed countries, however, water treatment and distribution methods have almost eradicated microbial contamination in drinking water. We now recognize that water pollution constitutes a much broader threat and continues to pose serious health risks to the public as well as aquatic life. In this unite we discuss the sources of water pollution and the effect of this pollution on streams, lakes and oceans.

1. Sources of Water Pollution

Water pollutants are categorized as point source or nonpoint source, the former being identified as all dry weather pollutants that enter watercourses through pipes or channels. Storm drainage, even though the water may enter watercourses by way of pipes or channels, is considered nonpoint source pollution. Other nonpoint source pollution comes from agricultural runoff, construction sites, and other land disturbances. Point source pollution comes mainly from industrial facilities and municipal wastewater treatment plants. The range of pollutants is vast, depending only on what gets "thrown down the drain."

Oxygen-Demanding Substances

Dissolved oxygen is a key element in water quality that is necessary to support aquatic life. A demand is placed on the natural supply of dissolved oxygen by many pollutants in wastewater. This is called biochemical oxygen demand, or BOD, and is used to measure how well a sewage treatment plant is working. If the effluent, the treated wastewater produced by a treatment plant, has a high content of organic pollutants or ammonia, it will demand more oxygen from the water and leave the water with less oxygen to support fish and other aquatic life.

Organic matter and ammonia are "oxygen-demanding" substances. Oxygen-demanding substances are contributed by domestic sewage and agricultural and industrial wastes of both plant and animal origin, such as those from food processing, paper mills, tanning, and other manufacturing processes. These substances are usually destroyed or converted to other compounds by bacteria if there is sufficient oxygen present in the water, but the dissolved oxygen needed to sustain fish life is used up in this break down process.

Pathogens

Disinfection of wastewater and chlorination of drinking water supplies has reduced the occurrence of waterborne diseases such as typhoid fever, cholera, and dysentery, which remain problems in underdeveloped countries while they have been virtually eliminated in the U. S.

Infectious micro-organisms, or pathogens, may be carried into surface and groundwater by sewage from cities and institutions, by certain kinds of industrial wastes, such as tanning and meat packing plants, and by the contamination of storm runoff with animal wastes from pets, livestock and wild animals, such as geese or deer. Humans may come in contact with these pathogens either by drinking contaminated water or through swimming, fishing, or other contact activities. Modern disinfection techniques have greatly reduced the danger of waterborne disease.

Nutrients

Carbon, nitrogen, and phosphorus are essential to living organisms and are the chief nutrients present in natural water. Large amounts of these nutrients are also present in sewage, certain industrial wastes, and drainage from fertilized land. Conventional secondary biological treatment processes do not remove the phosphorus and nitrogen to any substantial extent—in fact, they may convert the organic forms of these substances into mineral form, making them more usable by plant life. When an excess of these nutrients over stimulates the growth of water plants, the result causes unsightly conditions, interferes with drinking water treatment processes, and causes unpleasant and disagreeable tastes and odors in drinking water. The release of large amounts of nutrients, primarily phosphorus but occasionally nitrogen, causes nutrient enrichment which results in excessive growth of algae. Uncontrolled algae growth blocks out sunlight and chokes aquatic plants and animals by depleting dissolved oxygen in the water at night. The release of nutrients in quantities that exceed the affected water body's ability to assimilate them results in a condition called eutrophication or cultural enrichment.

Inorganic and Synthetic Organic Chemicals

A vast array of chemicals is included in this category. Examples include detergents, household cleaning aids, heavy metals, pharmaceuticals, synthetic organic pesticides and herbicides, industrial chemicals, and the wastes from their manufacture. Many of these substances are toxic to fish and aquatic life and many are harmful to humans. Some are known to be highly poisonous at very low concentrations. Others can cause taste and odor problems, and many are not effectively removed by conventional wastewater treatment.

Thermal

Heat reduces the capacity of water to retain oxygen. In some areas, water used for cooling is discharged to streams at elevated temperatures from power plants and industries. Even discharges from wastewater treatment plants and storm water retention ponds affected by summer heat can be released at temperatures above that of the receiving water, and elevate the stream temperature. Unchecked discharges of waste heat can seriously alter the ecology of a lake, a stream, or estuary.

2. Elements of Aquatic Ecology

The effects of water pollution can be best understood in the context of an aquatic ecosystem, by studying one or more specific interactions of pollutants with that ecosystem.

Plants and animals in their physical and chemical environment make up an ecosystem. The study of ecosystems is termed ecology. Although we often draw lines around a specific ecosystem to be able to study it more fully (e. g. , a farm pond) and thereby assume that the system is completely self-contained, this is obviously not me. One of the tenets of ecology is that "everything is connected with everything else".

Three categories of organisms make up an ecosystem. The producers use energy from the sun and nutrients like nitrogen and phosphorus from the soil to produce high energy chemical compounds by the process of photosynthesis. The energy from the sun is stored in the molecular structure of these compounds. Producers are often referred to as being in the first trophic (growth) level and are called autotrophs. The second category of organisms in an ecosystem includes the consumers, who use the energy stored during photosynthesis by ingesting the high-energy compounds. Consumers in the second trophic level use the energy of the producers directly. There may be several more trophic levels of consumers, each using the level below it as an energy source. The third category of organisms, the decomposers or decay organisms, use the energy in animal wastes, along with dead animals and plants, converting the organic compounds to stable inorganic compounds (e. g. , nitrate) that can be used as nutrients by the producers.

Ecosystems exhibit a flow of both energy and nutrients. The original energy source for nearly all ecosystems is the sun (the only notable exception is oceanic hydrothermal vent communities, which derive energy from geothermal activity). Energy flows in only one direction: from the sun and through each trophic level. Nutrient flow, on the other hand, is cyclic: nutrients are used by plants to make high-energy molecules that are eventually decomposed to the original inorganic nutrients, ready to be used again.

Most ecosystems are sufficiently complex that small changes in plant or animal populations will not result in long-term damage to the ecosystem. Ecosystems are constantly changing, even without human intervention, so ecosystem stability is best defined by its ability to return to its original rate of change following a disturbance. For example, it is unrealistic to expect to find the exact same numbers and species of aquatic invertebrates in a "restored" stream ecosystem as were present before any disturbance. Stream invertebrate populations vary markedly from year to year, even in undisturbed streams. Instead, we should look for the return of similar types of invertebrates, in about the same relative proportion as would be found in undisturbed streams.

The amount of perturbation that an ecosystem can absorb is called resistance. Communities dominated by large, long-lived plants (e.g., old growth forests) tend to be fairly resistant to perturbation (unless the perturbation is a chain saw!). Ecosystem resistance is partially based on which species are most sensitive to the particular disturbance.

Even relatively small changes in the populations of "top of the food chain" predators (including humans) or critical plant types (e.g., plants that provide irreplaceable habitat) can have a substantial impact on the structure of the ecosystem. The ongoing attempt to limit the logging of old-growth forests in the Pacific Northwest is an attempt to preserve critical habitat for species that depend on old growth, such as the spotted owl and the marbled murmlet. The rate at which the ecosystem recovers from perturbation is called resilience. Resilient ecosystems are usually populated with species that have rapid colonization and growth rates. Most aquatic ecosystems are very resilient (but not particularly resistant). For example, during storm events, the stream bottom is scoured, removing most of the attached algae that serve as food for small invertebrates. The algae grow quickly after the storm flow abates, so the invertebrates do not starve. In contrast, the deep oceanic ecosystem is extraordinarily fragile, not resilient, and not resistant to environmental disturbances. This must be considered before the oceans are used as waste repositories.

Although inland waters (streams, lakes, wetlands, etc.) tend to be fairly resilient ecosystems, they are certainly not totally immune to destruction by outside perturbations. In addition to the direct effect of toxic materials like metals, pesticides, and synthetic organic compounds, one of the most serious effects of pollutants on inland waters is depletion of dissolved oxygen. All higher forms of aquatic life exist only in the presence of oxygen, and most desirable microbiologic life also requires oxygen. Natural streams and lakes are usually aerobic. If a watercourse becomes anaerobic, the entire ecology changes and the water becomes unpleasant and unsafe. The dissolved oxygen concentration in waterways and the effect of pollutants are closely related to the concept of decomposition and biodegradation, part of the total energy transfer system that sustains life.

Vocabulary and Phrases

pathogenic　*adj.* 致病的，病原的，发病的
organism　*n.* 生物体，有机体
cholera　*n.* ［医］霍乱
brewery　*n.* 酿酒厂
watercourse　*n.* 水道，河道
runoff　*n.* 径流，未被土壤吸收的降雨
aquatic　*adj.* 水的，水上的，水生的，水栖的
effluent　*n.* 处理后出水
ammonia　*n.* ［化］氨，氨水
disinfection　*n.* 消毒
chlorination　*n.* ［化］氯化，用氯处理
typhoid fever　*n.* ［医］伤寒症
dysentery　*n.* ［医］痢疾
carbon　*n.* ［化］碳（元素符号 C）
nitrogen　*n.* ［化］氮
phosphorus　*n.* 磷
algae　*n.* 藻类，海藻
eutrophication　*n.* 超营养作用，富营养化
detergent　*n.* 洗涤剂
pharmaceutical　*n.* 药物 *adj.* 制药(学)上的
pesticide　*n.* 杀虫剂
herbicide　*n.* 除草剂
ecology　*n.* 生态学
estuary　*n.* 河口，江口
ecosystem　*n.* 生态系统
self-contained　*adj.* 设备齐全的，独立的
producer　*n.* 生产者
photosynthesis　*n.* 光合作用
trophic　*adj.* 营养的，有关营养的
autotroph　*n.* ［生］自养生物，靠无机物质生存的生物
consumer　*n.* 消费者
decomposer　*n.* 分解者
predator　*n.* 掠夺者，食肉动物
resilience　*n.* 弹性
inland　*adj.* 内陆的

Notes

the city council　市议会
water pollution was viewed primarily as a threat to human health because of the transmission of bacterial and viral waterborne diseases　水污染被看作对人类健康的主要威胁是因为细菌或病毒引起的与水相关的疾病传播
be categorized as　被分为……类
land disturbance　土地扰动
municipal wastewater treatment plant　城市污水处理厂
oxygen-demanding substance　耗氧物质
biochemical oxygen demand　生化需氧量
organic pollutant　有机污染物
domestic sewage　生活污水
surface water and groundwater　地表水和地下水
conventional secondary biological treatment process　传统二级生物处理过程

a vast array of chemicals are included in this category 这类物质包括一系列的化学物质

be toxic to 对……有毒

Question and Exercises

1. Describe briefly the main sources of water pollution.
2. Describe briefly the function of dissolved oxygen in the water.
3. What phenomena will happen when a large amounts of nutrients, especially phosphorus is released to the estuary?
4. What's aquatic ecology?

Unit 11   Wastewater Properties

1. Wastewater Sources

The principal sources of domestic wastewater in a community are the residential areas and commercial districts. Other important sources include institutional and recreational facilities and storm water (runoff) and groundwater (infiltration). Each source produces wastewater with specific characteristics. In this section wastewater sources and the specific characteristics of wastewater are described.

Generation of Wastewater

Wastewater is generated by five major sources: human and animal wastes, household wastes, industrial wastes, storm water runoff, and groundwater infiltration.

(1) Human and animal wastes—Contains the solid and liquid discharges of humans and animals and is considered by many to be the most dangerous from a human health viewpoint. The primary health hazard is presented by the millions of bacteria, viruses, and other microorganisms (some of which may be pathogenic) present in the waste stream.

(2) Household wastes—Consists of wastes, other than human and animal wastes, discharged from the home. Household wastes usually contain paper, household cleaners, detergents, trash, garbage, and other substances the homeowner discharges into the sewer system.

(3) Industrial wastes—Includes industry specific materials that can be discharged from industrial processes into the collection system. Typically contains chemicals, dyes, acids, alkalis, grit, detergents, and highly toxic materials.

(4) Storm water runoff—Many collection systems are designed to carry both the wastes of the community and storm water runoff. In this type of system when a storm event occurs, the wastestream can contain large amounts of sand, gravel, and other grit as well as excessive amounts of water.

(5) Groundwater infiltration—Groundwater will enter older improperly sealed collection systems through cracks or unsealed pipe joints. Not only can this add large amounts of water to wastewater flows, but also additional grit.

Classification of Wastewater

Wastewater can be classified according to the sources of flows: domestic, sanitary, industrial, combined, and storm water.

(1) Domestic (sewage) wastewater—Containing mainly human and animal wastes, household wastes, small amounts of groundwater infiltration and small amounts of industrial wastes.

(2) Sanitary wastewater—Consisting of domestic wastes and significant amounts of industrial wastes. In many cases, the industrial wastes can be treated without special precautions. However; in some cases, the industrial wastes will require special precautions or a pretreatment program to ensure the wastes do not cause compliance problems for the wastewater treatment plant.

(3) Industrial wastewater—Consisting of industrial wastes only. Often the industry will determine that it is safer and more economical to treat its waste independent of domestic waste.

(4) Combined wastewater—Consisting of a combination of sanitary wastewater and storm water runoff. All the wastewater and storm water of the community is transported through one system to the treatment plant.

(5) Storm water—Containing a separate collection system (no sanitary waste) that carries storm water runoff including street debris, road salt, and grit.

2. Wastewater Characteristics

Wastewater contains many different substances that can be used to characterize it. The specific substances and amounts or concentrations of each will vary, depending on the source. It is difficult to precisely characterize wastewater. Instead, wastewater characterization is usually based on and applied to an average domestic wastewater. Wastewater is characterized in terms of its physical, chemical, and biological characteristics.

Physical Characteristics

The physical characteristics of wastewater are based on color, odor, temperature, and flow.

(1) Color—Fresh wastewater is usually a light brownish-gray color. However, typical wastewater is gray and has a cloudy appearance. The color of the wastewater will change significantly if allowed to go septic (if travel time in the collection system increases). Typical septic wastewater will have a black color.

(2) Odor—Odors in domestic wastewater usually are caused by gases produced by the decomposition of organic matter or by other substances added to the wastewater. Fresh domestic wastewater has a musty odor. If the wastewater is allowed to go septic, this odor will significantly change to a rotten egg odor associated with the production of hydrogen sulfide (H_2S).

(3) Temperature—The temperature of wastewater is commonly higher than that of the water supply because of the addition of warm water from households and industrial plants. However, significant amounts of infiltration or storm water flow can cause major temperature fluctuations.

(4) Flow—The actual volume of wastewater is commonly used as a physical characterization of wastewater and is normally expressed in terms of gallons per person per day.

Most treatment plants are designed using an expected flow of 100 to 200gallons per person per day. This figure may have to be revised to reflect the degree of infiltration or storm flow the plant receives. Flow rates will vary throughout the day. This variation, which can be as much as 50% to 200% of the average daily flow is known as the diurnal flow variation.

Chemical Characteristics

In describing the chemical characteristics of wastewater, the discussion generally includes topics such as organic matter, the measurement of organic matter, inorganic matter, and gases. For the sake of simplicity, in this section we specifically describe chemical characteristics in terms of alkalinity, BOD, chemical oxygen demand (COD), dissolved gases, nitrogen compounds, pH, phosphorus, solids (organic, inorganic, suspended, and dissolved solids), and water.

(1) Alkalinity—this is a measure of the wastewater's capability to neutralize acids. It is measured in terms of bicarbonate, carbonate, and hydroxide alkalinity. Alkalinity is essential to buffer (hold the neutral pH) of the wastewater during the biological treatment processes.

(2) Biochemical oxygen demand—this is a measure of the amount of biodegradable matter in the wastewater, normally measured by a 5-d test conducted at 20℃. The BOD_5 domestic waste is normally in the range of 100 to 300mg/L.

(3) Chemical oxygen demand—this is a measure of the amount of oxidizable matter present in the sample. The COD is normally in the range of 200 to 500mg/L. The presence of industrial wastes can increase this significantly.

(4) Dissolved gases—these are gases that are dissolved in wastewater. The specific gases and normal concentrations are based upon the composition of the wastewater. Typical domestic wastewater contains oxygen in relatively low concentrations, carbon dioxide, and hydrogen sulfide (if septic conditions exist).

(5) Nitrogen compounds—the type and amount of nitrogen present will vary from the raw wastewater to the treated effluent. Nitrogen follows a cycle of oxidation and reduction. Most of the nitrogen in untreated wastewater will be in the forms of organic nitrogen and ammonia nitrogen. Laboratory tests exist for determination of both of these forms. The sum of these two forms of nitrogen is also measured and is known as total kjeldahl nitrogen (TKN). Wastewater will normally contain between 20 to 85mg/L of nitrogen. Organic nitrogen will normally be in the range of 8 to 35mg/L, and ammonia nitrogen will be in the range of 12 to 50mg/L.

(6) pH—this is a method of expressing the acid condition of the wastewater. pH is expressed on a scale of 1 to 14. For proper treatment, wastewater pH should normally be in the range of 6.5 to 9.0 (ideally 6.5 to 8.0).

(7) Phosphorus—this element is essential to biological activity and must be present in

at least minimum quantities or secondary treatment processes will not perform. Excessive amounts can cause stream damage and excessive algal growth. Phosphorus will normally be in the range of 6 to 20mg/L. The removal of phosphate compounds from detergents has had a significant impact on the amounts of phosphorus in wastewater.

(8) Solids—most pollutants found in wastewater can be classified as solids. Wastewater treatment is generally designed to remove solids or to convert solids to a form that is more stable or can be removed. Solids can be classified by their chemical composition (organic or inorganic) or by their physical characteristics (settleable, floatable, and colloidal). Concentration of total solids in wastewater is normally in the range of 350 to 1200mg/L.

(a) Organic solids—consists of carbon, hydrogen, oxygen, nitrogen and can be converted to carbon dioxide and water by ignition at 550℃, also known as fixed solids or loss on ignition.

(b) Inorganic solids—mineral solids that are unaffected by ignition, also known as fixed solids or ash.

(c) Suspended solids—these solids will not pass through a glass fiber filter pad and can be further classified as Total Suspended Solids (TSS), volatile suspended solids, and fixed suspended solids. Suspended solids can also be separated into three components based on settling characteristics: settleable solids, floatable solids, and colloidal solids. Total suspended solids in wastewater are normally in the range of 100 to 350mg/L.

(d). Dissolved solids—these solids will pass through a glass fiber filter pad and can also be classified as Total Dissolved Solids (TDS), volatile dissolved solids, and fixed dissolved solids. TDS are normally in the range of 250 to 850mg/L.

(9) Water—this is always the major constituent of wastewater. In most cases water makes up 99.5% to 99.9% of the wastewater. Even in the strongest wastewater, the total amount of contamination present is less than 0.5% of the total and in average strength wastes it is usually less than 0.1%.

Biological Characteristics and Processes

After undergoing physical aspects of treatment in preliminary and primary treatment, wastewater still contains some suspended solids and other solids that are dissolved in the water. In a natural stream, such substances are a source of food for protozoa, fungi, algae, and several varieties of bacteria. In secondary wastewater treatment, these same microscopic organisms (which are one of the main reasons for treating wastewater) are allowed to work as fast as they can to biologically convert the dissolved solids to suspended solids that will physically settle out at the end of secondary treatment.

Raw wastewater influent typically contains millions of organisms. The majority of these organisms are nonpathogenic, but several pathogenic organisms may also be present. (These may include the organisms responsible for diseases such as typhoid, tetanus, hepatitis, dysentery, gastroenteritis, and others.）

Many of the organisms found in wastewater are microscopic (microorganisms); they include algae, bacteria, protozoa (e. g. , amoeba, flagellates, free-swimming ciliates, and stalked ciliates), rotifers, and viruses. Table 11-1 is a summary of typical domestic wastewater characteristics.

Typical Domestic Wastewater Characteristics　　　　Table 11-1

Characteristic	Typical Characteristic		
Color	Gray	BOD	100~300mg/L
Odor	Musty	COD	200~500mg/L
TSS	100~350mg/L	Total nitrogen	20~85mg/L
DO	>1.0mg/L	Total phosphorus	6~20mg/L
pH	6.5~9.0	Fecal coliform	500000~3000000MPN/100mL
Flow	100~200gal/(person·d)		

3. Wastewater Collection Systems

Wastewater collection systems collect and convey wastewater to the treatment plant. The complexity of the system depends on the size of the community and the type of system selected. Methods of collection and conveyance of wastewater include gravity systems, force main systems, vacuum systems, and combinations of all three types of systems.

Gravity Collection System

In a gravity collection system, the collection lines are sloped to permit the flow to move through the system with as little pumping as possible. The slope of the lines must keep the wastewater moving at a velocity (speed) of 2 to 4ft/sec. Otherwise, at lower velocities, solids will settle out and cause clogged lines, overflows, and offensive odors. To keep collection systems lines at a reasonable depth, wastewater must be lifted (pumped) periodically so that it can continue flowing downhill to the treatment plant. Pump stations are installed at selected points within the system for this purpose.

Force Main Collection System

In a typical force main collection system, wastewater is collected to central points and pumped under pressure to the treatment plant. The system is normally used for conveying wastewater long distances. The use of the force main system allows the wastewater to flow to the treatment plant at the desired velocity without using sloped lines. It should be noted that the pump station discharge lines in a gravity system are considered to be force mains since the content of the lines is under pressure.

Vacuum System

In a vacuum collection system, wastewaters are collected to central points and then drawn toward the treatment plant under vacuum. The system consists of a large amount of mechanical equipment and requires a large amount of maintenance to perform properly. Generally, the vacuum type collection systems are not economically feasible.

Pumping Stations

Pumping stations provide the motive force (energy) to keep the wastewater moving at the desired velocity. They are used in both the force main and gravity systems. They are designed in several different configurations and may use different sources of energy to move the wastewater (i. e., pumps, air pressure or vacuum). One of the more commonly used types of pumping station designs is the wet well/dry well design.

Vocabulary and Phrases

waste　*n.* 废物，排泄物
infiltration　*n.* 渗透
discharge　*vt. n* 排放，放出
bacteria　*n.* 细菌
virus　*n.* [微] 病毒，滤过性微生物
microorganism　*n.* [微生] 微生物，微小动植物
household　*adj.* 家庭的，生活的
trash　*n.* 垃圾，废物
garbage　*n.* 垃圾，废物
chemical　*n.* 化学制品，化学药品
dye　*n.* 染料，染色
acid　*n.* [化] 酸，*adj.* 酸的
grit　*n.* 粗砂
gravel　*n.* 砂砾，砂砾层
brownish-gray　*n.* 褐灰色
septic　*adj.* 腐败的，*n.* 腐烂物
volume　*n.* 体积，量
nitrogen compound　氮化合物
organic　*adj.* 有机的
inorganic　*adj.* 无机的
suspended　*adj.* 悬浮的
dissolved　*adj.* 溶解的
neutralize　*v.* 中和，使（酸和碱）进行中和

bicarbonate　*n.* [化] 重碳酸盐
carbonate　*n.* [化] 碳酸盐；*vt.* 使变成碳酸盐
hydroxide　*n.* 氢氧化物，羟化物
buffer　*v.* 缓冲
biodegradable　*adj.* 能生物分解的
carbon dioxide　*n.* [化] 二氧化碳
oxidizable　*adj.* [化] 可氧化的
hydrogen sulfide　氢化硫
oxidation　*n.* [化] 氧化
reduction　*n.* [化] 还原
total kjeldahl nitrogen (TKN) 总凯氏氮
settleable　*adj.* 可沉的
floatable　*adj.* 可漂浮的
colloidal　*adj.* 胶体的
ignition　*n.* 点火，点燃，燃烧
mineral　*n.* 矿物，矿石
Total Suspended Solid (TSS) 总悬浮固体
Volatile Suspended Solid (VSS) 挥发性悬浮固体
Total Dissolved Solids (TDS) 总溶解性固体
protozoa　*n.* 原生动物
fungi　*n.* 真菌
influent　*adj.* 流入的；*n.* 进水
nonpathogenic　*adj.* [医] 非病原的，不

致病的
tetanus　*n*. [医] 破伤风
hepatitis　*n*. [医] 肝炎
dysentery　*n*. [医] 痢疾
gastroenteritis　*n*. [医] 肠胃炎

amoeba　*n*. 阿米巴，变形虫
flagellate　*n*. 鞭毛虫
rotifer　*n*. 轮虫
fecal coliform　粪大肠菌
overflow　*n*. 溢流

Notes

　　discharges into the sewer system.　排放到污水管网系统
　　be discharged from industrial processes into the collection system　从工业加工过程排放到污水收集系统
　　storm water runoff　暴雨径流
　　be characterized in terms of　根据……刻画
　　be caused by gases produced by the decomposition of organic matter　由有机物质分解产生的气体引起
　　the production of hydrogen sulfide (H_2S).　硫化氢的产生
　　is normally expressed in terms of gallons per person per day　通常表达为每人每天的加仑数
　　the average daily flow　日均流量
　　in the range of　在……范围
　　chemical oxygen demand　化学需氧量
　　are based upon the composition of the wastewater　基于污水的成分
　　vary from　不同
　　be in the forms of organic nitrogen and ammonia nitrogen　以有机氮和氨氮形式存在
　　is essential to biological activity　对生物活性是必要的
　　have a significant impact on　对……有重大影响.
　　be classified as　被分为……（类）
　　be classified by　通过……分类
　　be generally designed to remove　通常设计为去除……
　　can be converted to　可以转化为……
　　be separated into three components based on settling characteristics　根据沉淀性能被分为三种成分
　　can to biologically convert the dissolved solids to suspended solids　可以生物转化溶解态固体为悬浮固体
　　gravity collection system　重力收集系统
　　force main collection system　压力收集系统
　　vacuum collection system　真空收集系统
　　be economically feasible　是经济可行的
　　pump station　泵站

Question and Exercises

1. Please describe different kinds of wastewater according to the sources of flows.
2. How to describe characteristics of a wastewater?
3. What's the definition of BOD_5?
4. What are the main parts of a wastewater collection system?

Unit 12 Wastewater Treatment

Wastewater treatment is designed to use the natural purification processes (self-purification processes of streams and rivers) to the maximum level possible. It is also designed to complete these processes in a controlled environment rather than over many miles of a stream or river. Moreover, the treatment plant is also designed to remove other contaminants that are not normally subjected to natural processes, as well as treating the solids that are generated through the treatment unit steps. The typical wastewater treatment plant is designed to achieve many different purposes: protect public health, protect public, water supplies, protect aquatic life, preserve the best uses of the waters, protect adjacent lands.

Wastewater treatment is a series of steps. Each of the steps can be accomplished using one or more treatment processes or types of equipment. The major categories of treatment steps are:

(1) Preliminary treatment—removes materials that could damage plant equipment or would occupy treatment capacity without being treated.

(2) Primary treatment—removes settleable and floatable solids.

(3) Secondary treatment—removes BOD and dissolved and colloidal suspended organic matter by biological action. Organics are converted to stable solids, carbon dioxide and more organisms.

(4) Advanced waste treatment—uses physical, chemical, and biological processes to remove additional BOD, solids and nutrients (not present in all treatment plants).

(5) Disinfection—removes microorganisms to eliminate or reduce the possibility of disease when the flow is discharged.

(6) Sludge treatment—stabilizes the solids removed from wastewater during treatment, inactivates pathogenic organisms, and reduces the volume of the sludge by removing water.

The various treatment processes described above are discussed in detail later.

1. Preliminary Treatment

The initial stage in the wastewater treatment process (following collection and influent pumping) is preliminary treatment. Raw influent entering the treatment plant may contain many kinds of materials (trash). The purpose of preliminary treatment is to protect plant equipment by removing these materials that could cause clogs, jams, or excessive wear to plant machinery. In addition, the removal of various materials at the beginning of the treatment process saves valuable space within the treatment plant.

Preliminary treatment may include many different processes. Each is designed to remove a specific type of material—a potential problem for the treatment process.

Screening

The purpose of screening is to remove large solids, such as rags, cans, rocks, branches, leaves, roots, etc., from the flow before the flow moves on to downstream processes.

A bar screen traps debris as wastewater influent passes through. Typically, a bar screen consists of a series of parallel, evenly spaced bars or a perforated screen placed in a channel. The wastestream passes through the screen and the large solids (screenings) are trapped on the bars for removal.

The bar screen may be coarse (2 to 4-in. openings) or fine (0.75 to 2.0-in. openings). The bar screen may be manually cleaned or mechanically cleaned.

Grit Removal

The purpose of grit removal is to remove the heavy inorganic solids that could cause excessive mechanical wear. Grit is heavier than inorganic solids and includes, sand, gravel, clay, egg shells, coffee grounds, metal filings, seeds, and other similar materials.

There are several processes or devices used for grit removal. All of the processes are based on the fact that grit is heavier than the organic solids, which should be kept in suspension for treatment in following processes. Grit removal may be accomplished in grit chambers or by the centrifugal separation of sludge. Processes use gravity and velocity, aeration, or centrifugal force to separate the solids from the wastewater.

Gravity and velocity controlled grit removal is normally accomplished in a channel or tank where the speed or the velocity of the wastewater is controlled to about 1 foot per second (ideal), so that grit will settle while organic matter remains suspended. As long as the velocity is controlled in the range of 0.7 to 1.4 ft/sec the grit removal will remain effective. Velocity is controlled by the amount of water flowing through the channel, the depth of the water in the channel, the width of the channel, or the cumulative width of channels in service.

2. Primary Treatment (Sedimentation)

The purpose of primary treatment (primary sedimentation or primary clarification) is to remove settleable organic and floatable solids. Normally, each primary clarification unit can be expected to remove 90% to 95% settleable solids, 40% to 60% TSS, and 25% to 35% BOD.

Sedimentation may be used throughout the plant to remove settleable and floatable solids. It is used in primary treatment, secondary treatment, and advanced wastewater treatment processes. In this section, we focus on primary treatment or primary clarification, which uses large basins in which primary settling is achieved under relatively quiescent conditions. Within these basins, mechanical scrapers collect the primary settled solids into a hopper where they are pumped to a sludge-processing area. Oil, grease, and other floating materials (scum) are skimmed from the surface. The effluent is discharged over weirs into a collection trough.

Process Description

In primary sedimentation, wastewater enters a settling tank or basin. Velocity is reduced to approximately 1 ft/min.

Solids that are heavier than water settle to the bottom, while solids that are lighter than water float to the top. Settled solids are removed as sludge and floating solids are removed as scum. Wastewater leaves the sedimentation tank over an effluent weir and on to the next step in treatment. Detention time, temperature, tank design, and condition of the equipment control the efficiency of the process.

Overview of Primary Treatment

(1) Primary treatment reduces the organic loading on downstream treatment processes by removing a large amount of settleable, suspended, and floatable materials.

(2) Primary treatment reduces the velocity of the wastewater through a clarifier to approximately 1 to 2 ft/min, so that settling and floatation can take place. Slowing the flow enhances removal of suspended solids in wastewater.

(3) Primary settling tanks remove floated grease and scum, remove the settled sludge solids, and collect them for pumped transfer to disposal or further treatment.

(4) Clarifiers used may be rectangular or circular. In rectangular clarifiers, wastewater flows from one end to the other, and the settled sludge is moved to a hopper at the one end, either by flights set on parallel chains or by a single bottom scraper set on a traveling bridge. Floating material (mostly grease and oil) is collected by a surface skimmer.

(5) In circular tanks, the wastewater usually enters at the middle and flows outward. Settled sludge is pushed to a hopper in the middle of the tank bottom, and a surface skimmer removes floating material.

(6) Factors affecting primary clarifier performance include: rate of flow through the clarifier, wastewater characteristics (strength; temperature; amount and type of industrial waste; and the density, size, and shapes of particles), performance of pretreatment processes, nature and amount of any wastes recycled to the primary clarifier.

3. Secondary Treatment

The main purpose of secondary treatment (sometimes referred to as biological treatment) is to provide BOD removal beyond what is achievable by primary treatment.

There are three commonly used approaches, and all take advantage of the ability of microorganisms to convert organic wastes (via biological treatment) into stabilized, low-energy compounds. Two of these approaches, the trickling filter (and its variation, the RBC) and the activated sludge process, sequentially are followed the normal primary treatment.

The third, ponds (oxidation ponds or lagoons), can provide equivalent results without preliminary treatment.

In this section, we present a brief overview of the secondary treatment process firstly. Then shift focus to the trickling filter and the activated sludge process.

Secondary treatment refers to those treatment processes that use biological processes to convert dissolved, suspended, and colloidal organic wastes to more stable solids that can either be removed by settling or discharged to the environment without causing harm.

Exactly what is secondary treatment? As defined by the Clean Water Act (CWA), the secondary treatment will produce an effluent with nor more than 30mg/L BOD and 30mg/L TSS.

Most secondary treatment processes decompose solids aerobically, producing carbon dioxide, stable solids, and more organisms. Since solids are produced, all of the biological processes must include some form of solids removal (settling tank, filter, etc.). Secondary treatment processes can be separated into two large categories: fixed film systems and suspended growth systems.

Fixed film systems are processes that use a biological growth (biomass or slime) that is attached to some form of media. Wastewater passes over or around the media and the slime. When the wastewater and slime are in contact, the organisms remove and oxidize the organic solids. The media may be stone, redwood, synthetic materials, or any other substance that is durable (capable of withstanding weather conditions for many years), provides a large area for slime growth and an open space for ventilation, and is not toxic to the organisms in the biomass. Fixed film devices include trickling filters and RBCs.

Suspended growth systems are processes that use a biological growth that is mixed with the wastewater. Typical suspended growth systems consist of various modifications of the activated sludge process.

Trickling Filters

Trickling filters have been used to treat wastewater since the 1890s. It was found that if settled wastewater was passed over rock surfaces, slime grew on the rocks and the water became cleaner. Today we still use this principle, but in many installations we use plastic media instead of rocks.

In most wastewater treatment systems, the trickling filter follows primary treatment and includes a secondary settling tank or clarifier. Trickling filters are widely used for the treatment of domestic and industrial wastes. The process is a fixed film biological treatment method designed to remove BOD and suspended solids.

A trickling filter consists of a rotating distribution arm that sprays and evenly distributes liquid wastewater over a circular bed of fist-sized rocks, other coarse materials, or synthetic media. The spaces between the media allow air to circulate easily so that aerobic conditions can be maintained. The spaces also allow wastewater to trickle down through, around, and over the media. A layer of biological slime that absorbs and consumes the wastes trickling through the bed covers the media material. The organisms aerobically decompose the solids and produce more organisms and stable wastes that either become part of the slime or are discharged back into the wastewater flowing over the media. This slime consists mainly of bacteria, but it may also include algae, protozoa, worms, snails, fungi, and insect larvae. The accumulating slime occasionally sloughs off (sloughings) individual media materials and is collected at the bottom of the filter, along with the treated wastewater, and passed on to the secondary settling tank where it is removed.

The overall performance of the trickling filter is dependent on hydraulic and organic loading, temperature, and recirculation.

Activated Sludge Process

The activated sludge process is a treatment technique in which wastewater and reused biological sludge full of living microorganisms are mixed and aerated. The biological solids are then separated from the treated wastewater in a clarifier and are returned to the aeration process or wasted.

The microorganisms are mixed thoroughly with the incoming organic material, and they grow and reproduce by using the organic material as food. As they grow and are mixed with air, the individual organisms cling together (flocculate). Once flocculated, they more readily settle in the secondary clarifiers.

The wastewater being treated flows continuously into an aeration tank where air is injected to mix the wastewater with the returned activated sludge and to supply the oxygen needed by the microbes to live and feed on the organics. Aeration can be supplied by injection through air diffusers in the bottom of tank or by mechanical aerators located at the surface.

The mixture of activated sludge and wastewater in the aeration tank is called the mixed liquor. The mixed liquor flows to a secondary clarifier where the activated sludge is allowed to settle.

The activated sludge is constantly growing, and more is produced than can be returned for use in the aeration basin. Some of this sludge must be wasted to a sludge handling system for treatment and disposal. The volume of sludge returned to the aeration basins is normally 40% to 60% of the wastewater flow. The rest is wasted.

4. Advanced Wastewater Treatment

Advanced wastewater treatment is defined as the methods and processes that remove more contaminants (suspended and dissolved substances) from wastewater than are taken out by conventional biological treatment. In other words, advanced wastewater treatment is the application of a process or system that follows secondary treatment or that includes phosphorus removal or nitrification in conventional secondary treatment.

Advanced wastewater treatment is used to augment conventional secondary treatment because secondary treatment typically removes only between 85% and 95% of the BOD and TSS in raw sanitary sewage. Generally, this leaves 30mg/L or less of BOD and TSS in the secondary effluent. To meet stringent water-quality standards, this level of BOD and TSS in secondary effluent may not prevent violation of water—quality standards—the plant may not make permit. Thus, advanced wastewater treatment is often used to remove additional pollutants from treated wastewater.

In addition to meeting or exceeding the requirements of water-quality standards, treatment facilities use advanced wastewater treatment for other reasons as well. For example, conventional secondary wastewater treatment is sometimes not sufficient to protect the aquatic environment. This is the case when periodic flow events occur in a stream; the stream may not provide the amount of dilution of effluent needed to maintain the necessary DO levels for aquatic organism survival.

Secondary treatment has other limitations. It does not significantly reduce the effluent concentration of nitrogen and phosphorus (important plant nutrients) in sewage. If discharged into lakes, these nutrients contribute to algal blooms and accelerated eutrophication (lake aging). Also, the nitrogen in the sewage effluent may be present mostly in the form of ammonia compounds. If in high enough concentration, ammonia compounds are toxic to aquatic organisms. Yet another problem with these compounds is that they exert a nitrogenous oxygen demand in the receiving water as they convert to nitrates. This process is called nitrification.

Advanced wastewater treatment can remove more than 99% of the pollutants from raw sewage and can produce an effluent of almost potable (drinking) water quality. However, advanced treatment is not free. The cost of advanced treatment for operation and maintenance as well as for retrofit of present conventional processes is very high (sometimes doubling the cost of secondary treatment).

A plan to install advanced treatment technology calls for careful study; the benefit-to-cost ratio is not always big enough to justify the additional expense. The term tertiary treatment is commonly used as a synonym for advanced wastewater treatment. These two terms do not have precisely the same meaning. Tertiary suggests a third step that is applied after primary and secondary treatment.

Vocabulary and Phrases

contaminant *n.* 致污物，污染物
disinfection *n.* 消毒
inactivate *v.* 使不活泼，阻止活动，钝化
bar screen *n.* 格栅
downstream *adv.* 下游地；*adj.* 下游的

debris *n.* 碎片，残骸
aeration *n.* 曝气
sedimentation *n.* 沉淀
quiescent *adj.* 静止的
hopper *n.* 泥斗

scum　　n. 浮渣，
weir　　n. 堰，
trough　　n. 槽，水槽
clarifier　　n. 沉淀池
rectangular　　adj. 矩形的，成直角的
circular　　adj. 圆形的
scraper　　n. 刮泥机
lagoon　　n. 污水池；氧化塘
decompose　　v. 分解
aerobically　　adv. 好氧地
filter　　n. 滤池
film　　n. 膜，生物膜
media　　n. 载体

ventilation　　n. 通风，流通空气
biomass　　n. （单位面积或体积内）生物的数量
installation　　n. [计] 安装，装置
absorb　　vt. 吸收，吸引
worm　　n. 虫，蠕虫
larvae.　　n. 幼虫
slough　　vi. 剥落
flocculate　　v. 絮凝，絮结
microbe　　n. 微生物，细菌
nitrification　　n. [化] 硝化作用
dilution　　n. 稀释，稀释法
nitrate　　n. [化] 硝酸盐
potable　　adj. 适于饮用的

Notes

self-purification processes of rivers　　河流的自净过程
preliminary treatment　　预处理
primary treatment　　一(初)级处理
secondary treatment　　二级处理
biological action　　生物作用
advanced waste treatment　　废水深度处理，废水高级处理
manually cleaned or mechanically cleaned.　　人工清理或机械清理
grit removal may be accomplished in grit chambers or by the centrifugal separation of sludge　　粗砂可以通过污泥离心分离或者在沉砂池完成
primary sedimentation or primary clarification　　一级沉淀
secondary treatment　　二级沉淀
be skimmed from the surface　　从表面撇去
detention time　　停留时间
trickling filter and RBC　　滴滤池和生物转盘
a trickling filter consists of a rotating distribution arm　　滴滤池包括旋转布水臂
returned activated sludge　　回流污泥
air diffuser　　空气扩散器
mixed liquor　　混合液

Question and Exercises

1. What are the major categories of wastewater treatment steps?
2. What's the purpose of primary treatment?
3. What's the main purpose of secondary treatment?
4. Which categories can secondary treatment processes be separated into mainly?

5. What's the definition of advanced wastewater treatment?

Further Reading

Processes of Wastewater Treatment

1. Basic Wastewater Treatment Processes

Physical **Processes**

Physical processes were some of the earliest methods to remove solids from wastewater, usually by passing wastewater through screens to remove debris and solids. In addition, solids that are heavier than water will settle out from wastewater by gravity. Particles with entrapped air float to the top of water and can also be removed.

These physical processes are employed in many modern wastewater treatment facilities today.

Biological **Processes**

In nature, bacteria and other small organisms in water consume organic matter in sewage, turning it into new bacterial cells, carbon dioxide, and other by-products. The bacteria normally present in water must have oxygen to do their part in breaking down the sewage. In the 1920s, scientists observed that these natural processes could be contained and accelerated in systems to remove organic material from wastewater.

With the addition of oxygen to wastewater, masses of microorganisms grew and rapidly metabolized organic pollutants. Any excess microbiological growth could be removed from the wastewater by physical processes.

Chemical **Processes**

Chemicals can be used to create changes in pollutants that increase the removal of these new forms by physical processes. Simple chemicals such as alum, lime or iron salts can be added to wastewater to cause certain pollutants, such as phosphorus, to floc or bunch together into large, heavier masses which can be removed faster through physical processes.

Over the past 30years, the chemical industry has developed synthetic inert chemicals know as polymers to further improve the physical separation step in wastewater treatment. Polymers are often used at the later stages of treatment to improve the settling of excess microbiological growth or biosolids.

2. Stages of Wastewater Treatment

There are basically three types of stages or processes that take place to render wastewater for disposal. These processes are called primary, secondary, and tertiary treatment. Likewise, there are three types of treatment plants—primary, secondary, and tertiary—that reduce the pollutant load in wastewater and chlorinate it before discharging the effluent into outfall sewer.

Most treatment plants have primary treatment (physical removal of floatable and settleable solids) and secondary treatment (the biological removal of dissolved solids).

Primary Treatment

Primary treatment, essentially a physical process, includes the removal of settleable and floating residues. The process order is screening, followed by grinding or shredding if the facility or equipment is present (in this subcourse, we will assume that it is), and then grit removal, primary clarification of sedimentation, and sludge removal (Figure 12-1).

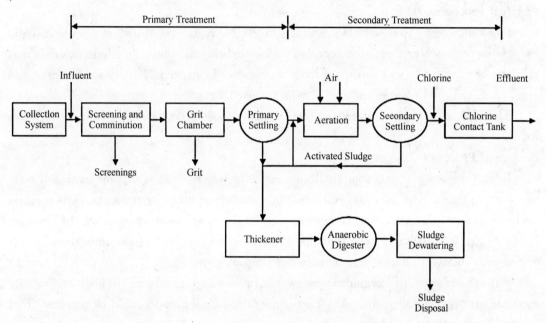

Figure 12-1 Schematic of conventional wastewater treatment processes.

(1) Screening and/or grinding. Screening and grinding prevent the entrance into the treatment plant of large objects such as rags, pieces of wood, dead animals, and other objects which may clog pipes, pumps, or other mechanical equipment. Types of equipment most commonly used are fixed-bar screens and communitors.

(a) Fixed-bar screens—a series of evenly spaced bars set in the wastewater flow.

(b) Communitors—machines for cutting or shredding solids and passing them to the wastewater flow.

(2) Grit removal. Grit removal is the preliminary removal of nonfiltrable, inorganic material (sand, gravel, cinders, etc.). This material, if not removed, will damage pumps and other equipment. It will also settle in digesters, thus reducing their capacity and treatment efficiency and necessitating frequent and costly cleaning. Grit removal is accomplished by passing wastewater through a grit chamber that retards the flow enough to permit the grit to settle. There are two types of grit removal units commonly used—the flow rate controlled grit removal unit and the aerated grit removal unit.

(3) Sedimentation. When fresh domestic wastewater stands quiescent or flows very slowly, a considerable portion of the nonfiltrable residue will settle fairly rapidly. Under

average conditions, most of the settling occurs within 1 hour. The quantity and rate of sedimentation that occurs after 2 hours is almost negligible.

Sedimentation may be accomplished in any of the following devices.

(a) Septic tank. A septic tank is a device that receives raw wastewater from a single residence, several residences, a hotel, or an institution. The septic tank retains the wastewater long enough for the settleable residue to accumulate on the bottom. The effluent is discharged through an overflow pipe.

(b) Imhoff tank. An Imhoff tank is a two-story tank that serves a small installation or community. It is a further refinement of the septic tank principle whereby the wastewater enters the upper chamber and, as it passes slowly through, solids settle to the lower chamber. A detention period of 2 to 3 hours in an Imhoff tank will reduce nonfiltrable residue by 45 to 60 percent.

(c) Separate settling tank. The separate settling tank is the most common device used for primary settling at large installations and municipal wastewater systems. It may be circular or rectangular, having a sloped hopper bottom for the accumulation of solids. A 2-hour sedimentation period will usually effect removal of 50 to 60 percent of the suspended matter and 30 to 40 percent of the BOD. With the addition of chemicals to effect coagulation, as in water treatment, sedimentation may remove as much as 75 to 85 percent of the suspended matter and 50 to 70 percent of the total BOD.

(4) Sludge removal. Removal of sludge is required to ensure continued efficient operation of primary treatment equipment. Sludge from septic tanks is normally pumped into commercially operated tank trucks and disposed of by burial or discharge into municipal wastewater systems. Sludge from the bottom compartments of Imhoff tanks is pumped on a periodic basis, as digestion is completed, into sludge drying beds, tank sludge lagoons, or some similar type of disposal facility. Sludge from primary settling tanks is removed continuously, or frequently, to prevent septic conditions from developing in the tank. It is pumped into sludge digesters for complete digestion.

Secondary Treatment

The secondary wastewater treatment plant performs, essentially, a biochemical process, although some of the same physical processes employed in primary treatment are also involved. Wastewater that received only primary treatment usually still contains 50 to 60 percent of the original wastewater material, including all of the dissolved matter, most of the colloidal and finely divided organic matter, and approximately as high a bacterial content as the raw wastewater.

This partly treated wastewater is highly putrescible and will pollute large bodies of receiving water unless further treated. The objectives of secondary treatment are to bring about a rapid oxidation and stabilization of this remaining organic matter, to remove the resulting settleable solids, and to reduce the bacterial content. The following processes are considered secondary treatment processes.

(1) **Filtration.** Filtration effects biological oxidation of organic material on and in beds of stone or sand.

(a) Absorption trenches (subsurface tile fields). These trenches are perforated tile lines laid in gravel beds in loose, porous soil into which the effluent from a septic tank is continuously or intermittently dosed. Oxidation takes place as the effluent percolates through the gravel and soil.

(b) Intermittent sand filters. These are beds of underdrained sand on which settled wastewater is applied. Oxidation takes place in the bed.

(c) Trickling filters. These filters are beds of stone where effluent from primary treatment is intermittently or continuously distributed. Films of organisms that form on the surfaces of the stones stabilize the solids by aerobic methods.

(2) **Aeration.**

(a) Activated sludge process. This process accelerates aeration whereby the effluent from primary settling is mixed with return sludge and agitated continuously in the presence of oxygen. The activated sludge thus formed has the property of absorbing dissolved organic material and converting it into stable substances that will settle.

(b) Oxidation pond. The pond is a relatively large, shallow pond, either natural or artificial, into which settled wastewater is discharged for natural purification under the influence of sunlight and air.

(3) Secondary settling. Sedimentation is an essential step following biochemical processes such as the trickling filter and the activated sludge process. These biochemical processes do not remove organic material; they convert it to a stable form which will settle out by sedimentation. The sludge from final settling tanks is pumped into digestion tanks. In the activated sludge process, a portion is returned and mixed with the settled wastewater entering the aeration tanks.

(4) Sludge digestion. The sludge which settles out during primary and secondary sedimentation is about 95 percent water. The remaining 5 percent is highly putrescible organic matter. It is normally pumped directly into covered tanks and permitted to digest by anaerobic bacterial action. In the Imhoff tank and the septic tank, the sludge settles to the bottom where it is digested anaerobically. Digested sludge is withdrawn and discharged into sludge drying beds. These beds are usually provided with underdrains to facilitate drying. Dried, well-digested sludge is quite stable and is an excellent low-grade fertilizer.

Unit 13 Treatment and Disposal of Solids and Biosolids in the Wastewater

The field of wastewater treatment and engineering is littered with unique and imaginative processes for achieving high degrees of waste stabilization at attractive costs. In practice, few of these "wonder plants" have met expectations, often because they fail to pay sufficient attention to sludge treatment and disposal problems. Currently sludge treatment and disposal accounts for over 50% of the treatment costs in a typical secondary plant, prompting renewed interest in this none-too-glamorous, but essential aspect, of wastewater treatment. This chapter is devoted to the problem of sludge treatment and disposal. The sources and quantities of sludge from various types of wastewater treatment systems are examined, followed by a definition of sludge characteristics. Such solids concentration techniques as thickening and dewatering are discussed next, concluding with considerations for ultimate disposal.

1. Sources of Sludge

Sludges are generated from nearly all phases of wastewater treatment. The first source of sludge in a wastewater treatment facility is the suspended solids from the primary settling tank or clarifier. Ordinarily about 60% of the suspended solids entering the treatment facility become raw primary sludge, which is highly putrescible and very wet (about 96% water).

The removal of BOD is basically a method of wasting energy, and secondary wastewater treatment plants are designed to reduce the high-energy organic material that enters the treatment plant to low-energy chemicals. This process is typically accomplished by biological means, using microorganisms (the "decomposers" in ecological terms) that use the energy for their own life and procreation. Secondary treatment processes such as the popular activated sludge system are almost perfect systems. Their major fault is that the microorganisms convert too little of the high-energy organics to CO_2 and H_2O and too much of it to new organisms. Thus the system operates with an excess of these microorganisms, or waste activated sludge. As defined in the previous chapter, the mass of waste activated sludge per mass of BOD removed in secondary treatment is known as the yield, expressed as kilograms of suspended solids produced per kilogram of BOD removed.

Phosphorus removal processes also invariably end up with excess solids. If lime is used, calcium carbonates and calcium hydroxyapatites solids are formed. Similarly, aluminm sulfate produces solids in the form of aluminum hydroxides and aluminum phosphates. Even so-called "totally biological processes" for phosphorus removal end up with solids. The use of an oxidation pond or marsh for phosphorus removal is possible only if some organics (algae, water hyacinths, fish, etc.) are harvested periodically. The quantities of sludge obtained from various wastewater treatment processes may be calculated.

2. Characteristics of Sludge

The important or relevant characteristics of sludge depend on what is to be done to the sludge. For example, if the sludge is to be thickened by gravity, its settling and compaction rates are important. On the other hand, if the sludge is to be digested anaerobically, the concentrations of volatile compounds, other organic solids, and heavy metals are important. Variability of the sludge is immensely important in the design of sludge handling and disposal operations. In fact, this variability may be stated in terms of three "laws":

(1) No two wastewater sludges are alike in all respects.

(2) Sludge characteristics change with time.

(3) There is no "average sludge".

The first "law" of sludge reflects the fact that no two wastewaters are alike and that if the variable of treatment is added, the sludge produced will have significantly different characteristics. The second "law" is often overlooked. For example, the settling characteristics of chemical sludge from the treatment of plating wastes (e.g., $Pb(OH)_2$, $Zn(OH)_2$, or $Cr(OH)_3$) vary with time simply because of uncontrolled pH changes. Biological sludge is, of course, continually changing, with the greatest change occurring when the sludge changes from aerobic to anaerobic (or vice versa). Not surprisingly, it is quite difficult to design sludge-handling equipment because the sludge may change in some significant characteristic in only a few hours. The third "law" is constantly violated.

Tables showing "average values" for "average sludge" are useful for illustrative and comparative purposes, and as such, are included in this chapter for general information; however, they should not be used for treatment design. Instead, you need to determine the specific and unique characteristics of the sludge that needs treatment. For illustrative purposes only, Table 13-1 shows characteristics of hypothetical, "average sludge". The first characteristic, solids concentration, is perhaps the most important variable, defining the volume of sludge to be handled and determining whether the sludge behaves as a liquid or a solid. The second characteristic, volatile solids, is also important to sludge disposal. Disposal is difficult if the sludge contains high concentrations of volatile solids because gases and odors are produced as the sludge outgases and the volatile substances are degraded. The volatile solids parameter is often interpreted as a biological rather than a physical characteristic, the assumption being that volatile suspended solids are a gross measure of the viable biomass. Another important parameter, especially in regard to the ultimate disposal, is the concentration of pathogens, both bacteriological and viral. The primary clarifier seems to act as a viral and bacteriological concentrator, with a substantial fraction of these microorganisms existing in the sludge instead of the liquid effluent.

Sludge Characteristics   Table 13-1

Sludge type	Physical				Chemical		
	Solids conc. (mg/L)①	Volatile solids (%)	Yield strength (dyne/cm²)	Plastic viscosity (g/cm-s)	N as N (%)	P as P_2O_5 (%)	K as K_2O (%)
Water	—	—	0	0.01	—	—	—
Raw primary	60000	60	40	0.3	2.5	1.5	0.4
Mixed digested	80000	40	15	0.9	4.0	1.4	0.2
Waste activated	15000	70	0.1	0.06	4.0	3.0	0.5
Alum. ppt	20000	40	—	—	2.0	2.0	—
Lime, ppt	200000	18	—	—	2.0	3.0	—

① Note that sludge with 10000mg/L solids is approximately 1% solids.

Rheological characteristics (degree of plasticity) are one of only a few truly fundamental physical parameters of sludges. Two-phase mixtures like sludges, however, are almost without exception non-Newtonian and thixotropic. Sludges tend to act as pseudoplastics, with an apparent yield **stress** and a plastic viscosity.

3. Sludge Treatment

A great deal of money could be saved, and troubles averted, if sludge disposal could be done as the sludge is drawn off the main process train. Unfortunately, sludges have three characteristics that make such a simple solution unlikely: they are aesthetically displeasing, they are potentially harmful, and they have too much water.

The first two problems are often solved by stabilization, which may involve anaerobic or aerobic digestion. The third problem requires the removal of water, either by thickening or dewatering. The next three sections cover the topics of stabilization, thickening, and dewatering, followed by the considerations of ultimate sludge disposal.

Sludge Stabilization

The objective of sludge stabilization is to reduce the problems associated with sludge odor and putrescence, as well as reducing the hazard presented by pathogenic organisms. Sludge may be stabilized using lime, aerobic digestion, or anaerobic digestion. Lime stabilization is achieved by adding lime, either as hydrated lime ($Ca(OH)_2$) or as quicklime (CaO) to the sludge, which raises the pH to 11 or above. This significantly reduces the odor and helps in the destruction of pathogens. The major disadvantage of lime stabilization is that its odor reduction is temporary. Within days the pH drops and the sludge once again becomes putrescible.

Aerobic digestion is a logical extension of the activated sludge system. Waste activated sludge is placed in dedicated tanks for a very long time, and the concentrated solids are allowed to progress well into the endogenous respiration phase, in which food is obtained only by the destruction of other viable organisms. Both total and volatile solids are thereby reduced. One drawback to this process is that aerobically digested sludges are more difficult to dewater than anaerobic sludges.

Sludge can also be stabilized by anaerobic digestion, as illustrated in Figure 13-1. The biochemistry of anaerobic digestion is a staged process: solution of organic compounds by extracellular enzymes is followed by the production of organic acids by a large and hearty group of anaerobic microorganisms known, appropriately enough, as the acid formers. The organic acids are in turn degraded further by a group of strict anaerobes called methane formers. These microorganisms are the prima donnas of wastewater treatment, getting upset at the least change in their environment. The success of anaerobic treatment depends on maintenance of suitable conditions for the methane formers. Since they are strict anaerobes, they are unable to function in the presence of oxygen and are very sensitive to environmental conditions like pH, temperature, and the presence of toxins. A digester goes "sour" when the methane formers have been inhibited in some way. The acid formers keep chugging away, making more organic acids, thus further lowering the pH and making conditions even worse for the methane formers. Curing a sick digester requires suspension of feeding and, often, massive doses of lime or other antacids.

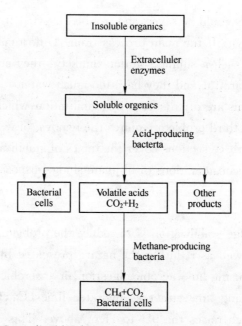

Figure 13-1 Generalized biochemical reactions in anaerobic sludge digestion.

Sludge Thickening

Sludge thickening is a process in which the concentration of solids is increased and the total sludge volume is decreased, but the sludge still behaves like a liquid instead of a solid. Thickening commonly produces sludge solids concentrations in the 3% to 5% range, whereas the point at which sludge begins to have the properties of a solid is between 15 and 20% solids.

The sludge thickening process is gravitational, using the difference between particle and fluid densities to achieve greater compacting of solids.

The advantages of sludge thickening are substantial. When sludge with 1% solid is thickened to 5%, the result is an 80% reduction in volume (Figure 13-2). Reducing from 1% solid to 20% solid, which might be achieved by sludge dewatering, would result in a 95% reduction in volume. The volume reduction translates into considerable savings in treatment, handling, and disposal costs.

Two types of non-mechanical thickening processes are presently in use: the gravity thickener and the flotation thickener.

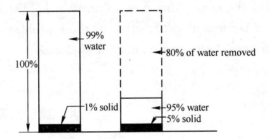

Figure 13-2 Volume reduction owing to sludge thickening

Sludge Dewatering

Unlike sludge thickening, where the treated sludge continues to behave as a liquid, dewatered sludge will behave like a solid after treatment. Dewatering is seldom used as an intermediate process unless the sludge is to be incinerated. Most wastewater plants use dewatering as a final method of volume reduction before ultimate disposal. In the United States, the usual dewatering techniques are sand beds, vacuum filters, pressure filters, belt filters, and centrifuges.

4. Ultimate Disposal

Even after treatment, we are left with a large volume of sludge that needs a final resting place. The choices for ultimate disposal of sludge are limited to air, water, and land. Until quite recently, incineration ("air disposal") was viewed as an effective sludge reduction method, if not exactly an ultimate sludge disposal method (the residual ash still required disposal). However, strict controls on air pollution and increasing concern over global warming are making incineration an increasingly unlikely option. Disposal of sludge in deep water (such as oceans) is decreasing owing to adverse or unknown detrimental effects on aquatic ecology. Land disposal, particularly the use of sludge as fertilizer or soil conditioner, has historically been a favored disposal method, and is currently growing in popularity as other options become more problematic.

Incineration is actually not a method of disposal at all, but rather a sludge treatment step in which the organics are converted to H_2O and CO_2 and the inorganic drop out as a nonputrescible residue. Two types of incinerators have found use in sludge treatment: multiple hearth and fluid bed. The multiple-hearth incinerator, as the name implies, has several hearths stacked vertically, with rabble arms pushing the sludge progressively downward through the hottest layers and finally into the ash pit.

Sludge disposal represents a major headache for many municipalities. Sludge represents the true residues of our civilization, and its composition reflects our style of living, our technological development, and our ethical concerns. "Pouring things down the drain" is our way of getting rid of all manner of unwanted materials, not recognizing that these materials often become part of the sludge that must, ultimately, be disposed of in the environment. All of us need to become more sensitive to these problems and keep potentially harmful materials out of our sewage system and out of sludge.

Vocabulary and Phrases

stabilization n. 稳定化
disposal n. 处理，处置
thickening n. 浓缩
dewatering n. 脱水
putrescible adj. 易腐烂的
yield n. 产量
lime n. 石灰
Calcium n. ［化］钙（元素符号 Ca）
aluminum n. ［化］铝
hyacinth n. ［植］风信子，水葫芦
compaction n. 压缩
digest vi. 消化
anaerobic adj. 厌氧的
aerobic adj. 依靠氧气的，与需氧菌有关的
parameter n. 参数，参量
bacteriological adj. 细菌学的，细菌学上的

viral adj. 滤过性毒菌的，滤过性毒菌引起的
rheological adj. ［物］流变学的
putrescence n. 腐败，腐烂
hydrated lime n. 熟石灰
quicklime n. 生石灰
extracellular adj. ［生］（位于或发生于）细胞外的
enzyme n. ［生化］酶
anaerobes n. ［微］厌氧性生物
toxin n. ［生化］［生］毒素
antacid adj. 中和酸性的，抗酸性的；n. 抗酸剂
gravitational adj. 重力的
incinerate vi. 把…烧成灰，焚烧
detrimental adj. 有害的
hearth n. 炉膛

Notes

waste activated sludge 剩余活性污泥
expressed as kilograms of suspended solids produced per kilogram of BOD removed 表达为去除每千克 BOD 产生的 SS 千克数
the settling characteristics of chemical sludge 化学污泥的沉淀特征
plastic viscosity 塑黏性，塑性黏度
the endogenous respiration phase 内源呼吸阶段

The organic acids are in turn degraded further by a group of strict anaerobes called methane formers. 有机酸被一群严格的厌氧菌即产甲烷菌进一步降解。

A digester goes "sour" when the methane formers have been inhibited in some way. 当产甲烷菌在某种程度上被抑制，消化池就发生酸化。

dewatered sludge 脱水污泥

be viewed as an effective sludge reduction method 被看作有效的污泥减量方法

Questions

1. Which parameters can be used to describe the Characteristics of sludge?
2. What's the objective of sludge stabilization process?
3. Describe the general biochemical reactions in anaerobic sludge digestion.
4. What's the situation of the sludge ultimate disposal at present?

Further Reading

Wastewater Management

The idea of wastewater management is as old as man himself. Simply put, man has struggled through the ages with the problem of what to do with his waste. The painstaking effort of plumber past is evidenced by the ancient drains, grandiose palaces, and bath houses of the Minoan civilization some 4000 years ago.

Man knew instinctively, even in his earliest existence, the importance of allowing animal and human waste to go downstream, yielding to the natural flow of things. He may not have known all of the consequences, but he surely found the prospect of harvesting drinking water from the same area of the stream used for waste distasteful. Now, in the dawning of a new century, wastewater management is still an issue in the forefront. As our ancestors sought to answer that eternal question, we, in a more sophisticated manner today, are still trying to figure out the best way to manage our waste.

1. What is Wastewater Management?

Imagine that you are opening a new business. It is a considerable investment. You have put a lot of time and hard-earned money into it. Would you open your new store without a long-term plan, having no control over future sales or purchasing? Although this question may seem rudimentary, in many parts of the country the onsite wastewater treatment industry has been functioning just this way, without a long-term plan or management program. The dictionary defines management as "the act, manner, or practice of managing, supervising, or controlling." Whether you spend millions or thousands of dollars, or whether the system is part of a public works project or an individual septic tank, there should be some entity responsible for the overall consequences and direction. Most communities already manage their onsite systems to some extent through regulation. But the term "management" as it is used today implies a broader definition. In other words, wastewater systems, particularly onsite systems, need to be managed or controlled, not just technologically, but with a broad concept connecting individuals, communities, local officials, and regulatory agencies if failures and malfunctions are to be avoided.

2. Why is Management Important?

Trends and numbers speak volumes about the need for onsite wastewater management today. As we enter the new millennium, population growth is moving more and more homeowners into suburban areas, many relying on onsite wastewater treatment and disposal. The majority of homes in rural America already rely solely on onsite systems. Approximately one fourth of the estimated 109 million housing units in the United States are served with septic tanks or cesspools, according to a 1995 American Housing Survey (AHS). During that year alone, more than 2.5 million septic tanks in America were reported as malfunctioning (or as having a total breakdown of the system). Graham Knowles of the National Small Flows Clearinghouse's (NSFC) National Onsite Demonstration Project (NODP) authored a report titled "Septic Stats, An Overview," In the report, he combines U.S. Department of Commerce Bureau of Census statistics with the AHS data to establish septic tank trends. Knowles' report projects that by the year 2025 there will be 40million housing units with septic tanks. If the current trend continues, that could mean as many as 4million septic systems could be malfunctioning by 2025. If this projection becomes a reality, the necessity for greater control through management programs should be self-evident.

As In Olden Times turning the clock back to study how wastewater systems evolved and how management programs have fared throughout the years can be a useful tool. Close to 4000 years ago, approximately 1700 B.C., the Minoan Palace of Knossos on the isle of Crete featured four separate drainage systems that emptied into great sewers constructed of stone. The palace latrine was the world's first flushing toilet with a wooden seat and a small reservoir of water. From 3000 to 1500 B.C., early plumbers laid sewage and drainage systems. Archaeologists have discovered underground channels that remained virtually unchanged for centuries. Ancient gravity sewers were developed in response to the density of populations living in close proximity or in cities, according to Peter Casey, program coordinator for the NSFC. These large central systems were actually analagous to sewers developed in the 1800s in London and other large cities. During these times, there were many outbreaks of various diseases, such as dysentery, cholera, infectious hepatitis, typhoid and paratyphoid, and various other types of diarrhea. "The biggest health benefit of the 20th century was brought about by the purification of drinking water and treatment of wastewater," Casey added. "It increased life expectancies and had a tremendous impact on man's health and survival." Casey said in 1870, the average person could expect to live to be 40 years old. By 1900, that age climbed to 47 with steady increases throughout the decades since. Today, the average person in a developed country can expect to live into his or her 70s or beyond. According to a 1997 U.S. Environmental Protection Agency (EPA) document, onsite wastewater systems have been around since the mid-1800s. "In rural areas during the mid-1800s to the early 1900s, sanitation was not a problem because the water supply was hand carried or pumped. No water was required for the privy," Casey said. "If it became full, they would simply cover it and dig a new privy. "Many people are familiar with the early 1900 image of a splintered, wooden shed, usually with one door and a hole in the floor, as the rural family's outhouse. Chamber pots were dumped outside or in the privy. After the 1930s when electricity and gas became available in rural areas, the need for onsite treatment arose because of the increased volume of liquids in the wastewater, Casey said.

"Once farmhouses got electricity and indoor plumbing and the conveniences of the large cities, the flows became too great and caused problems," Casey added. Suddenly, there was running water in the house, making way for baths, showers, and flush toilets. Cesspools were the earliest form of onsite system in response to increased water use. They were usually just a large, covered hole dug in an inaccessible area. "With the move to the suburbs in the 1940s, we saw dense housing units trying to use all this water on half-acre lots. There was no place for all of the water to go," Casey said. Septic tank systems, specifically, have been used for wastewater treatment since the turn of the century, according to a report from a 1994 University of Waterloo, Ontario, conference. The report, by Richard J. Otis and Damann L. Anderson, adds that the use of septic tanks did not become widespread until after World War II when the suburban housing boom outgrew the rate of sewer construction.

3. Regulation Begins

The Otis and Anderson report notes that in the 1950s, states began to adopt regulations to provide a universal basis for the design and installation of septic tank systems. These early codes did not, however, provide much in the way of broad management or prevention of system failure. The programs regulating the installation and use of onsite systems could not keep up with the increasing demand. The report adds, "Today, it is generally recognized that past approaches to managing onsite wastewater treatment systems use are no longer adequate... The failure of these systems to gain acceptance as effective and permanent facilities is due primarily to shortcomings in management programs." The report notes that the biggest assumption at that time was that onsite systems would ultimately be replaced by central sewerage. Despite this, some early management programs did arise. In 1954, Fairfax County, Virginia, established an onsite wastewater management program when the board of supervisors there directed the health department to develop a program that would prevent future septic system failures. The management plan focused on the planning, design, and construction review of septic tank systems through an extensive permit program. Under this early management plan, the county was in charge of site evaluation, design review, installation supervision, monitoring, and public education while the homeowners were responsible for the operation, maintenance, and repair of the systems. With the establishment of the wastewater treatment construction grant program under the Federal Water Pollution Control Act Amendments in 1956, the focus continued on construction of centralized sewers. Throughout the 1960s, the concept of septic tank systems being a temporary solution continued.

4. Onsite Systems Are Recognized

In the 1970s, millions of dollars were still being spent on constructing sewers and centralized wastewater treatment facilities, while at the same time, many federal and state agencies started to consider regulating and managing onsite systems as part of environmental pollution control issues. Throughout the 1970s, management programs sprang up across the country.

By 1974, many states had identified the need for better managed individual onsite systems through studies conducted under the Federal Water Pollution Control Act, Section 208. More importantly for the onsite system industry, EPA regulations required the inclusion of a cost-effectiveness analysis of alternatives by all applications initiated after April 30, 1974, under the Federal Construction Grants Program. In fact, the 1977 Clean Water Act (CWA) Amendments required communities to examine alternatives to conventional systems. In addition, the NSFC was established by Congress as part of the amendments to provide technical information and assistance to small communities across the country. In the 1978"Report to the Congress,"the Comptroller General of the U.S. stated that septic systems can function as effectively and permanently as central facilities and are a cost-effective alternative to sewage treatment plants, adding that "EPA and other federal agencies should increase the acceptance of septic systems by requiring established public management entities to control their design, installation, and operation." If the 1970s could be remembered as the decade septic systems became recognized as permanent wastewater treatment options, then the 1980s might be remembered as the decade of onsite exploration. During the 1980s, the field progressed significantly, and many onsite management system models were developed. In the 1990s, the issue of management has been tweaked further, focusing on the development of adequate monitoring and comprehensive management systems. In the 1997"Response to Congress on Use of Decentralized Wastewater Treatment Systems,"EPA stated that "adequately managed decentralized wastewater systems are a cost-effective and long-term option for meeting public health and water quality goals, particularly in less densely populated areas." Since then, septic tank systems and other alternative systems generally have been recognized not only as environmentally and technologically sound treatment methods, but have been viewed as viable, permanent methods of treatment. In the 1997 report, EPA noted that one of the barriers to implementing decentralized systems is a lack of management programs. To overcome this, EPA recommended development of management programs "on state, regional, or local levels, as appropriate, to ensure that decentralized wastewater systems are sited, designed, installed, operated, and maintained properly and that they continue to meet health and water quality performance standards." As one of the responses to these 1997 recommendations, EPA launched Phase IV of the NODP in 1998.

5. Enter Phase IV

Phase IV of the NODP is a three-year program, focusing on establishing the necessary processes to help small communities develop a broad concept of management for onsite systems. Knowles, program coordinator of Phase IV, has been studying management issues. He commented, "Management programs are imperative today because they will enable communities to control the effectiveness of wastewater treatment and can help ensure public health, improve water quality, and sustain the environment."Knowles said Phase IV's mission has three components: To gather data, information, knowledge and insights concerning all aspects of onsite management systems nationwide. Under this component, Knowles said objectives will be to establish a repository of information and expertise on the topic of onsite management, forming a national database of management systems complete with case studies addressing issues of management approaches, compliance, improvement, and prevention perspectives. Knowles explained that this component seeks to"develop a framework of guiding ideas to assist communities through the process of moving from their current reality toward increasingly effective onsite wastewater management."He added,"The aim is to establish strategies for change, create"a network of interested publics to partner with, for disseminating the onsite management idea and delivering products, tools, and services."To analyze, evaluate, review, and refine onsite management models, methods, and materials at strategically selected sites. Knowles said this component of the project is designed to select suitable sites to pilot onsite management systems. It also will " provide NODP expertise, materials, mentors, management insights, tools, and techniques to communities interested in adopting an onsite management systems approach."This component will document and track management strategies, products, tools, and services to meet differing community needs. Once these components have been met, Knowles said Phase IV ultimately will provide interested communities with practical, hands-on technological and management expertise facilitating community onsite system management programs tailor-made to meet a particular local community's needs. To help the project succeed, NODP IV has enlisted an expert panel, made up of talented individuals in the wastewater field who have made and are continuing to make significant contributions to the evolution and development of onsite management plans.

6. Cranberry Lake's Success

One panel member, Jane Schautz, vice president and director of the Small Towns Environment Program at The Rensselaerville Institute in New York, is working with several communities, studying their onsite management programs. Schautz's role is that of an observer, documenting the progress and noting the plan's assets and possible defects. She defines onsite management as"the systematic monitoring and maintenance of onsite systems to anticipate and/or correct malfunction in order to preserve the life of the system and prevent environmental degradation."The challenging task, she said, is to make a management program work in existing communities that have onsite systems and do not automatically see the benefit of adopting a management system with all of its associated costs that the residents have not paid previously. She cited Cranberry Lake, New Jersey, as an excellent case study of this scenario. One important lesson she has learned is that residents have to be shown there are innumerable benefits to whatever costs might be incurred.

The Cranberry Lake Septic Management System was established in 1990. It is a relatively affluent area. Most of the houses around the lake were built in the 1950s and intended for seasonal use. Because of this, some of the lots are small, approximately 50 by 50 feet. "Year-round occupancy was not expected, but with retirement increasing, more and more people are living there year-round," she said. "With retirees you have to be sensitive to their limited incomes. One fear was that they would be thrown out of their houses if a malfunction were discovered. That made the problem more intense. "Prior to the establishment of the management system, Cranberry Lake had a nitrate problem from failing septic systems and was overgrown by weeds. Schautz said that as a bonus, having the management system for wastewater in place helped the township to successfully secure funding for treating unwanted plants. Schautz believes Cranberry Lake's success should be credited largely to Margaret McGarrity, the "spark plug," or local person who took the initiative to get things moving there. She added that Township Manager Ronald Gatti also played a major role in their success. "Trying to establish a management district is going to be controversial, and people have to be willing to deal with controversy without being damaged," said Schautz. "You have to have savvy people who have the guts to stick with it. A plan is inert until somebody believes in it. You have to have a champion to give any plan a life. "McGarrity, a member of the environmental township commission, was that person for Cranberry Lake. Schautz said McGarrity felt that sewers were inappropriate for the area. "Sewers just take wastewater from one area and move it to someplace else," said Schautz. "McGarrity felt they couldn't afford that for the wells or lake. She looked at all of the components and decided it made no sense to spend money installing septic systems and allowing them to malfunction. "Under the management plan, residents pay a flat $15 fee that covers a three-year period, that extends from one date of pumping to another. By paying the fee, they update their permit by showing proof of pumping. Municipal officers oversee the process. The township board of health is responsible for enforcement. "They have astonishing compliance," said Schautz. Schautz added that the township residents' drinking water is provided by privately owned wells. "People understand that this is all related to maintaining the purity of their lake, as well as preserving their drinking wells. "It took awhile to convince people that this was in their best interest, but they now see that this is an improvement of their relationship with the township government," said Schautz. In fact, Cranberry Lake's management plan has been the model for other communities in the area. She said, "To me one of the most persuasive evidences of the plan's success is that others in the area have seen the results and are taking steps to follow that model. "In the beginning, people were saying, 'why us?' and now after the evidence, not only are they seeing many upgrades to systems made voluntarily, but people are saying, 'why not us?' and have petitioned the township board to include other areas," she added. Schautz cautioned that management plans cannot be established overnight. She quipped, "Starting a wastewater management district is like planting asparagus—the first rule is the ground should have been prepared three years ago. Cranberry Lake did it faster, but very intensively. "Cranberry Lake's first step was education, including presentations at local meetings, seminars, articles in the local newspaper, information booths at community meetings, and insert fliers. Schautz said this process took approximately one year. "McGarrity and Gatti said their success depended on persistence—getting the word out and allowing it to take hold, giving people time to come to their own conclusions," Schautz added. "In this case, their commitment and belief eventually became infectious. "Another helpful aspect of gaining acceptance was that the ordinance was relatively mild. "That way, there was less opposition," she said. "There's no reason to make this harder than it has to be. In fact, they went out of their way to accommodate people. "The management plan gives the township the authority to fine residents $1000 per day or order them to do 90 days of community service for noncompliance, but Schautz said there has rarely been a need to impose those punishments. In addition to having a spark plug, educating the public, and persistence, Schautz believes humor is an imperative component to the key to success. "McGarrity and Gatti livened up their material with graphics and energy," she said. "It really worked for the community. "In the end, Schautz said a management plan must be based on the local culture and philosophy. "Some of the purists say it isn't a management system unless you have inspectors there all the time, tearing up the soil. I'm not saying that doesn't work; but for an older established community, it seems that moving in areas of environmental sensitivity makes sense when people come to the understanding that they are at risk."

7. Getting Utilities into the Plan

Another expert panel member, Bridget Chard, is a Small Communities Project coordinator and a township supervisor from Pillager, Minnesota. She works as a consultant for many townships in Minnesota, helping them implement a management model, called an "Environmental" Subordinate Service District, that can be tailored to meet differing needs. In essence, the model allows local township boards, usually lacking the expertise, time, and experience needed, to develop and maintain a management plan by partnering with the local rural utilities. "The rural utilities are already in place and providing electric power to the rural residents. These residents are already part of the rural electric co-op. Therefore, these utilities are usually more than willing to provide this management service," she explained. "They have the needed assets to oversee the systems. They do the billing, administrative work, and actually manage the wastewater system and therefore relieve the work that the town board would have to do. Essentially this becomes a public-private management system. It's a good set of checks and balances. The townships have the authority to levy onto the residents property taxes for any unpaid service charges. "In this model, enforcement issues are taken care of through a partnership with the township or county. "This is a choice situation," said Chard. "We continually are building new partnerships and better ways to do things. As homeowners and township board people become introduced to this new model, it's always an education process. We do a great deal of informational work up front before we create the districts."

The model allows for different methods of funding, including service charges and/or a property charge. She added, however, that a township and its residents sometimes find other alternative and equitable methods for financing their projects. Chard, who is an independent contractor and chairman of her township board, said this model works well for old and new systems. "This model is usually used to retrofit and replace old groups of nonconforming wastewater systems as well as being used for new conservation-based designed subdivisions. It's a fluid, dynamic model that can change and adapt to the local homeowner's needs. You can come up with different ways of handling old systems versus new systems. We want everybody's environment to be protected," she added. "It also stabilizes the local economy and protects the landowners real estate investment." Like Cranberry Lake, the area has water sources to protect. As a result, Chard said lake associations are very active in Minnesota with education programs as well as performing lake monitoring. They have been very supportive of projects that protect their lake quality and well supplies. Most of the areas where Chard works as a consultant are served by individual well systems. Also like Cranberry Lake, many of the lots were platted years ago and originally may have been set up for seasonal homes and have very small lot sizes. These lots are now seeing a need to replace a failing system with nowhere to place it on the property. Chard has helped township projects, ranging from as small as 8 to 200 homeowners, set up management districts. "We continually learn from the evolution of these and older districts," she said. "You should always be improving on the models." Chard was involved with the establishment of Cass County's first management district model, which was included in EPA's 1997 "Response to Congress on Use of Decentralized Wastewater Treatment Systems." The statute used in Minnesota for the framework of this management model is Minnesota Statute 365A for townships. This statute is used to provide many services that residents need within a township including road paving, animal control, and many other services. The statute was used to develop "Environmental" Subordinate Service Districts, which manage a water or wastewater projects or both at the same time. Under that plan, the Rural Utilities Services, formerly the Rural Electrification Association, was a major player. Cass County sought out the local utility, Crow Wing Power and Light of Brainerd, Minnesota, and asked them to help with the management program, including monitoring, monthly inspections, pumping, record keeping, and billing administration. Chard said this type of plan is typical of the model. She added that there are currently four known wastewater management districts operating in Cass County today with many others being implemented around the state by townships and counties. The county usually partners with the townships to do all of the enforcement, permitting, and sitting of treatment sites as well as implementation of a Geographic Information System (GIS) database for the districts. "They don't want to micromanage small groupings of wastewater systems, but would rather partner with townships. This method keeps them informed about the smaller wastewater management systems. Now they have started doing planning and zoning, road work, and many other ideas have evolved from this original partnership and dialogue," she added.

"The county attends yearly audit meetings with the township boards, residents, and rural utility representatives. They physically review the system and look at the management logs to see how the wastewater system could be improved." The keys to success, in Chard's view, are education and the ability to keep an open mind. "It all goes back to working with your neighbor, building a trust base. From there, you are challenged to find answers," she said. "You sit down with the property owners in a meeting and say here is the problem, now what can we do. I have yet to come up against a group that can't find their own solutions." Like Schautz, Chard recommends keeping education material and any documents homeowner friendly and humorous. "Try to make it fun. Homeowners always think of the government as being very imposing, but there is a lot of flexibility in this model that can be used to help the homeowners and town board work together and find solutions they need. When it's done, all feel that they have ownership in their project," she added. Chard believes the definition for onsite management depends on a person's perspective. "Onsite management from the homeowner's perspective is new," she said. "It means taking care of their system, which is something they have never done before. By taking responsibility of your system, you are also protecting your neighbor. Further, we are managing and protecting a considerable investment and not wasting anyone's money to replace it sooner than is necessary." Chard added that onsite systems will always need some type of management tool, from the simplest "tank management" tools to the more sophisticated technologies that homeowners would not understand. "To me, it is doing it right from cradle to grave. It has gone beyond knowing that there are problems, that central piping is no longer a necessary evil because it's so costly," she added. "The NODP IV theory is truly fourth-generation thinking regarding the evolution of wastewater management for the new millennium," Chard said. "Now we know we not only have the technology, but the tools to manage and maintain any onsite-and cluster-designed wastewater system."

Unit 14 Wastewater Reuse

Once freshwater has been used for an economic or beneficial purpose, it is generally discarded as waste. In many countries, these wastewaters are discharged, either as untreated waste or as treated effluent, into natural watercourses, from which they are abstracted for further use after undergoing "self-purification" within the stream. Through this system of indirect reuse, wastewater may be reused up to a dozen times or more before being discharged to the sea. Such indirect reuse is common in the larger river systems of Latin America. However, more direct reuse is also possible: the technology to reclaim wastewaters as potable or process water is a technically feasible option for agricultural and some industrial purposes (such as for cooling water or sanitary flushing), and a largely experimental option for the supply of domestic water. Wastewater reuse for drinking raises public health, and possibly religious, concerns among consumers. The adoption of wastewater treatment and subsequent reuse as a means of supplying freshwater is also determined by economic factors.

In many countries, water quality standards have been developed governing the discharge of wastewater into the environment. Wastewater, in this context, includes sewage effluent, stormwater runoff, and industrial discharges. The necessity to protect the natural environment from wastewater-related pollution has led to much improved treatment techniques. Extending these technologies to the treatment of wastewaters to potable standards was a logical extension of this protection and augmentation process.

1. Technical Description

One of the most critical steps in any reuse program is to protect the public health, especially that of workers and consumers. To this end, it is most important to neutralize or eliminate any infectious agents or pathogenic organisms that may be present in the wastewater. For some reuse applications, such as irrigation of non-food crop plants, secondary treatment may be acceptable. For other applications, further disinfection, by such methods as chlorination or ozonation, may be necessary.

A typical example of wastewater reuse is the system at the Sam Lords Castle Hotel in Barbados. Effluent consisting of kitchen, laundry, and domestic sewage ("gray water") is collected in a sump, from which it is pumped, through a comminutor, to an aeration chamber. No primary sedimentation is provided in this system, although it is often desirable to do so. The aerated mixed liquor flows out of the aeration chamber to a clarifier for gravity separation. The effluent from the clarifier is then passed through a 16-foot-deep chlorine disinfection chamber before it is pumped to an automatic sprinkler irrigation system. The irrigated areas are divided into sixteen zones; each zone has twelve sprinklers. Some areas are also provided with a drip irrigation system. Sludge from the clarifier is pumped, without thickening, as a slurry to suckwells, where it is disposed of. Previously the sludge was pumped out and sent to the Bridgetown Sewage Treatment Plant for further treatment and additional desludging.

2. Extent of Use

For health and aesthetic reasons, reuse of treated sewage effluent is presently limited to non-potable applications such as irrigation of non-food crops and provision of industrial cooling water. There are no known direct reuse schemes using treated wastewater from sewerage systems for drinking. Indeed, the only known systems of this type are experimental in nature, although in some cases treated wastewater is reused indirectly, as a source of aquifer recharge. Table 14-1 presents some guidelines for the utilization of wastewater, indicating the type of treatment required, resultant water quality specifications, and appropriate setback distances. In general, wastewater reuse is a technology that has had limited use, primarily in small-scale projects in the region, owing to concerns about potential public health hazards.

Guidelines for Water Reuse Table 14-1

Type of Reuse	Treatment Required	Reclaimed Quality	Water Recommended Monitoring	Setback Distances
Agricultural	Secondary Disinfection	$pH=6\sim9$	pH weekly	300 ft from potable water supply wells
Food crops commercially processed		$BOD<30mg/L$ $SS<30mg/L$	BOD weekly SS daily	
Orchards and Vinerds		$FC<200/100mL$ $Cl_2\ residual=1mg/(L\cdot min.)$	FC daily Cl_2 residual continuous	100 ft from areas accessible to public
Pasturage	Secondary Disinfection	$pH=6\sim9$	pH weekly	300 ft from potable water supply wells
Pasture for milking animals		$BOD<30mg/L$ $SS<30mg/L$	BOD weekly SS daily	
		$FC<200/100mL$	FC daily	
Pasture for livestock		$Cl_2\ residual=1mg/(L\cdot min.)$	Cl_2 residual continuous	100 ft from areas accessible to the public
Forestation	Secondary Disinfection	$pH=6\sim9$	pH weekly	300 ft from potable water supply wells
		$BOD<30mg/L$	BOD weekly	
		$SS<30mg/L$	SS daily	
		$FC<200/100mL$	FC daily	
		$Cl_2\ residual=1mg/(L\cdot min.)$	Cl_2 residual continuous	100 ft from areas accessible to the public

Continued

Type of Reuse	Treatment Required	Reclaimed Quality	Water Recommended Monitoring	Setback Distances
Agricultural	Secondary Filtration Disinfection	pH=6~9	pH weekly	50 ft from potable water supply wells
Food crops not commercially processed		BOD<30mg/L	BOD weekly	
		Turbidity<1NTU	Turbidity daily	
		FC=0/100mL	FC daily	
		Cl_2 residual=1mg/L·min.	Cl_2 residual continuous	
Groundwater recharge	Site-specific use-dependent	and Site-specific and use-dependent	Depends on treatment Site-specific and use	

In Latin America, treated wastewater is used in small-scale agricultural projects and, particularly by hotels, for lawn irrigation. In Chile, up to 220 L/s of wastewater is used for irrigation purposes in the desert region of Antofagasta. In Brazil, wastewater has been extensively reused for agriculture. Treated wastewaters have also been used for human consumption after proper disinfection, for industrial processes as a source of cooling water, and for aquaculture. Wastewater reuse for aquacultural and agricultural irrigation purposes is also practiced in Lima, Peru. In Argentina, natural systems are used for wastewater treatment. In such cases, there is an economic incentive for reusing wastewater for reforestation, agricultural, pasturage, and water conservation purposes, where sufficient land is available to do so. Perhaps the most extensive reuse of wastewater occurs in Mexico, where there is large-scale use of raw sewage for the irrigation of parks and the creation of recreational lakes.

In the United States, the use of reclaimed water for irrigation of food crops is prohibited in some states, while others allow it only if the crop is to be processed and not eaten raw. Some states may hold, for example, that if a food crop is irrigated in such a way that there is no contact between the edible portion and the reclaimed water, a disinfected, secondary-treated effluent is acceptable. For crops that are eaten raw and not commercially processed, wastewater reuse is more restricted and less economically attractive. Less stringent requirements are set for irrigation of non-food crops.

International water quality guidelines for wastewater reuse have been issued by the World Health Organization (WHO). Guidelines should also be established at national level and at the local/project level, taking into account the international guidelines. Some national standards that have been developed are more stringent than the WHO guidelines. In general, however, wastewater reuse regulations should be strict enough to permit irrigation use without undue health risks, but not so strict as to prevent its use. When using treated wastewater for irrigation, for example, regulations should be written so that attention is paid to the interaction between the effluent, the soil, and the topography of the receiving area, particularly if there are aquifers nearby.

3. Operation and Maintenance

The operation and maintenance required in the implementation of this technology is related to the previously discussed operation and maintenance of the wastewater treatment processes, and to the chlorination and disinfection technologies used to ensure that pathogenic organisms will not present a health hazard to humans. Additional maintenance includes the periodic cleaning of the water distribution system conveying the effluent from the treatment plant to the area of reuse; periodic cleaning of pipes, pumps, and filters to avoid the deposition of solids that can reduce the distribution efficiency; and inspection of pipes to avoid clogging throughout the collection, treatment, and distribution system, which can be a potential problem. Further, it must be emphasized that, in order for a water reuse program to be successful, stringent regulations, monitoring, and control of water quality must be exercised in order to protect both workers and the consumers.

4. Level of Involvement

The private sector, particularly the hotel industry and the agricultural sector, are becoming involved in wastewater treatment and reuse. However, to ensure the public health and protect the environment, governments need to exercise oversight of projects in order to minimize the deleterious impacts of wastewater discharges. One element of this oversight should include the sharing of information on the effectiveness of wastewater reuse. Government oversight also includes licensing and monitoring the performance of the wastewater treatment plants to ensure that the effluent does not create environmental or health problems.

5. Costs

Cost data for this technology are very limited. Most of the data relate to the cost of treating the wastewater prior to reuse. Additional costs are associated with the construction of a dual or parallel distribution system. In many cases, these costs can be recovered out of the savings derived from the reduced use of potable freshwater (i.e., from not having to treat raw water to potable standards when the intended use does not require such extensive treatment). The feasibility of wastewater reuse ultimately depends on the cost of recycled or reclaimed water relative to alternative supplies of potable water, and on public acceptance of the reclaimed water. Costs of effluent treatment vary widely according to location and level of treatment (see the previous section on wastewater treatment technologies). The degree of public acceptance also varies widely depending on water availability, religious and cultural beliefs, and previous experience with the reuse of wastewaters.

6. Effectiveness of the Technology

The effectiveness of the technology, while difficult to quantify, is seen in terms of the diminished demand for potable-quality freshwater and, in the Caribbean islands, in the diminished degree of degradation of water quality in the near-shore coastal marine environment, the area where untreated and unreclaimed wastewaters were previously disposed. The analysis of beach waters in Jamaica indicates that the water quality is better near the hotels with wastewater reuse projects than in beach areas where reuse is not practiced: Beach #1 in Table 14-2 is near a hotel with a wastewater reuse project, while Beach #2 is not. From an aesthetic point of view, also, the presence of lush vegetation in the areas where lawns and plants are irrigated with reclaimed wastewater is further evidence of the effectiveness of this technology.

Water Quality of Beach Water in Wastewater Reuse Project in Jamaica   Table 14-2

Site	BOD	TC	FC	NO_3
Beach #1	0.30	<2	<2	0.01
Beach #2	1.10	2.40000	280.00	0.01

7. Suitability

This technology has generally been applied to a small-scale projects, primarily in areas where there is a shortage of water for supply purposes. However, this technology can be applied to larger-scale projects. In many developing countries, especially where there is a water deficit for several months of the year, implementation of wastewater recycling or reuse by industries can reduce demands for water of potable quality, and also reduce impacts on the environment.

Large-scale wastewater reuse can only be contemplated in areas where there are reticulated sewerage and/or stormwater systems. (Micro-scale wastewater reuse at the household or farmstead level is a traditional practice in many agricultural communities that use night soils and manures as fertilizers.) Urban areas generally have sewerage systems, and, while not all have stormwater systems, those that do are ideal localities for wastewater reuse schemes. Wastewater for reuse must be adequately treated, biologically and chemically, to ensure the public health and environmental safety. The primary concerns associated with the use of sewage effluents in reuse schemes are the presence of pathogenic bacteria and viruses, parasite eggs, worms, and helminths (all biological concerns) and of nitrates, phosphates, salts, and toxic chemicals, including heavy metals (all chemical concerns) in the water destined for reuse.

8. Advantages

(1) This technology reduces the demands on potable sources of freshwater. It may reduce the need for large wastewater treatment systems, if significant portions of the waste stream are reused or recycled.

(2) The technology may diminish the volume of wastewater discharged, resulting in a

beneficial impact on the aquatic environment.

(3) Capital costs are low to medium, for most systems, and are recoverable in a very short time; this excludes systems designed for direct reuse of sewage water.

(4) Operation and maintenance are relatively simple except in direct reuse systems, where more extensive technology and quality control are required.

(5) Provision of nutrient-rich wastewaters can increase agricultural production in water-poor areas.

(6) Pollution of seawater, rivers, and ground waters may be reduced.

(7) Lawn maintenance and golf course irrigation is facilitated in resort areas.

(8) In most cases, the quality of the wastewater, as an irrigation water supply, is superior to that of well water.

9. Disadvantages

(1) If implemented on a large scale, revenues to water supply and wastewater utilities may fall as the demand for potable water for non-potable uses and the discharge of wastewaters is reduced.

(2) Reuse of wastewater may be seasonal in nature, resulting in the overloading of treatment and disposal facilities during the rainy season; if the wet season is of long duration and/or high intensity, the seasonal discharge of raw wastewaters may occur.

(3) Health problems, such as water-borne diseases and skin irritations, may occur in people coming into direct contact with reused wastewater.

(4) Gases, such as sulfuric acid, produced during the treatment process can result in chronic health problems.

(5) In some cases, reuse of wastewater is not economically feasible because of the requirement for an additional distribution system.

(6) Application of untreated wastewater as irrigation water or as injected recharge water may result in groundwater contamination.

10. Cultural Acceptability

A large percentage of domestic water users are afraid to use this technology to supply of potable water (direct reuse) because of the potential presence of pathogenic organisms. However, most people are willing to accept reused wastewater for golf course and lawn irrigation and for cooling purposes in industrial processes. On the household scale, reuse of wastewaters and manures as fertilizer are a traditional technology.

11. Further Development of the Technology

Expansion of this technology to large-scale applications should be encouraged. Cities and towns that now use mechanical treatment plants that are difficult to operate, expensive to maintain, and require a high skill level can replace these plants with the simpler systems;

treated wastewater can be reused to irrigate crops, pastures, and lawns. In new buildings, plumbing fixtures can be designed to reuse wastewater, as in the case of using gray water from washing machines and kitchen sinks to flush toilets and irrigate lawns. Improved public education to ensure awareness of the technology and its benefits, both environmental and economic, is recommended.

Vocabulary and Phrases

watercourse   n. 水道，河道
reclaim   vt. 收回
freshwater   n. 淡水(不是海洋水的)
ozonation   n. [化] 臭氧
kitchen   n. 厨房
laundry   n. 洗衣店
sump   n. 污水坑
sprinkler   n. 洒水车，洒水装置
slurry   n. 泥浆
aesthetic   adj. 美学的
aquifer   n. 含水土层，蓄水层
recharge   vt. 再充电，再控告，再袭击 vi. 再袭击 n. 再袭击，再装填
daily   adj. 每日的，日常的；adv. 每日
residual   adj. 剩余的，残留的
turbidity   n. 混浊
filtration   n. 过滤，筛选
pasture   n. 牧地，草原，牧场

livestock   n. 家畜，牲畜
forestation   n. 造林
aquaculture   n. 水产业
recreational   adj. 休养的，娱乐的
edible   adj. 可食用的
topography   n. 地形学
deposition   n. 沉积作用
deleterious   adj. 有害的，有毒的
quantify   vt. 确定数量；v. 量化
reticulated   adj. 网状的
parasite   n. 寄生虫
phosphate   n. 磷酸盐
revenue   n. 收入，国家的收入，税收
overload   vt. 使超载，超过负荷；n. 超载，负荷过多
sink   n. 水槽，水池
toilet   n. 盥洗室，梳洗

Notes

    gray water   灰水
    chlorine disinfection chamber   氯消毒间
    the periodic cleaning of the water distribution system   周期性的给水布水管道清洗
    recycled or reclaimed water   回用水
    near-shore coastal marine environment   近海岸环境
    From an aesthetic point of view   从美学观点上看

Question and Exercises

    1. What are the main disadvantages of wastewater reuse?
    2. What are the main advantages of wastewater reuse?
    3. What's your opinion to use reclaimed wastewater?

Further Reading

Advanced Methods of Wastewater Treatment

As our country and the demand for clean water have grown, it has become more important to produce cleaner wastewater effluents, yet some contaminants are more difficult to remove than others. The demand for cleaner discharges has been met through better and more complete methods of removing pollutants at wastewater treatment plants, in addition to pretreatment and pollution prevention which helps limit types of wastes discharged to the sanitary sewer system.

Currently, nearly all WWTPs provide a minimum of secondary treatment. In some receiving waters, the discharge of secondary treatment effluent would still degrade water quality and inhibit aquatic life.

Further treatment is needed. Treatment levels beyond secondary are called advanced treatment. Advanced treatment technologies can be extensions of conventional secondary biological treatment to further stabilize oxygen-demanding substances in the wastewater, or to remove nitrogen and phosphorus. Advanced treatment may also involve physical-chemical separation techniques such as adsorption, flocculation/precipitation, membranes for advanced filtration, ion exchange, and reverse osmosis.

In various combinations, these processes can achieve any degree of pollution control desired. As wastewater is purified to higher and higher degrees by such advanced treatment processes, the treated effluents can be reused for urban, landscape, and agricultural irrigation, industrial cooling and processing, recreational uses and water recharge, and even indirect augmentation of drinking water supplies.

1. Nitrogen Control

Nitrogen in one form or another is present in municipal wastewater and is usually not removed by secondary treatment. If discharged into lakes and streams or estuary waters, nitrogen in the form of ammonia can exert a direct demand on oxygen or stimulate the excessive growth of algae. Ammonia in wastewater effluent can be toxic to aquatic life in certain instances.

By providing additional biological treatment beyond the secondary stage, nitrifying bacteria present in wastewater treatment can biologically convert ammonia to the non-toxic nitrate through a process known as nitrification. The nitrification process is normally sufficient to remove the toxicity associated with ammonia in the effluent. Since nitrate is also a nutrient, excess amounts can contribute to the uncontrolled growth of algae. In situations where nitrogen must be completely removed from effluent, an additional biological process can be added to the system to convert the nitrate to nitrogen gas. The conversion of nitrate to nitrogen gas is accomplished by bacteria in a process known as denitrification. Effluent with nitrogen in the form of nitrate is placed into a tank devoid of oxygen, where carbon-containing chemicals, such as methanol, are added or a small stream of raw wastewater is mixed in with the nitrified effluent. In this oxygen free environment, bacteria use the oxygen attached to the nitrogen in the nitrate form releasing nitrogen gas. Because nitrogen comprises almost 80 percent of the air in the earth's atmosphere, the release of nitrogen into the atmosphere does not cause any environmental harm.

2. Biological Phosphorus Control

Like nitrogen, phosphorus is also a necessary nutrient for the growth of algae. Phosphorus reduction is often needed to prevent excessive algal growth before discharging effluent into lakes, reservoirs and estuaries. Phosphorus removal can be achieved through chemical addition and a coagulation sedimentation process discussed in the following section.

Some biological treatment processes called biological nutrient removal (BNR) can also achieve nutrient reduction, removing both nitrogen and phosphorus. Most of the BNR processes involve modifications of suspended growth treatment systems so that the bacteria in these systems also convert nitrate nitrogen to inert nitrogen gas and trap phosphorus in the solids that are removed from the effluent.

3. Coagulation Sedimentation

A process known as chemical coagulation-sedimentation is used to increase the removal of solids from effluent after primary and secondary treatment.

Solids heavier than water settle out of wastewater by gravity. With the addition of specific chemicals, solids can become heavier than water and will settle.

Alum, lime, or iron salts are chemicals added to the wastewater to remove phosphorus. With these chemicals, the smaller particles "floc" or clump together into large masses.

The larger masses of particles will settle faster when the effluent reaches the next step—the sedimentation tank. This process can reduce the concentration of phosphate by more than 95 percent.

Although used for years in the treatment of industrial wastes and in water treatment, coagulation sedimentation is considered an advanced process because it is not routinely applied to the treatment of municipal wastewater. In some cases, the process is used as a necessary pretreatment step for other advanced techniques. This process produces a chemical sludge, and the cost of disposing this material can be significant.

4. Carbon Adsorption

Carbon adsorption technology can remove organic materials from wastewater that resist removal by biological treatment. These resistant, trace organic substances can contribute to taste and odor problems in water, taint fish flesh, and cause foaming and fish kills.

Carbon adsorption consists of passing the wastewater effluent through a bed or canister of activated carbon granules or powder which remove more than 98 percent of the trace organic substances. The substances adhere to the carbon surface and are removed from the water. To help reduce the cost of the procedure, the carbon granules can be cleaned by heating and used again.

Part C Urban Air Environment

Unit 15 Type and Sources of Air Pollutants

1. What Is Air Pollution

The composition of "unpolluted" air is unknown to us. Humans have lived on the planet thousands of years and influenced the composition of the air through their many activities before it was possible to measure the constituents of the air. Air is a complex mixture made up of many chemical components. The primary components of air are nitrogen (N_2), oxygen (O_2), and water vapor (H_2O). About 99 percent of air is nitrogen (78%) and oxygen (21%). The remaining percent includes trace quantities of substances such as carbon dioxide (CO_2), methane (CH_4), hydrogen (H_2), argon (Ar) and helium (He).

In theory, the air has always been polluted to some degree. Natural phenomena such as volcanoes, wind storms, the decomposition of plants and animals, and even the aerosols emitted by the ocean "pollute" the air. However, the pollutants we usually refer to when we talk about air pollution are those generated as a result of human activity.

Air pollution can be defined as the presence in the outdoor atmosphere of one or more contaminants (pollutants) in such quantities and of such duration as may be (or may tend to be) injurious to human, plant, or animal life, or to property (materials), or which may unreasonably interfere with the comfortable enjoyment of life or property, or the conduct of business. It should be stressed that the attention in this definition is on the outdoor, or ambient air as opposed to the indoor, or work environment air. This definition mentions the quantity or concentration of the contaminant in the atmosphere, and its associated duration or period of occurrence. This is an important concept in that pollutants that are present at extremely low concentration and for short time periods may be insignificant in terms of damage effect.

2. Type of Air Pollutants

An air pollutant can be considered as a substance in the air that, in high enough concentrations, produces a detrimental environmental effect. These effects can be either health effects or welfare effects. A pollutant can affect the health of humans, as well as the health of plants and animals. Pollutants can also affect non-living materials such as paints, metals, and fabrics. An environmental effect is defined as a measurable or perceivable detrimental change resulting from contact with an air pollutant.

Human activities have had a detrimental effect on the makeup of air. Activities such as driving cars and trucks, burning of coal, oil and other fossil fuels, and manufacturing chemicals have changed the composition of air by introducing many pollutants. There are hundreds of pollutants in the ambient air. Ambient air is the air to which the general public has access, i. e. any unconfined portion of the atmosphere. The two basic physical forms of air pollutants are particulate matter and gases. Particulate matter includes small solid and liquid particles such as dust, smoke, sand, pollen, mist, and fly ash. Gases include substances such as carbon monoxide (CO), sulfur dioxide (SO_2), nitrogen oxide (NO_2), and volatile organic compounds (VOCs).

Pollutants can also be classified as either primary pollutants or secondary pollutants. A primary pollutant is one that is emitted into the atmosphere directly from the source of the pollutant and retains the same chemical form. An example of a primary pollutant is the ash produced by the burning of solid waste. A secondary pollutant is one that is formed by atmospheric reactions of precursor or primary emissions. Secondary pollutants undergo a chemical change once they reach the atmosphere. An example of a secondary pollutant is ozone created from organic vapors given off at a gasoline station. The organic vapors react with sunlight in the atmosphere to produce the ozone, the primary component of smog. Control of secondary pollutants is generally more problematic than that of primary pollutants, because mitigation of secondary pollutants requires the identification of the precursor compounds and their sources as well as an understanding of the specific chemical reactions that result in the formation of the secondary pollutants.

The Environmental Protection Agency (EPA) has further classified ambient air pollutants for regulatory purposes as hazardous air pollutants (HAPs) and criteria pollutants. Criteria pollutants are pollutants that have been identified as being both common and detrimental to human welfare and are found over all the United States (ubiquitous pollutants). EPA currently designates six pollutants as criteria pollutants. These criteria pollutants are: carbon monoxide (CO), sulfur oxides (SO_x), nitrogen oxides (NO_x), ozone (O_3), lead (Pb), and particulate matter (PM). On the other hand, EPA refers to chemicals that cause serious health and environmental hazards as hazardous air pollutants (HAPs) or air toxics. Hazardous air pollutants are those pollutants that are known or suspected to cause cancer or other serious health effects, such as reproductive effects or birth defects, or adverse environmental effects.

3. Sources of Air Pollutants

A source of air pollution is any activity that causes pollutants to be emitted into the air. Air pollutant sources can be categorized from several perspectives, including the type of source, their frequency of occurrence and spatial distribution, and the types of emissions.

Characterization by source can be delineated as arising from natural sources or from man-made sources. There have always been natural sources of air pollution, also known as biogenic sources. For example, volcanoes have spewed particulate matter and gases into our atmosphere for millions of years. Lightening strikes have caused forest fires, with their resulting contribution of gases and particles, for as long as storms and forests have existed. Organic matter in swamps decay and wind storms whip up dust. Trees and other vegetation contribute large amounts of pollen and spores to our atmosphere. These natural pollutants can be problematic at times, but generally are not as much of a problem as human-generated pollutants or anthropogenic sources.

The quality of daily life depends on many modern conveniences. People enjoy the freedom to drive cars and travel in airplanes for business and pleasure. They expect their homes to have electricity and their water to be heated for bathing and cooking. They use a variety of products such as clothing, pharmaceuticals, and furniture made of synthetic materials. At times, they rely on services that use chemical solvents, such as the local dry cleaner and print shop. Yet the availability of these everyday conveniences comes at a price, because they all contribute to air pollution.

Human-generated sources of air pollution or anthropogenic sources are categorized in two ways: mobile and stationary sources. Mobile sources of air pollution include most forms of transportation such as automobiles, trucks, and airplanes. Stationary sources of air pollution consist of non-moving sources such as power plants and industrial facilities.

Source characterization according to spatial distribution can be categorized as station sources and mobile sources. Stationary sources are classified as point source or area source (types of emissions). A point source refers to a source at a fixed point, such as a smokestack or storage tank, that emits air pollutants. An area source refers to a series of small sources that together can affect air quality in a region. For example, a community of homes using woodstoves for heating would be considered as an area source, even though each individual home is contributing small amounts of various pollutants.

4. Sources of Pollutants-Stationary Sources

Stationary sources are non-moving sources, fixed-site producers of pollution such as power plants, chemical plants, oil refineries, manufacturing facilities, and other industrial facilities. There are hundreds of thousands of stationary sources of air pollution in the United States. Stationary sources emit both criteria pollutants and hazardous air pollutants (HAPs). Air pollution from stationary sources is produced by two primary activities. These activities are stationary combustion of fuel such as coal and oil at power generating facilities, and the pollutant losses from industrial processes. Industrial processes include refineries, chemical manufacturing facilities, and smelters.

Stationary sources have many possible emission points. An emission point is the specific place or piece of equipment from which a pollutant is emitted. Air pollutants can be emitted from smokestacks, storage tanks, equipment leaks, process wastewater handling/ treatment area, loading and unloading facilities, and process vents. A process vent is basically an opening where substances (mostly in gaseous form) are "vented" into the atmosphere. Common process vents in a chemical plant are distillation columns and oxidation vents. Emissions from storage tanks are due to pollutants that can leak through the roofs, and can leak through tank openings when liquids expand or cool because of outdoor temperature changes. Also, air pollutants can escape during the filling and emptying of a storage tank. Air pollution produced from wastewater occurs when wastewater containing volatile chemicals comes in contact with the air. Volatile means that it can be evaporated, or pass from a liquid state to a gaseous state.

Large, stationary sources of emissions that have specific locations and release pollutants in quantities above an emission threshold are known as point sources. Those facilities or activities whose individual emissions do not qualify them as point sources are called area sources. Area sources represent numerous facilities or activities that individually release small amounts of a given pollutant, but collectively can release significant amounts of a pollutant. For example, dry cleaners, vehicle refinishing and gasoline dispensing facilities, and residential heating will not typically qualify as point sources, but collectively the various emissions from these sources are classified as area sources.

Stationary sources are also classified as major and minor sources. A major source is one that emits, or has the potential to emit, pollutants over a major source threshold. A minor source is any source which emits fewer pollutants than the major source threshold.

5. Sources of Pollutants-Mobile Sources

"Mobile sources" is a term used to describe a wide variety of vehicles, engines, and equipment that generate air pollution and that move, or can be moved, from place to place. Mobile sources are classified as on-road and nonroad sources. "On-road" or highway sources include vehicles used on roads for transportation of passengers or freight. On-road sources include light-duty vehicles (LDVs, also referred to as passenger cars), heavy-duty vehicles (HDVs), and motorcycles that are used for transportation on the road. On-road vehicles may be fueled with gasoline, diesel fuel, or alternative fuels, such as alcohol or natural gas. "Nonroad" sources include gasoline and diesel powered vehicles, engines, and equipment used for construction, agriculture, transportation, recreation, and many other purposes. These sources emit both criteria pollutants and other hazardous air pollutants.

Mobile sources account for more than half of all the air pollution in the United States. The primary mobile source of air pollution is the automobile. EPA studies show that today's cars emit 75 to 90 percent less pollution (for each mile driven) than their 1970 counterparts, thanks largely to advancements in vehicle and fuel technology. But today's motor vehicles are still responsible for up to half of all the emissions released into the air. The specific pollutant categories include 45 percent of the volatile organic compound (VOC) emissions, 50 percent of the nitrogen oxides (NO_x) emissions, approximately 60 percent of the carbon monoxide (CO) emissions and 50 percent of the hazardous air pollutants in urban areas.

Mobile sources pollute the air through combustion and fuel evaporation. These emissions contribute greatly to air pollution nationwide and are the primary causes of air pollution in many urban areas. Combustion is the process of burning. Motor vehicles and equipment typically burn fuel in an engine to create power. Gasoline and diesel fuels are mixtures of hydrocarbons, which are compounds that contain hydrogen and carbon atoms.

In "perfect" combustion, oxygen in the air would combine with all the hydrogen in the fuel to form water and with all the carbon in the fuel to form carbon dioxide. Nitrogen in the air would remain unaffected. In reality, the combustion process is not "perfect," and engines emit several types of pollutants as combustion byproducts. Evaporation is the process by which a substance is converted from a liquid into a vapor. "Evaporative emissions" occur when a liquid fuel evaporates and fuel molecules escape into the atmosphere. A considerable amount of hydrocarbon pollution results from evaporative emissions that occur when gasoline leaks or spills, or when gasoline gets hot and evaporates from the fuel tank or engine.

Perfect Combustion

Fuel (hydrocarbons) + Air (oxygen and nitrogen) → Carbon dioxide (CO_2) + water (H_2O) + unaffected nitrogen

Typical Engine Combustion

Fuel + Air → Unburned Hydrocarbons + Nitrogen Oxides (NOx) + Carbon monoxide (CO) + Carbon dioxide (CO_2) + water (H_2O)

Vocabulary and Phrases

trace quantity 微量
aerosol n. 浮质(气体中的悬浮微粒，如烟，雾等)，[化] 气溶胶，气雾剂，烟雾剂
birth defect 出生缺陷
pharmaceutical n. 药剂，药品；
ubiquitous adj. 到处存在的，(同时)普遍存在的
ozone n. [化] 臭氧
dry cleaner n. 干洗店，干洗工

vent [vent] n. 通风孔，出烟孔，出口
distillation column 分馏塔
volatile adj. 飞行的，挥发性的
threshold n. 上限，下限，阀值
freight n. 货物，货运
evaporation n. 蒸发(作用)
hydrocarbon n. 烃，碳氢化合物

Notes

1. Air pollution can be defined as the presence in the outdoor atmosphere of one or more contaminants (pollutants) in such quantities and of such duration as may be (or may tend to be) injurious to human, plant, or animal life, or to property (materials), or which may unreasonably interfere with the comfortable enjoyment of life or property, or the conduct of business.

空气污染可以定义为存在于室外大气中的一种或多种污染物，其数量和持续时间已达到（或将会）危害人类、动、植物及财产的程度，或妨碍（人们）对生活财产的舒适享受或影响商业活动。

2. fossil fuel 化石燃料

3. The organic vapors react with sunlight in the atmosphere to produce the ozone, the primary component of smog.
有机蒸气在光照下反应生成臭氧，烟雾（光化学烟雾）的主要成分。

4. power plants $n.$ 发电厂，发电站

Question and Exercises

1. Put the following into Chinese
   natural sources   primary pollutant   secondary pollutant
   human-generated pollutants   anthropogenic sources   criteria pollutants
   volatile organic compound   hazardous air pollutants

2. Put the following into English
   颗粒物   自然来源   流动源   固定源   点源
   面源   二氧化硫   氮氧化物   一氧化碳   完全燃烧

3. Translate the following paragraphs into Chinese

(1) A primary air pollutant is a chemical added directly to the air that occurs in a harmful concentration. It can be a natural air component, such as carbon dioxide, that rises above its normal concentration, or something not usually found in the air, such as a lead compound emitted by cars burning leaded gasoline.

(2) Mobile sources account for more than half of all the air pollution in the United States. The primary mobile source of air pollution is the automobile. EPA studies show that today's cars emit 75 to 90 percent less pollution (for each mile driven) than their 1970 counterparts, thanks largely to advancements in vehicle and fuel technology. But today's motor vehicles are still responsible for up to half of all the emissions released into the air. The specific pollutant categories include 45 percent of the volatile organic compound (VOC) emissions, 50 percent of the nitrogen oxides (NO_x) emissions, approximately 60 percent of the carbon monoxide (CO) emissions and 50 percent of the hazardous air pollutants in urban areas.

Further Reading

Major Air Pollutants

Sulfur Dioxide

Sulfur dioxide (SO_2) is a colorless and odorless gas normally present at earth's surface at low concentrations. One of the significant features of SO_2 is that once it is emitted into the atmosphere it may be converted through complex reactions to fine particulate sulfate (MSO_4). The major anthropogenic source of sulfur dioxide is the burning of fossil fuels, mostly coal in power plants. Another major source comprises a variety of industrial processes, raging from petroleum refining to the production of paper, cement, and aluminum.

Adverse effects associated with sulfur dioxide depend on the dose or concentration present and include corrosion of paint and metals and injury or death to animals and plants. Crops such as alfalfa, cotton, and barely are especially susceptible. Sulfur dioxide is capable of causing severe damage to human and other animal lungs, particularly in the sulfate form. It is also an important precursor to acid rain, as are nitrogen oxides.

Nitrogen Oxides

Nitrogen Oxides (NO_x), is the generic term used to describe the sum of NO, NO_2 and other oxides of nitrogen. Although nitrogen oxides occur in many forms in the atmosphere, only NO and NO_2 are subject to emission regulations. NO_x is a group of highly reactive gases that play a major role in the formation of ozone. Many of the nitrogen oxides are colorless and odorless. However, one common pollutant, nitrogen dioxide (NO_2) along with particles in the air can often be seen as a reddish-brown layer over many urban areas. Nitrogen oxides form when fuel is burned at high temperatures, as in a combustion process. The primary sources of NO_x are motor vehicles, electric utilities, and other industrial, commercial, and residential sources that burn fuels.

The environmental effects of nitrogen oxides on humans are variable but include the irritation of eyes, nose, throat, and lungs and increased susceptibility to viral infections, including influenza (which can cause bronchitis and pneumonia). Nitrogen oxides suppress plant growth and damage leaf tissue. When the oxides are converted to their nitrate form in the atmosphere, they impair visibility. However, when nitrate is deposited on the soil, it can promote plant growth through nitrogen fertilization.

Carbon Monoxide

Carbon momoxide (CO), is a colorless, odorless gas formed when carbon in fuel is not burned completely. In the United States, motor vehicle exhaust contributes about 60 percent of all CO emissions nationwide (Latest Finding on National Air Quality 2002). Other non-road engines and vehicles (such as construction equipment and boats) contribute about 22 percent of all CO emissions nationwide. Higher levels of CO generally occur in areas with heavy traffic congestion. In cities, 95 percent of all CO emissions may come from motor vehicle exhaust. Other sources of CO emissions include industrial processes (such as metals processing and chemical manufacturing), residential wood burning, and natural sources such as forest fires. Woodstoves, gas stoves, cigarette smoke, and unvented gas and kerosene space heaters are sources of CO indoors. The highest levels of CO in the outside air typically occur during the colder months of the year when CO automotive emissions are greater and nighttime inversion conditions are more frequent. In inversion conditions the air pollution becomes trapped near the ground beneath a layer of warm air.

The high toxicity results from a striking physiological effect, namely, that carbon monoxide and hemoglobin in blood have a strong natural attraction for one another. Hemoglobin in our blood will take up carbon monoxide nearly 250 times more rapidly than it will oxygen. Therefore, if there is any carbon monoxide in the vicinity, a person will take it in very readily, with potentially dire effects. Many people have been accidentally asphyxiated by carbon monoxide produced from incomplete combustion of fuels in campers, tents, and houses. The effects depend on the dose or concentration of exposure and range from dizziness and headaches to death. Carbon Monoxide is particularly hazardous to people with known heart disease, anemia, or respiration disease. In addition, it may cause birth defects, including mental retardation and impairment of growth of fetus.

Photochemical oxidants

Photochemical oxidants result from atmospheric interactions of nitrogen dioxide and sunlight. The most common photochemical oxidant is ozone (O_3). Ozone is a gas composed of three oxygen atoms. It is a colorless compound that has an electric-discharge-type odor. It is a unique pollutant in that it is exclusively a secondary pollutant. It is not usually emitted directly into the air, but at ground level is created by a chemical reaction between oxides of nitrogen (NO_x) and volatile organic compounds (VOCs) in the presence of heat and sunlight. The concentration of ozone in a given locality is influenced by many factors, including the concentration of NO_2 and VOCs in the area, the intensity of the sunlight, and the local weather conditions. Ozone and the chemicals that react to form it can be carried hundreds of miles from their origins, causing air pollution over wide regions.

Ozone has the same chemical structure whether it occurs miles above the earth or at ground level and can be "good" or "bad," depending on its location in the atmosphere. "Good" ozone occurs naturally in the stratosphere and forms a layer that protects life on earth from the sun's harmful rays or ultraviolet radiation. In the earth's lower atmosphere, or troposphere, ground-level ozone is considered "bad". Ozone is the most prevalent chemical found in photochemical air pollution, or smog.

Hydrocarbons

Hydrocarbons are compounds composed of hydrogen and carbon. There are thousands of such compounds, including natural gas or methane (CH_4), butane (C_4H_{10}) and propane (C_3H_8). Analysis of urban air has identified many different hydrocarbons, some of which are asverse effects of hydrocarbons are numerous, many at a specific dose or concentration are toxic to plants and animals or may be converted to harmful compounds through complex chemical changes that occur in the atmosphere. Over 80% of hydrocarbons (which are primary pollutants) that enter the atmosphere are emitted from natural sources. The most important anthropogenic source is the automobile. Hydrocarbons may also escape to the atmosphere when a car's tank is being filled with gasoline to the tank are now required in many urban areas and are helping to reduce the problem of hydrocarbons (vapors) escaping while tanks are being filled.

Hydrogen Sulfide

Hydrogen Sulfide (H_2S) is a highly toxic and corrosive gas, easily identified by its rotten egg odor. Hydrogen sulfide is produced from natural sources, such as geysers, swamps, and bogs, as well as from human sources, such as petroleum processing-refining and metal smelting. The potential effects of hydrogen sulfide include functional damage to plants and health problems ranging from toxicity to death for humans and other animals.

Hydrogen Fluoride

Hydrogen fluoride (HF) is a gaseous pollutant that is released primarily by aluminum production, coal gasification and the burning of coal in power plants. Hydrogen fluoride is extremely toxic, and even a small concentration (as low as 1×10^{-9}) may cause problems for plants and animals.

Other HazardousGases

Pollutants or air toxics are those pollutants that are known or suspected to cause cancer or other serious health effects, such as reproductive effects or birth defects, or adverse environmental effects. Air toxics may also cause adverse environmental and ecological effects. The presence of hazardous air pollutants in the air is more localized than are the criteria pollutants, and they are usually found at highest levels close to their sources. Examples of air toxic pollutants include benzene, found in gasoline; mercury, from coal combustion; perchloroethylene, emitted from some dry cleaning facilities; and methylene chloride, used as a solvent by a number of industries. Most air toxics originate from man-made sources, including mobile sources (e. g. , cars, trucks, construction equipment), stationary sources (e. g. , factories, refineries, power plants), and indoor sources (e. g. , some buildings materials and cleaning solvents).

Particulate Matter

Particulate matter is the general term used for a heterogeneous mixture of solid particles and liquid droplets found in the air, including dust, dirt, soot, smoke, and liquid droplets. Particles can be suspended in the air for long periods of time. Some particles are large or dark enough to be seen as soot or smoke. Others are so small that individually they can only be detected with an electron microscope. PM can be a primary or secondary pollutant. "Primary"particles, such as dust or black carbon (soot) are directly emitted into the air. They come from a variety of sources such as cars, trucks, buses, factories, construction sites, tilled fields, unpaved roads, stone crushing, and burning of wood. "Secondary"particles are formed in the air from the chemical change of primary gaseous emissions. They are indirectly formed when gases from burning fuels react with sunlight and water vapor. These can result from fuel combustion in motor vehicles, at power plants, and in other industrial processes. PM2. 5 describes the"fine"particles that are less than or equal to 2. 5μm in diameter. PM10 refers to all particles less than or equal to 10μm in diameter (about one-seventh the diameter of a human hair).

Asbestos

Asbestos is the term for several minerals that have the form of small elongated particles. In the past, asbestos was treated rather casually, and people working in asbestos plants were not protected from dust. Asbestos was used in building insulation, electrical insulation, roofing material, and in brake pads for automobiles, trucks, and other vehicles. As a result, a considerable amount of asbestos fibers have been spread throughout industrialized countries, especially in urban environments of Europe and North American. In one cases, the products containing asbestos were sold in burlap bags that were recycled by nurseries and other secondary businesses, thus further spreading the pollutant. Some types of asbestos particles are believed to be carcinogenic or carry with them carcinogenic materials, and so must be carefully controlled.

Lead (Pb)

Lead is a metal found naturally in the environment as well as in manufactured products. Because of unique physical properties that allow it to be easily formed and molded, lead has been used in many applications. The major sources of lead emissions have historically been motor vehicles (such as cars and trucks) and industrial sources. Due to the phase out of leaded gasoline, metals processing is the major source of lead emissions to the air today. The highest levels of lead in air are generally found near lead smelters. Other stationary sources are waste incinerators, utilities, and lead-acid battery manufacturers.

Unit 16 Effects of Air Pollution

Air pollution has a detrimental effect on nearly all phases of our lives. Along with the health effects there are many welfare effects of air pollution. These include effects on vegetation, soil, water, manmade materials, climate, and visibility. In this unit we will discuss the effects of air pollution.

Health Effects

Exposure to air pollution is associated with numerous effects on human health, including pulmonary, cardiac, vascular, and neurological impairments. The health effects vary greatly from person to person. High-risk groups such as the elderly, infa-nts, pregnant women, and sufferers from chronic heart and lung diseases are more susceptible to air pollution. Children are at greater risk because they are generally more active outdoors and their lungs are still developing. Exposure to air pollution can cause both acute (short-term) and chronic (long-term) health effects. Acute effects are usually immediate and often reversible when exposure to the pollutant ends. Some acute health effects include eye irritation, headaches, and nausea. Chronic effects are usually not immediate and tend not to be reversible when exposure to the pollutant ends. Some chronic health effects include decreased lung capacity and lung cancer resulting from long-term exposure to toxic air pollutants. The scientific techniques for assessing health impacts of air pollution include air pollutant monitoring, exposure assessment, dosimetry, toxicology, and epidemiology.

Acid Rain

Acid rain is a broad term used to describe several ways that acids fall out of the atmosphere. A more precise term is acid deposition, which has two parts: wet and dry. Acid deposition occurs when emissions of sulfur dioxide and nitrogen oxides in the atmosphere react with water, oxygen, and oxidants to form acidic compounds. These compounds fall to the earth in either dry form (gas and particles) or wet form (rain, snow, and fog). In the United States, about 63 percent of annual SO_x emissions and 22 percent of NO_x emissions are produced by burning fossil fuels for electricity generation. Because it typically takes days to weeks for atmospheric SO_2 and NO_x to be converted to acids and deposited on the earth's surface, acid deposition occurs in a multistate scale hundreds of miles away from its sources.

Acidity is measured in terms of pH on a logarithmic scale from 1.0 to 14.0. A pH of 1.0 indicates high acidity, whereas a pH of 14.0 indicates high alkalinity; a pH of 7.0 indicates a neutral solution. Precipitation falling through a "clean" atmosphere is normally somewhat acidic, with a pH of about 5.6. Acid rain, however, can have a pH values below 4.0.

In the environment, acid deposition causes soil and water bodies to acidify (making the water unsuitable for some fish and other wildlife) and damages some trees, particularly at high elevations. It also speeds the decay of buildings, statues, and sculptures that are part of our national heritage. The nitrogen portion of acid deposition contributes to eutrophication (oxygen depletion) of water bodies, the symptoms of which include algal blooms (some of which may be toxic), fish kills, and loss of plant and animal diversity. These ecological changes impact human populations by changing the availability of seafood and creating a risk of consuming contaminated fish or shellfish, reducing our ability to use and enjoy our coastal ecosystems, and causing economic impact on people who rely on healthy coastal ecosystems, such as fishermen and those who cater to tourists.

Stratospheric Ozone Depletion

The stratosphere, located about 6 to 31 miles above the earth, contains a layer of ozone gas that protects living organisms from harmful ultraviolet-B radiation (UV-B) from the Sun. UV-B (280 to 315 nanometer wavelength) has been linked to many harmful effects including various types of skin cancer, cataracts, and harm to some crops, certain materials, and some forms of marine life. In the mid-1970s, it was discovered that some human-produced gases could cause stratospheric ozone depletion. Gases containing chlorine and bromine accumulate in the lower atmosphere, are eventually transported to the stratosphere and then converted to more reactive gases that participate in reactions that destroy ozone. Ozone depletion allows additional UV-B radiation to pass through the atmosphere and reach the earth's surface, leading to increases in UV-related health and environmental effects.

Several substances have been associated with the stratospheric ozone depletion, including chlorofluorocarbons (CFCs), halons, carbon tetrachloride, methyl bromide, and methyl chloroform. One example of ozone depletion is the annual ozone "hole" over Antarctica that has occurred during the Antarctic spring since the early 1980s. Rather than being a literal hole, the ozone hole is a large area of the stratosphere with extremely low amounts of ozone. Ozone levels fall by over 60% during the worst years. Even over the United States, ozone levels are about 3 percent below normal in the summer and 5 percent below normal in the winter.

Smog

"Smog" is a term used in our daily language. It is the mixing of smoke particles from industrial plumes with fog that produces a yellow-black color near ground level. Under the right conditions, the smoke and sulfur dioxide produced from the burning of coal can combine with fog to create industrial smog. The burning of fossil fuels like gasoline can create another atmospheric pollution problem known as photochemical smog. Photochemical smog is a condition that develops when primary pollutants (oxides of nitrogen and volatile organic compounds created from fossil fuel combustion) interact under the influence of sunlight to produce a mixture of hundreds of different and hazardous chemicals known as secondary pollutants. Smog is the brownish haze that pollutes our air, particularly over cities in the summertime. Smog can make it difficult for some people to breathe and it greatly reduces how far we can see through the air.

Smog is a mixture of pollutants with ground-level ozone being the main culprit. Increased levels of ground level-ozone are generally harmful to living systems because ozone reacts strongly to destroy or alter many other molecules. Excessive ozone exposure reduces crop yield and forest growth. It interferes with the ability of plants to produce and store food, reducing overall plant health and the ability to grow and reproduce. The weakened plants are more susceptible to harsh weather, disease, and pests. In addition, increases in tropospheric ozone lead to a warming of earth's surface.

Visibility

Air pollution also has an effect on visibility. Visibility is a measure of aesthetic value and the ability to enjoy scenic vistas, but it also can be an indicator of general air quality. Visibility degradation results when light encounters tiny pollution particles (sulfates, nitrates, organic carbon, soot, and soil dust) and some gases (nitrogen dioxide) in the air. Some light is absorbed by the particles and other light is scattered away before it reaches the observer. More pollutants mean more absorption and scattering of light, resulting in more haze. Haze obscures the clarity, color, texture, and form of what we see. Humidity magnifies the haze problem because some particles, such as sulfates, attract water and grow in size, scattering more light. In the United States' scenic areas, the visual range has been substantially reduced by air pollution. In eastern parks, average visual range has decreased from 90 miles to 15～25 miles. In the west, average visual range has decreased from 140 miles to 35～90 miles.

Vocabulary and Phrases

aesthetic *adj.* 美的
algal *n.* 藻类
bromine *n.* ［化］溴
cataract *n.* ［医］白内障
chlorine *n.* ［化］氯
culprit *n.* 事故起因，罪魁祸首
eutrophication *n.* 富营养化
halon *n.* ［化］哈龙
haze *n.* 阴霾，薄雾
multistate *n.* 多态

photochemical smog 光化学烟雾
plume *n.* 烟囱
primary pollutant 一次污染物
sculpture *n.* 雕塑
secondary pollutant 二次污染物
soot *n.* 烟尘
stratosphere *n.* ［地］平流层
tropospheric *n.* ［地］对流层
ultraviolet *n.* ［地］紫外线

Notes

  eutrophication of water bodies 水体富营养化

methyl bromide　甲基溴，溴化甲烷

methyl chloroform　甲基氯仿，三氯乙烷

Question and Exercises

1. Put the following into Chinese

acid rain　greenhouse effect　stratospheric　global warming　nitrous oxide　Stratospheric Ozone Depletion　ozone depletion

2. Put the following into English

酸沉降　温室气体　生物多样性　光化学烟雾　可见度

3. Translate the following paragraphs into Chinese

(1) Acid deposition occurs when emissions of sulfur dioxide and nitrogen oxides in the atmosphere react with water, oxygen, and oxidants to form acidic compounds.

(2) Gases containing chlorine and bromine accumulate in the lower atmosphere, are eventually transported to the stratosphere and then converted to more reactive gases that participate in reactions that destroy ozone.

(3) In the environment, acid deposition causes soil and water bodies to acidify (making the water unsuitable for some fish and other wildlife) and damages some trees, particularly at high elevations. It also speeds the decay of buildings, statues, and sculptures that are part of our national heritage.

4. We release greenhouse gases as a result of using energy to drive, using electricity to light and heat your home, and through other activities that support our quality of life like growing food, raising livestock and throwing away garbage. What can we do to reduce our greenhouse gases emissions?

Further Reading

Environmental Effects of Carbon Oxidation

The role of carbon dioxide in the global heat balance is well recognized. Because CO_2 can absorb thermal energy, it decreases the atmosphere's emissivity, that is, its ability to radiate long-wave infrared and transparency to visible light is characterized as the "greenhouse effect". This "greenhouse effect" is due primarily to water vapor and carbon dioxide, which have strong infrared absorption bands. Changes in atmospheric levels of either or both of these gases would, of course, affect the amount of heat or long-wave infrared trapped within the greenhouse and thus determine global climates. Since 1890 a atmospheric CO_2 levels have increased from about 290 to 322 ppm. Of this increase 25% has occurred in the past decade. Since 1958 the average annual increase has been approximately 0.7 ppm/yr. If this trend continues, atmospheric CO_2 levels are expected to double by the year 2035 AD. Based on some widely accepted climatic models, this doubling could result in the warming of surface temperatures in the mid-latitudes on the order of 2.4 degree, with greater potential warming in the polar regions.

The augmentation of the atmosphere's greenhouse effect from anthropogenically derived CO_2 has the potential for significantly warming global climate. By increasing the length of the growing season, the consequences of this warming may in the short term be beneficial. In the long term, however, continued global warming may have deleterious effects on global agriculture. According to a National Research Council report, climatic warming associated with increased atmosphere CO_2 level could be expected to cause significant changes in agriculture zones. These changes could include: (1) an expansion of arid and semiarid regions; (2) poleward advance of arid and arable zones; and (3) summer temperature too high for full productivity of important middle latitude crops such as corn and soybeans. The net effect of these changes would be to reduce areas suitable for agriculture and disrupt existing agricultural patterns in many regions in the northern hemisphere. This could, of course, result in severe economic disruption and untold misery for billions of people.

The most severe consequence of CO_2 induced global warming would be melting of the polar ice caps and the subsequent rise of ocean levels and inundation of coastal plains. Such a scenario would have disastrous consequences for many of the largest population centers on the planet. Based on current and projected CO_2 additions to the atmosphere, such extreme consequences would not, however, be expected for at least several centuries. Within the next fifty years, changes in agricultural zones are more likely.

Although global warming from the already elevated levels of atmospheric CO_2 has been predicted by most climatic models, such a warming has not been empirically observed. Indeed, since the early 1940s global surface temperatures have actually declined. Initially this observation led some climatologists to discount the effect of fossil fuel CO_2 on global climate. However, expanded understanding of the mature of climatic changes suggests that over the next half century increases in atmospheric CO_2 may in fact result in global warming.

Confirmation of the effect of CO_2 on climate has been confounded by the natural fluctuations in climate and average global surface temperatures. Meteorological records are only adequate for the past century of less. Although they indicate trends in average surface temperatures and climate, they unfortunately coincide with increased industrialization and pollution. Consequently, they cannot adequately indicate the nature of natural climatic fluctuations. A more useful record of climatic fluctuations for the past millennium has been established from studies of oxygen isotope rations in Greenland ice cores. These oxygen isotope studies have indicated cyclic variations in global surface temperatures. Cycles of 80 and 180 years are apparent. The amplitude of the first half cycle from 1900 to 1940 is similar to that observed in meteorological records. Based on the observations of climatic changes as recorded in oxygen isotope rations in Greenland ice, the present cooling trend is expected to bottom out in the next decade or so and will be followed by a global warming trend. It is during this natural warming period that CO_2-induced warming is expected to have a significant effect on global climate. During the early 21st century the earth may experience surface temperature warmer than any time in the past 100 years.

Although CO_2 is expected to cause climate warming, uncertainties remain about the world carbon budget, the fate of anthropogenic CO_2, and the relative important of various sinks and sources. Many students of the problem contend that all of the observed increase in atmospheric CO_2 is due to fossil fuel composition. Of the release in this way only about 50% remains in the atmosphere, with the excess believed to be removed by the oceans and terrestrial biota. The calculations of some scientists suggest that the oceans are the principal CO_2 sinks, with land biota storing anywhere from 5% to 20% of fossil fuel CO_2.

Other scientists, however, suggest that the land biota is in fact a major source of CO_2. They contend that the widespread deforestation of tropical forests in this century, and subsequent land conversion to agricultural production, has resulted in a decrease of CO_2 storage in terrestrial biota. In addition, deforestation speeds the decay of soil humus, releasing CO_2 rather than a major sink as suggested, then all of he excess CO_2 must be taken up by the oceans. This conclusion has been disputed by many oceanographers, who believe that over the short term the oceans are too sluggish to exchange CO_2 fast enough to remove the excess CO_2. The uncertainties associated with sink mechanisms considerably complicate the problem of determining the world carbon budget and prediction of future atmospheric CO_2 levels.

Several climatic models predict that a doubling of atmospheric CO_2 will result in a 2.4℃ rise in surface temperature in the middle latitudes. These models are still relatively primitive, as they do not take into account a variety of possible feedback mechanisms. For example, global warming may result in increased cloudiness, which would reduce incoming radiation and counteract warming. On the other hand, increased sea surface temperature would result in decreased CO_2 absorption and retention. As a consequence, even greater warming might occur.

In addition to these feedback mechanisms, most climatic models do not take into account the potential augmentation of the greenhouse effect by other absorption bands and therefore have the potential to significantly contribute to global warming. The contribution of chlorofluorocarbons may be half that of CO_2. Trace gases such as N_2O, CH_4 and NH_3 also have infrared absorption bands and may also contribute to the earth's greenhouse effect. Atmospheric concentrations of these gases may increases as a result of human activities. For example, the use of agricultural fertilizer may result in significant quantities of N_2O and NH_3 are released from applied fertilizers. Methane is released from fossil fuel combustion.

At the present time the worldwide annual emission rate of fossil carbon is about 4.3%. Based on these emissions, atmospheric levels of CO_2 would be expected to double by the year 2035. The potential perturbation of world climate by the release of fossil fuel CO_2. and other anthropogenically derived gases poses a real dilemma for world governments as energy requirements continue to increase and the options to reduce emissions are very limited. The continued augmentation of CO_2 in the atmosphere may cause significant climatic changes that may have adverse consequences for the human species. Many climatologists and other scientists believe that the threat is real. However, as is the case for potential of the ozone layer, climatic changes resulting from increase atmospheric CO_2 levels are theoretical. They are based on predictions from climatic models which many agree are still rather primitive.

Unit 17 Control of Particulate Pollutants

Control techniques for particles focus on capturing the particles emitted by a pollution source. Several factors must be considered before choosing a particulate control device. Typically, particles are collected and channeled through a duct or stack. The characteristics of the particulate exhaust stream affect the choice of the control device. These characteristics include the range of particle sizes, the exhaust flow rate, the temperature, the moisture content, and various chemical properties such as explosiveness, acidity, alkalinity, and flammability.

The most commonly used control devices for controlling particulate emissions include: electrostatic precipitators, fabric filters, venture scrubbers, cyclones, and settling chambers. In many cases, more than one of these devices is used in a series to obtain the desired removal efficiencies. For example, a settling chamber can be used to remove larger particles before a pollutant stream enters an electrostatic precipitator.

1. Gravity Settling Chambers

As the name implies, this category of control devices relies upon gravity settling to remove particles from the gas stream. The gas stream enters a chamber where the velocity of the gas is reduced. Large particles drop out of the gas and are recollected in hoppers. Because settling chambers are effective in removing only larger particles (approximately 75 micrometers and larger), they are used in conjunction with a more efficient control device.

The stringent control requirements adopted in the late 1960s through early 1970s have resulted in a sharp decline in the use of this type of collector. There is very few gravity settling chambers still in commercial use.

2. Cyclones

Operating Principles

Cyclones use the inertia of the particles for collection. Cyclonic collectors are round conically shaped vessels in which the gas stream enters tangentially and followers a spiral path to the outlet. The spiral motion produces the centrifugal forces that cause the particulate matter to move toward the periphery of the vessel and collect on the walls and fall to the bottom of the vessel. The centrifugal force is the major force causing separation of the particulate in a cyclone separator. A typical large-diameter cyclone system is shown in Figure 17-1.

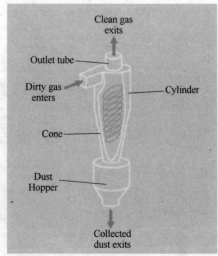

Figure 17-1 Top-Inlet Large-Diameter Cyclone

Types of Cyclones

The cyclonic collectors are generally of two types: the larger diameter, lower efficiency cyclones, and the small diameter, multitube high-efficiency units (a small-diameter cyclone tube is shown in Figure 17-2). The larger cyclones have lower efficiencies especially on particles less than $50 \mu m$. However, they have low initial cost and usually operate at pressure drops of 1 to 3 inches of water. The multitube cyclones are capable of efficiencies exceeding 90% but the cost is higher and their pressure drops is usually 3 to 5 inches of water. They are also more susceptible to plugging and erosion.

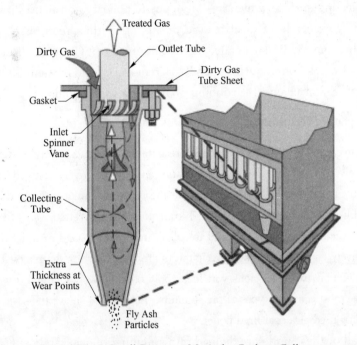

Figure 17-2   Small-Diameter Mutitube-Cyclone Collector

Advantages and Disadvantages of Cyclones

Cyclones provide a low-cost, low-maintenance method of removing larger particulates from a gas stream. They are used whenever the particle size distributions generated by the process are relatively large (greater than 5 micrometers) and/or the control efficiency requirements are in the low-to-moderate range of 50% to 90%.

They are also used as the pre-collector of large-diameter embers generated in some combustion systems. Removal of the embers is necessary to protect high-efficiency particulate control systems downstream from the mechanical collectors.

Most mechanical collectors are not applicable to industrial sources that generate sticky and/or wet particulate matter. These materials can accumulate on the cyclone body wall or the inlet spinner vanes of conventional multi-cyclone collectors.

3. Electrostatic Precipitators

Operating Principles

Gas cleaning by electrostatic precipitation is particularly suited for gas streams which can be easily ionized and which contain either liquid or particulate matter. The method of removal consists of passing the particle-laden gas through an electrostatic field produced by a high voltage electrode and grounded collection surface. The gas is ionized by the high voltage discharge and the particulate matter is charged by the interaction of the gas ions. The particles migrate to the collecting surface, which has an opposite polarity and neutralized.

The operating principle of electrostatic precipitation thus requires three basic steps: (1) electrical charging of the suspended particulate matter; (2) collection of the charged particulate matter on a ground surface; and (3) removal of the particulate matter from the collecting surfaces by mechanical scrubbing or flushing with liquids.

Types of Electrostatic Precipitators

There are three main styles of electrostatic precipitators: (1) negatively charged dry precipitators, (2) negatively charged wetted-wall precipitators, and (3) positively charged two-stage precipitators. The negatively charged dry precipitators are the type most frequently used on large applications such as coal-fired boilers, cement kilns, and kraft pulp mills. Wetted-wall precipitators (sometimes called wet precipitators) are often used to collect mist and/or solid material that is moderately sticky. The positively charged two-stage precipitators are used only for the removal of mists. In the remainder of this section, the discussions will focus only on negatively charged dry precipitators because these are the most common types of precipitators. Figure 17-3 shows the scale of a typical electrostatic precipitator used at a coal-fired boiler.

Essentially all of these units are divided into a number of separately energized areas that are termed fields (see Figure 17-4). Most precipitators have between three and ten fields

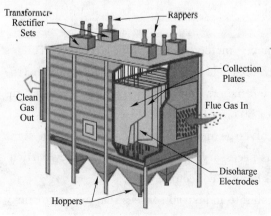

Figure 17-3 Conventional Electrostatic Precipitator

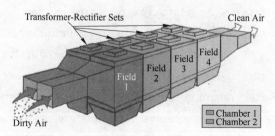

Figure 17-4 Arrangements of Fields and Chambers in an ESP

in series along the gas flow path. On large units, the precipitators are divided into a number of separate, parallel chambers, each of which has an equal number of fields in series. There is a solid partition or physical separation between the 2 to 8 chambers that are present on the large systems.

Advantages and Disadvantages of ESPs

Electrostatic precipitators can have very high efficiencies due to the strong electrical forces applied to the small particles. These types of collectors can be used when the gas stream is not explosive and does not contain entrained droplets or other sticky material.

The composition of the particulate matter is very important because it influences the electrical conductivity within the dust layers on the collection plate. Resistivity, an important concept associated with electrostatic precipitators, is a measure of the ability of the particulate matter to conduct electricity and is expressed in units of ohm-cm. As the resistivity increases, the ability of the particulate matter to conduct electricity decreases. Precipitators can be designed to work in any resistivity range; however, they usually work best when the resistivity is in the moderate range (10^8 to 10^{10} ohms/cm).

4. Fabric Filters

Operating Principles

Fabric filters collect particulate matter on the surfaces of filter bags. Most of the particles are captured by inertial impaction, interception, Brownian diffusion, and sieving on already collected particles that have formed a dust layer on the bags. The fabric material itself can capture particles that have penetrated through the dust layers. Electrostatic attraction may also contribute to particle capture in the dust layer and in the fabric itself. Due to the multiple mechanisms of particle capture possible, fabric filters can be highly efficient for the entire particle size range of interest in air pollution control.

Fabric filters, or baghouses, remove dust from a gas stream by passing the stream through a porous fabric. The fabric filter is efficient at removing fine particles and can exceed efficiencies of 99 percent in most applications.

The selection of the fiber material and fabric construction is important to baghouse performance. The fiber material from which the fabric is made must have adequate strength characteristics at the maximum gas temperature expected and adequate chemical compatibility with both the gas and the collected dust. One disadvantage of the fabric filter is that high-temperature gases often have to be cooled before contacting the filter medium.

Types of Fabric Filters

A reverse-air-type fabric filter, shown in Figure 17-5, is one of the major categories of fabric filters. It is used mainly for large industrial sources. In this type of unit, the particle-laden gas stream enters from the bottom and passes into the inside of the bags. The dust cake accumulates on the inside surfaces of the bags. Filtered gas passes through the bags and is exhausted from the unit.

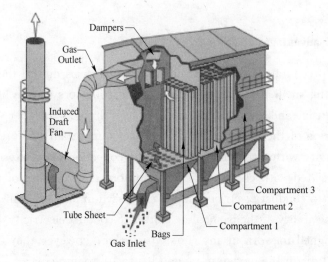

Figure 17-5 Reverse Air Fabric Filter

When cleaning is necessary, dampers are used to isolate a compartment of bags from the inlet gas flow. Then, some of the filtered gas passes in the reverse direction (from the outside of the bag to the inside) in order to remove some of the dust cake. The gas used for reverse air cleaning is re-filtered and released.

Another common type of fabric filter is the pulse jet shown in Figure 17-6. In this type of unit, the bags are supported on metal wire cages that are suspended from the top of the unit. Particulate-laden gas flows around the outside of the bags, and a dust cake accumulates on the exterior surfaces. When cleaning is needed, a very-short-duration pulse of compressed air is injected at the top inside part of each bag in the row of bags being cleaned. The compressed air pulse generates a pressure wave that moves down each bag and, in the process, dislodges some of the dust cake from the bag.

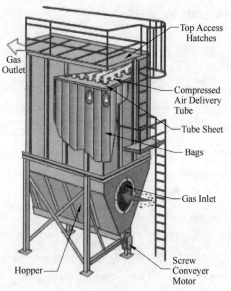

Figure 17-6 Pulse Jet Fabric Filter

Advantages and Disadvantages of Fabric Filters

Fabric filters are used in a wide variety of applications where high efficiency particulate collection is needed. The control efficiencies usually range from 99% to greater than

99.5% depending on the characteristics of the particulate matter and the fabric filter design. As mentioned earlier, fabric filters can be very efficient at collecting particles in the entire size range of interest in air pollution control.

The performance of fabric filters is usually independent of the chemical composition of the particulate matter. However, they are not used when the gas stream generated by the process equipment includes corrosive materials that could chemically attack the filter media. Fabric filters are also not used when there are sticky or wet particles in the gas stream. These materials accumulate on the filter media surface and block gas movement.

Fabric filters must be designed carefully if there are potentially combustible or explosive particulate matters, gases, or vapors in the gas stream being treated. If these conditions are severe, alternative control techniques, such as wet scrubbers, are often used.

5. Wet scrubbers

Operating Principles

Wet collectors use water "sprays" to collect and remove particulate matter. The wet scrubbers can provide high collection efficiency but may involve treatment of liquid wastes with settling ponds. They also saturate the gas stream and produce a resultant steam plume. The principal mechanisms involves in wet scrubbing are: (1) increasing the size of the particles by combination with liquid droplets thereby increasing their size so they may be collected more easily, and or (2) trapping them in liquid film and washing them away.

Types of Wet Scrubber

There are a number of major categories of particulate wet scrubbers: Venturis, spray towers, impingement plate towers, packed powers, etc. This lesson discusses only three of the above types of scrubbers: venturis, impingement plate scrubbers, and spray towers.

There are many variations of wet collectors but they may generally be classified as low or high energy scrubbers. Low energy scrubbers of 1 to 6 inches of pressure drop may consist of simple spray towers, packed towers or impingent plate towers. The water requirements may be 3 to 6gal/1000ft^3 of gas and collection efficiencies can exceed 90% to 95%. The lower energy scrubbers find frequent application in incinerators, fertilizer manufacturing, lime kilns, and iron foundries.

The high energy scrubbers, or Venturi scrubbers use a liquid stream to remove solid particles. In the venturi scrubber, gas laden with particulate matter passes through a short tube with flared ends and a constricted middle. This constriction causes the gas stream to speed up when the pressure is increased. A water spray is directed into the gas stream either prior to or at the constriction in the tube. The difference in velocity and pressure resulting from the constriction causes the particles and water to mix and combine. The reduced velocity at the expanded section of the throat allows the droplets of water containing the particles to drop out of the gas stream. Venturi scrubbers are effective in removing small particles, with removal efficiencies of up to 99 percent. One drawback of this device, however, is the production of wastewater.

Impingement Plate Scrubber

An impingement plate scrubber is shown in Figure 17-7. These scrubbers usually have one to three horizontal plates, each of which has a large number of small holes. The gas stream accelerating through the holes atomizes some water droplets in the water layer above the plate. Particles impact into these water droplets.

Spray Tower Scrubber

A typical spray tower scrubber is shown in Figure 17-8. This is the simplest type of particulate wet scrubber in commercial service. Sets of spray nozzles located near the top of the scrubber vessel generate water droplets that impact with particles in the gas stream as the gas stream moves upwards.

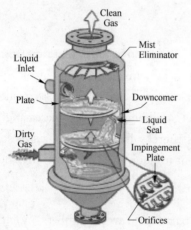

Figure 17-7　Impingement Plate Scrubber

Figure 17-8　Spray Tower Scrubber

Venturi Scrubbers

A typical venturi throat is shown in Figure 17-9. Particulate matter, which accelerates as it enters the throat, is driven into the slow moving, large water droplets that are introduced near the high velocity point at the inlet of the venturi throat. The adjustable dampers in the unit illustrated are used to adjust the open cross-sectional area and thereby affect the speed of the particles entrained in the inlet gas stream.

Advantages and Disadvantages of Scrubbers

Many types of particulate wet scrubbers can provide high efficiency control of particulate matter. One of the

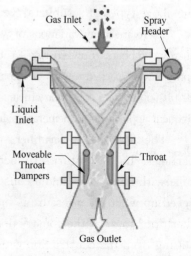

Figure 17-9　Adjustable-Throat Venturi Scrubber

main advantages of particulate wet scrubbers is that they are often able to simultaneously collect particulate matter and gaseous pollutants. Also, wet scrubbers can often be used on sources that have potentially explosive gases or particulate matter. They are compact and can often be retrofitted into existing plants with very limited space.

One of the main disadvantages of particulate wet scrubbers is that they require make-up water to replace the water vaporized into the gas stream and lost to purge liquid and sludge removed from the scrubber system. Wet scrubbers generate a waste stream that must be treated properly.

6. General Applicability of Particulate Control Systems

Particulate matter control systems are often selected based on the general criteria listed in Figure 17-10.

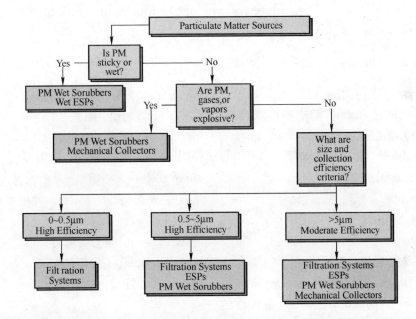

Figure 17-10 General Applicability of Particulate Control Systems

If there is a high concentration of wet and/or sticky particulate matter, either a particulate wet scrubber or a wet electrostatic precipitator is used. If wet or sticky materials are present with combustible materials or explosive gases or vapors, the particulate wet scrubber is most appropriate.

If the particulate matter is primarily dry, mechanical collectors, particulate wet scrubbers, conventional electrostatic precipitators, and fabric filters can be used. The next step in the selection process is to determine if the particulate matter and/or gases and vapors in the gas stream are combustible or explosive. If so, then mechanical collectors or particulate wet scrubbers can be used because both of these categories of systems can be designed to minimize the risks of ignition. In some cases, a fabric filter can also be used if it includes the appropriate safety equipment. An electrostatic precipitator is not used due to the risk of ignition caused by electrical sparking in the precipitator fields. When selecting between mechanical collectors and wet scrubbers, mechanical collectors are the more economical choice. They have a lower purchase cost and a lower operating cost than wet scrubbers.

If the dry particulate matter is present in a gas stream that is not combustible or explosive, the selection depends on the particle size range and the control efficiency requirements. If a significant portion of the gas stream is in the less than 0.5-micrometer size range, and high efficiency control is needed, a fabric filter is the most common choice. If a significant portion of the particulate matter is in the 0.5-to 5-micrometer size range, and high efficiency control is needed, fabric filters, electrostatic precipitators, or particulate wet scrubbers (certain types) could be used. If most of the particulate matter is larger than 5 micrometers, any of the four main types of particulate control systems could be used.

Vocabulary and Phrases

tangentially　　*ad.* 切向的
electrostatic precipitator　　静电除尘器
fabric filter　　布袋过滤器
venturi scrubber　　文丘里洗涤器
cyclone　　*n.* 旋风分离器
settling chamber　　沉降室
hopper　　*n.* 漏斗
interception　　*n.* 拦截
penetrate　　*vi.* 穿透，渗透
compatibility　　*n.* 兼容性
corrosive　　*adj.* 腐蚀的，蚀坏的，腐蚀性的

sticky　　*adj.* 黏的，黏性的
block　　*vt.* 妨碍，阻塞
inertia　　*n.* 惯性，惯量
resistivity　　*n.* 抵抗力，电阻系数
spray towers　　喷雾塔
impingement plate tower　　冲击式板式塔
packed power　　填料塔
drawback　　*n.* 缺点，障碍，
horizontal plates　　水平盘
nozzle　　*n.* 管口，喷嘴

Notes

1. The performance of fabric filters is usually independent of the chemical composition of the particulate matter. However, they are not used when the gas stream generated by the process equipment includes corrosive materials that could chemically attack the filter media.

布袋过滤器的性能通常和颗粒物的化学组成无关。但是，当气体中含有腐蚀性物质对滤料有腐蚀时，不能使用布袋过滤器。

2. In this type of unit, the particle-laden gas stream enters from the bottom and passes into the inside of the bags. The dust cake accumulates on the inside surfaces of the bags. Filtered gas passes through the bags and is exhausted from the unit.

对于逆气流清灰布袋除尘器，含尘气体从底部进入，通过袋的内部流动，粉尘聚集在

袋的内表面，被过滤的气体（清洁气体）穿过布袋从除尘器排出。

3. The spiral motion produces the centrifugal forces that cause the particulate matter to move toward the periphery of the vessel and collect on the walls and fall to the bottom of the vessel.

螺旋运动产生的离心力使得颗粒物向容器的周边运动，在容器壁上被收集然后降落至容器的底部。

4. The method of removal consists of passing the particle-laden gas through an electrostatic field produced by a high voltage electrode and grounded collection surface.

这种方法包括使含尘气体通过由一个高压电极和接地捕集表面产生的静电区域。

Question and Exercises

1. Put the following into Chinese

electrostatic attraction mutitube-Cyclone Collector Brownian diffusion reverse-air-type fabric filter compressed air pulse jet fabric filter high voltage discharge electrical forces

2. Put the following into English

重力沉降室 燃煤锅炉 导电性 比电阻 粉尘层 滤料 静电引力 离心力 压力降 高能湿式除尘器 文丘里（喉）管 喷雾嘴

3. Translate the following paragraphs or sentences into Chinese

(1) Cyclonic collectors are round conically shaped vessels in which the gas stream enters tangentially and followers a spiral path to the outlet.

(2) The operating principle of electrostatic precipitation thus requires three basic steps: 1) electrical charging of the suspended particulate matter; 2) collection of the charged particulate matter on a ground surface; and 3) removal of the particulate matter from the collecting surfaces by mechanical scrubbing or flushing with liquids.

(3) The composition of the particulate matter is very important because it influences the electrical conductivity within the dust layers on the collection plate. Resistivity, an important concept associated with electrostatic precipitators, is a measure of the ability of the particulate matter to conduct electricity and is expressed in units of ohm-cm.

(4) The principal mechanisms involves in wet scrubbing are: 1) increasing the size of the particles by combination with liquid droplets thereby increasing their size so they may be collected more easily, and or 2) trapping them in liquid film and washing them away.

4. Describe the basic stricture of five different categories of particle control devices and the collection mechanisms they use.

Further Reading

Mobile Sources Emission Control

Motor vehicles are an important source of air pollution. The Environmental Protection Agencies estimates that in 1990, on-road motor vehicles accounted for 29% of the total volatile organic compounds (VOCs) emissions, 33% of NO_x emissions, and 65% of all carbon monoxide emissions nationwide. In urban areas, motor vehicle emissions typically account for even larger fractions of total anthropogenic air pollutant emissions.

To achieve on-road mobile source emission control an integrated approach has been used. This integrated approach includes technological advances in vehicle and engine design together with cleaner, high-quality fuels plus the addition of vapor and particulate recovery systems and the development of auto inspection and maintenance (I/M) programs.

Mobile Sources-Cleaner Fuels

One way to reduce air pollution from cars and trucks is to use a gasoline that is designed to burn clean. This cleaner burning gasoline, called reformulated gasoline or RFG, is required in cities with the worst smog pollution, but other cities with smog problems may choose to use RFG. Reformulated gasoline contains less volatile organic compounds (VOCs), and will contain oxygen additives to make the fuel burn more efficiently. Also, all gasolines will have to contain detergents, which, by preventing build-up of engine deposits, keep engines working smoothly and burning fuel cleanly.

Methyl tertiary-butyl ether (MTBE) is the oxygen additive most commonly used by the petroleum industry to satisfy the two percent oxygen mandate in the RFG program. MTBE is used in approximately 87 percent of RFG, with ethanol being the second most commonly used additive. Oxygenates increase the combustion efficiency of gasoline, thereby reducing vehicle emissions of carbon monoxide. On the other hand, EPA encourages the development and sale of alternative fuels such as alcohols, liquefied petroleum gas (LPG) and natural gas in order to lower fuel emissions.

Mobile Sources-Cleaner Cars

In response to tighter standards, manufacturers equipped new cars with even more sophisticated emission control systems. These systems generally include a "three-way" catalyst (which converts carbon monoxide and hydrocarbons to carbon dioxide and water, and also helps reduce nitrogen oxides to elemental nitrogen and oxygen), plus an on-board computer and oxygen sensor. This equipment helps optimize the efficiency of the catalytic converter. The catalytic converter is an anti-pollution device located between a vehicle's engine and tailpipe. Catalytic converters work by facilitating chemical reactions that convert exhaust pollutants such as carbon monoxide and nitrogen oxides to normal atmospheric gases such as nitrogen, carbon dioxide, and water.

In diesel exhaust, the addition of a particulate filter as an anti-pollution device traps particles in the exhaust before they can escape into the atmosphere. A vapor recovery system, also an anti-pollution system, captures gasoline vapors that would otherwise escape into the atmosphere from hot vehicle engines and fuel tanks.

Finally, auto makers must build some cars that use clean fuels, including alcohol, and that release less pollution from the tailpipe through advanced engine design. Electric cars, which are low-pollution vehicles, are currently being used as clean cars. Electric cars

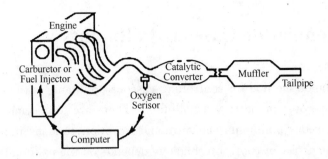

Figure 17-11   Typical Catalyst System for Exhaust Emissions

produce essentially no pollution from the tailpipe or through fuel evaporation. Car manufacturers are also beginning to sell "hybrid" vehicles that combine an electric motor with a separate gasoline or diesel engine. Hybrid vehicles can more than double the gas mileage of conventional gasoline.

Mobile Sources-Auto Inspection and Maintenance (I/M) Programs

Under the 1990 Clean Air Act, auto manufacturers will build cleaner cars, and cars will use cleaner fuels. However, to get air pollution down and keep it down, a third program is needed; vehicle inspection and maintenance (I/M), which makes sure cars are being maintained adequately to keep pollution emissions (releases) low. The 1990 Clean Air Act includes very specific requirements for inspection and maintenance programs.

Auto inspection and maintenance programs are designed to make sure that the cars we buy and drive perform their best and live up to their pollution control promise. Auto inspection and maintenance programs require the testing of motor vehicles in parts of the country with unhealthy air and the repair of those that do not meet standards. I/M tests use special equipment to measure the pollution in car exhaust. These tests check that the car's key emission controls are installed as designed and then analyze the exhaust to check acceptable control of carbon monoxide and hydrocarbons. Advanced tests also check nitrogen oxide emissions.

Unit 18　Control of Gaseous Pollutants

The most common method for controlling gaseous pollutants is the addition of add-on control devices to recover or destroy a pollutant. There are four commonly used control technologies for gaseous pollutants: absorption, adsorption, condensation, and incineration (combustion). The choice of control technology depends on the pollutant (s) that must be removed, the removal efficiency required, pollutant gas stream characteristics, and specific characteristics of the site. Absorption, adsorption, and condensation all are recovery techniques while incineration involves the destruction of the pollutant.

1. Absorption

The removal of one or more selected components from a gas mixture by absorption is probably the most important operation in the control of gaseous pollutant emissions. Absorption is a process in which a gaseous pollutant is dissolved in a liquid. Water is the most commonly used absorbent liquid. As the gas stream passes through the liquid, the liquid absorbs the gas, in much the same way that sugar is absorbed in a glass of water when stirred. Absorption is commonly used to recover products or to purify gas streams that have high concentrations of organic compounds. Absorption equipment is designed to get as much mixing between the gas and liquid as possible.

Absorbers are often referred to as scrubbers, and there are various types of absorption equipment. The principal types of gas absorption equipment include spray towers, packed columns, spray chambers, and venturi scrubbers. The packed column is by far the most commonly used for the absorption of gaseous pollutants. The packed column absorber has a column filled with an inert (non-reactive) substance, such as plastic or ceramic, which increases the liquid surface area for the liquid/gas interface. The inert material helps to maximize the absorption capability of the column. In addition, the introduction of the gas and liquid at opposite ends of the column causes mixing to be more efficient because of the counter-current flow through the column. In general, absorbers can achieve removal efficiencies grater than 95 percent. One potential problem with absorption is the generation of waste-water, which converts an air pollution problem to a water pollution problem.

2. Adsorption

When a gas or vapor is brought into contact with a solid, part of it is taken up by the solid. The molecules that disappear from the gas either enter the inside of the solid, or remain on the outside attached to the surface. The former phenomenon is termed absorption (or dissolution) and the latter adsorption. Adsorption is the binding of molecules or particles to a surface. In this phenomenon molecules from a gas or liquid will be attached in a physical way to a surface. The binding to the surface is usually weak and reversible. The most common industrial adsorbents are activated carbon, silica gel, and alumina, because they have enormous surface areas per unit weight.

Activated carbon is the universal standard for purification and removal of trace organic contaminants from liquid and vapor streams. Carbon adsorption uses activated carbon to control and/or recover gaseous pollutant emissions. In carbon adsorption, the gas is attracted and adheres to the porous surface of the activated carbon. Removal efficiencies of 95 percent to 99 percent can be achieved by using this process. Carbon adsorption is used in cases where the recovered organics are valuable. For example, carbon adsorption is often used to recover perchloroethylene, a compound used in the dry cleaning process.

Carbon adsorption systems are either regenerative or non-regenerative (Seen in Figure 18-1). A regenerative system usually contains more than one carbon bed. As one bed actively removes pollutants, another bed is being regenerated for future use. Steam is used to purge captured pollutants from the bed to a pollutant recovery device. By a "regenerating" the carbon bed, the same activated carbon particles can be used again and again. Regenerative systems are used when concentration of the pollutant in the gas stream is relatively high. Non-regenerative systems have thinner beds of activated carbon. In a non-regenerative adsorber, the spent carbon is disposed of when it becomes saturated with the pollutant. Because of the solid waste problem generated by this type of system, non-regenerative carbon adsorbers are usually used when the pollutant concentration is extremely low.

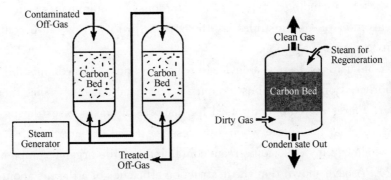

Figure 18-1 Regenerative carbon adsorption system and non-regenerative carbon adsorption system

3. Condensation

Condensation is the process of converting a gas or vapor to liquid. Any gas can be reduced to a liquid by lowering its temperature and/or increasing its pressure. The most common approach is to reduce the temperature of the gas stream, since increasing the pressure of a gas can be expensive. A simple example of the condensation process is droplets of water forming on the outside of a glass of cold water. The cold temperature of the glass causes water vapor from the surrounding air to pass into the liquid state on the surface of the glass.

Condensers are widely used to recover valuable products in a waste stream. Condensers are simple, relatively inexpensive devices that normally use water or air to cool and condense a vapor stream. Condensers are typically used as pretreatment devices. They can be used ahead of adsorbers, absorbers, and incinerators to reduce the total gas volume to be treated by more expensive control equipment. Condensers used for pollution control are contact condensers and surface condensers. In a contact condenser, the gas comes into contact with cold liquid. In a surface condenser, the gas contacts a cooled surface in which cooled liquid or gas is circulated, such as the outside of the tube. Removal efficiencies of condensers typically range from 50 percent to more than 95 percent, depending on design and applications.

4. Incineration

Incineration, also known as combustion, is most used to control the emissions of organic compounds from process industries. This control technique refers to the rapid oxidation of a substance through the combination of oxygen with a combustible material in the presence of heat. When combustion is complete, the gaseous stream is converted to carbon dioxide and water vapor. Incomplete combustion will result in some pollutants being released into the atmosphere. Smoke is one indication of incomplete combustion. Equipment used to control waste gases by combustion can be divided in three categories: direct combustion or flaring, thermal incineration and catalytic incineration. Choosing the proper device depends on many factors, including type of hazardous contaminants in the waste stream, concentration of combustibles in the stream, process flow rate, control requirements, and an economic evaluation.

A direct combustor or flare is a device in which air and all the combustible waste gases react at the burner. Complete combustion must occur instantaneously since there is no residence chamber. Flares are commonly used for disposal of waste gases during process upsets, such as those that take place when a process is started or shut down. A flare can be used to control almost any emission stream containing volatile organic compounds. Studies conducted by EPA have shown that the destruction efficiency of a flare is about 98 percent.

In thermal incinerators the combustible waste gases pass over or around a burner flame into a residence chamber where oxidation of the waste gases is completed. For thermal incineration, it is important that the vapor stream directed to the thermal incinerator have a constant combustible gas concentration and flow rate. These devices are not well-suited to vapor streams that fluctuate, because the efficiency of the combustion process depends on the proper mixing of vapors, and a specific residence time in the combustion chamber. Residence time is the amount of time the fuel mixture remains in the combustion chamber. Often, supplementary fuel is added to a thermal incinerator to supplement the quantity of pollutant gases being burned by the incinerator. Energy and heat produced by the incineration process can be recovered and put to beneficial uses at a facility. Thermal incinerators can destroy gaseous pollutants at efficiencies of greater than 99 percent when operated correctly.

Catalytic incinerators are very similar to thermal incinerators. The main difference is that after passing through the flame area, the gases pass over a catalyst bed. A catalyst is a substance that enhances a chemical reaction without being changed or consumed by the reaction. A catalyst promotes oxidation at lower temperatures, thereby reducing fuel costs. Destruction efficiencies greater than 95 percent are possible using a catalytic incinerator. Higher efficiencies are possible if larger catalyst volumes or higher temperatures are used. Catalytic incinerators are best suited for emission streams with low VOCs content.

Vocabulary and Phrases

absorption   n. 吸收
adsorption   n. 吸附
condensation   n. 浓缩，冷凝
incineration   n. 烧成灰，焚化，燃烧
absorbent   adj. 能吸收的，n. 吸收剂
organic compounds   有机化合物
spray tower   喷淋塔
reversible   adj. 可逆的

activated carbon   活性炭
silica gel   n. 硅胶
alumina   n. 氧化铝，矾土
enormous   adj. 巨大的，庞大的，
porous   adj. 多孔的，有孔的
perchloroethylene   n. [化] 四氯乙烯
regenerate   vt. 使新生，重建，改革，革新

Notes

1. by far   到目前为止
2. The packed column absorber has a column filled with an inert (non-reactive) substance, such as plastic or ceramic, which increases the liquid surface area for the liquid/gas interface.
填料塔是装填有惰性(非反应性)物质的圆柱形设备，这些惰性物质是塑料或陶瓷，可以增加气液相界面。
3. counter-current flow   逆流
4. Residence time is the amount of time the fuel mixture remains in the combustion chamber.
停留时间是燃烧混合物在燃烧室的逗留时间。
5. Often, supplementary fuel is added to a thermal incinerator to supplement the quantity of pollutant gases being burned by the incinerator.
通常，在热力燃烧炉中要加入辅助燃料补充燃烧所需污染气体的量(辅助燃料燃烧供热，将废气温度提高到热力燃烧所需的温度，使其进行氧化分解)。
6. VOCs: volatile organic compounds   挥发性有机化合物

Question and Exercises

1. Put the following into Chinese
removal efficiency   absorbent liquid   absorption equipment   adsorbent   absorber   organic

contaminants　catalytic incinerator　flame area　residence time　packed columns

2. Put the following into English

文丘里吸收器　填料塔　再生系统　接触冷凝　表面冷凝　直接燃烧　热力燃烧　催化燃烧　燃烧室

3. Translate the following paragraphs into Chinese

(1) Condensation is the process of converting a gas or vapor to liquid. Any gas can be reduced to a liquid by lowering its temperature and/or increasing its pressure. The most common approach is to reduce the temperature of the gas stream, since increasing the pressure of a gas can be expensive.

(2) Absorption is a process in which a gaseous pollutant is dissolved in a liquid. Water is the most commonly used absorbent liquid. As the gas stream passes through the liquid, the liquid absorbs the gas, in much the same way that sugar is absorbed in a glass of water when stirred.

(3) Absorption is commonly used to recover products or to purify gas streams that have high concentrations of organic compounds.

(4) In carbon adsorption, the gas is attracted and adheres to the porous surface of the activated carbon. Removal efficiencies of 95 percent to 99 percent can be achieved by using this process. carbon adsorption is used in cases where the recovered organics are valuable.

Further Reading (1)

Sulfur Oxides Control Techniques

1. Characteristics of Sulfur Oxides

Sulfur oxides include sulfur dioxide (SO_2), sulfur trioxide (SO_3), and sulfuric acid (H_2SO_4). Combustion of fossil fuels for generation of electric power is clearly the primary contributor of sulfur dioxide emissions. Industrial processes, such as nonferrous metal smelting, also contribute to sulfur dioxide emissions.

Sulfur dioxide is a colorless gas, which is moderately soluble in water and aqueous liquids. It is formed primarily during the combustion of sulfur-containing fuel or waste. Once released to the atmosphere, sulfur dioxide reacts slowly to form sulfuric acid (H_2SO_4), inorganic sulfate compounds, and organic sulfate compounds.

Some of the sulfur dioxide in high temperature processes is oxidized to form sulfur trioxide as shown below.

$$SO_2 + 1/2 O_2 \rightarrow SO_3$$

Sulfur trioxide remains in the vapor state while the combustion gases are very hot. As the gases cool, sulfur trioxide adds a water molecule and forms sulfuric acid as indicated by the reaction below.

$$SO_3 + H_2O \rightarrow H_2SO_4$$

Below 500 to 600°F, most of the sulfur trioxide, which is extremely hygroscopic, reacts with water molecules to form sulfuric acid.

Sulfuric acid is a strong acid. In addition to the sources mentioned above, sulfuric acid can also be released from plants that manufacture batteries.

Sulfuric acid vapor in moderate concentrations (2 to 8 ppm) is very beneficial to electrostatic precipitators because it adsorbs onto particle surfaces and creates a moderate resistivity. High concentrations can be detrimental to precipitator performance. High sulfuric acid levels can also cause significant corrosion problems for precipitators, fabric filters, and other control devices. The temperature of flue gases should be kept well above the dew point for sulfuric acid to prevent condensation on ductwork surfaces and components in the air pollution control system.

2. Formation Mechanisms

Sulfur dioxide and sulfuric acid are formed during the combustion of fuel or waste that contains sulfur compounds. Sulfur oxides can also be released from chemical reactors and sulfuric acid plants.

The sulfur in the fuel or waste being fired enters the combustion process in a variety of chemical forms including but not limited to inorganic sulfates, organic sulfur compounds, and pyrites.

A small fraction of the fuel or waste sulfur (usually less than five percent) remains in the bottom ash leaving the combustion processes. The remaining 95 percent is converted to sulfur dioxide, which remains in the gaseous form throughout the combustion system.

A small fraction of the sulfur dioxide generated in the combustion zone is oxidized further to form sulfur trioxide. The reaction mechanisms that could contribute to the formation of this pollutant are not entirely known; however, they probably include the following:

- Free radical reaction of sulfur dioxide with atomic oxygen in the high temperature zones.
- Catalytic oxidation of sulfur dioxide on the surfaces of particles entrained in the gas stream.
- Thermal reactions between sulfur dioxide and other inorganic gases generated during combustion.

The concentration of sulfur trioxide generated during combustion varies widely from unit to unit for reasons that have not been determined entirely; however, sulfur trioxide concentrations are generally related directly to the concentration of sulfur in the fuel and the concentration of oxygen in the combustion zone. The sulfur trioxide concentrations are usually 0.5 to 2 percent of the sulfur dioxide concentration. Sulfur trioxide quickly converts to sulfuric acid upon cooling in the gas stream or atmosphere.

3. Control Techniques

Air pollution control systems for sulfur dioxide removal are large and sophisticated. Sulfur dioxide is controlled by three different techniques: absorption, adsorption, and the use of low-sulfur fuels.

The control systems used for sulfur dioxide are usually not designed to remove sulfuric acid. The sulfuric acid concentrations are usually below the levels where it is economically feasible or environmentally necessary to install control systems.

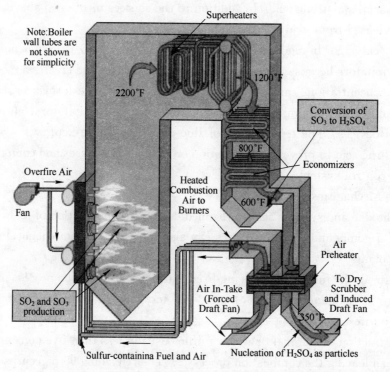

Figure 18-2 Location of Sulfur Oxides Formation in a Boiler

Absorption

Absorption processes use the solubility of sulfur dioxide in aqueous solutions to remove it from the gas stream. Once sulfur dioxide has dissolved in solution to form sulfurous acid (H_2SO_3), it reacts with oxidizers to form inorganic sulfites (SO_3^{2-}) and sulfates (SO_4^{2-}). This process prevents the dissolved sulfur dioxide from diffusing out of solution and being re emitted. The most common type of sulfur dioxide absorber is the limestone wet scrubber. An example flowchart is shown in Figure 18-3.

Limestone is the alkali most often used to react with the dissolved sulfur dioxide. Limestone slurry is sprayed into the sulfur dioxide-containing gas stream. The chemical reactions in the recirculating limestone slurry and reaction products must be carefully controlled in order to maintain the desired sulfur dioxide removal efficiency and to prevent operating problems.

Wet scrubbers used for sulfur dioxide control usually operate at liquid pH levels between 5 to 9 to maintain high efficiency removal. Typical removal efficiencies for sulfur dioxide in wet scrubbers range from 80% to 95%.

The wet scrubber (absorber) vessels do not efficiently remove particulate matter smaller than approximately 5 micrometers. However, as in the case with low-efficiency particu-

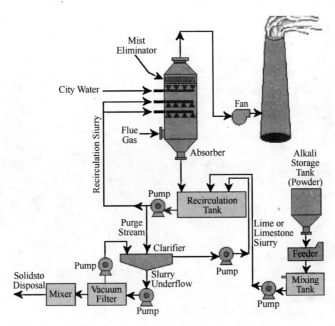

Figure 18-3 Example Flowchart of Limestone-Based SO₂ Scrubbing System

late wet scrubbers, the particulate removal efficiency increases rapidly with particle size above 5 micrometers. Usually, a moderate-to-high efficiency particulate control system is used upstream from the sulfur dioxide absorber to reduce the particulate matter emissions in the less than 3 micrometer size range. These upstream collectors also reduce the quantity of particulate matter that is captured in the absorber.

The evaporation of water that occurs in wet scrubber vessels can keep gas temperatures relatively cold, in the range of 110 to 140°F. These gas temperatures are well below the typical operating temperatures of other air pollution control systems used on sources that generate sulfur dioxide emissions.

Another type of absorption system is called a spray atomizer dry scrubber, which belongs to a group of scrubbers called spray-dryer-type dry scrubbers. In this case, an alkaline slurry is sprayed into the hot gas stream at a point upstream from the particulate control device. As the slurry droplets are evaporating, sulfur dioxide absorbs into the droplet and reacts with the dissolved and suspended alkaline material.

Large spray dryer chambers are used to ensure that all of the slurry droplets evaporate to dryness prior to going to a high efficiency particulate control system. The term "dry scrubber" refers to the condition of the dried particles approaching the particulate control system. Fabric filters or electrostatic precipitators are often used for high efficiency particulate control. The system shown in Figure 18-4 has a fabric filter.

Spray-dryer-type absorption systems have efficiencies that are similar to those for wet-scrubber-type absorption systems. These generate a waste stream that is dry and, therefore, easier to handle than the sludge generated in a wet scrubber. However, the equipment

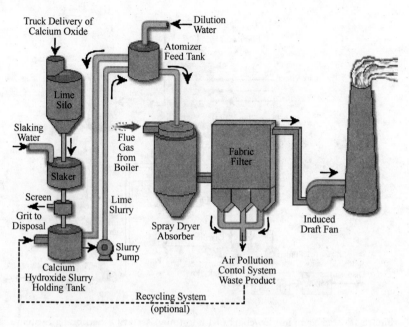

Figure 18-4 Spray-Dryer-Type Dry Scrubber

used to atomize the alkaline slurry is complicated and can require considerably more maintenance than the wet scrubber systems. Spray-dryer-type absorption systems operate at higher gas temperatures than wet scrubbers do and are less effective for the removal of other pollutants in the gas stream such as condensable particulate matter.

The choice between a wet-scrubber absorption system and a spray-dryer absorption system depends primarily on site-specific costs. The options available for environmentally sound disposal of the waste products are also an important consideration in selecting the type of system for a specific application. Both types of systems are capable of providing high efficiency sulfur dioxide removal.

Adsorption

Sulfur dioxide can be collected by adsorption systems. In this type of control system, a dry alkaline powder is injected into the gas stream. Sulfur dioxide adsorbs to the surface of the alkaline particles and reacts to form compounds that cannot be re-emitted to the gas stream. Hydrated lime (calcium hydroxide) is the most commonly used alkali. However, a variety of alkalis can be used effectively. A flowchart for a dry-injection-type dry scrubber (adsorber) is shown in Figure 18-5.

A dry-injection-type dry scrubber can be used on smaller systems as opposed to using the larger, more complicated spray-dryer-type dry scrubber. However, the dry injection system is slightly less efficient, and requires more alkali per unit of sulfur dioxide (or other acid gas) collected. Accordingly, the waste disposal requirements and costs are higher for adsorption systems than absorption systems.

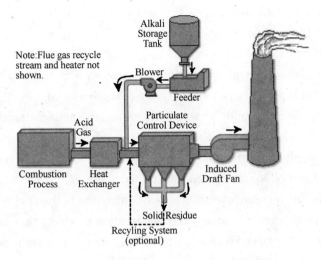

Figure 18-5　Dry-Injection-Type Dry Scrubber

Alternative Fuels

Other techniques used for limiting the emissions of sulfur dioxide are simply to switch to fuels that have less sulfur or to convert to synthetic (processed) fuels that have low sulfur levels. The sulfur dioxide emission rate is directly related to the sulfur levels in coal, oil, and synthetic fuels. However, not all boilers can use these types of fuels. Each type of boiler has a number of very specific and important fuel characteristic requirements and not all low sulfur fuels meet these fuel-burning characteristics.

Further Reading (2)

Nitrogen Oxides Control Techniques

1. Characteristics of Nitrogen Oxides

Due to cleaner fuels and more efficient engines, total emissions of nitrogen oxides from mobile sources has decreased over the past twenty years. However mobile sources, along with electric utilities, continue to be the leading contributor of nitrogen oxide emissions as shown in Figure 18-6.

Nitrogen oxides (often abbreviated NO_x) are nitric oxide (NO) and nitrogen dioxide (NO_2). They are formed simultaneously in combustion processes and other high temperature operations such as metallurgical furnaces, blast furnaces, plasma furnaces, and kilns. Nitrogen oxides can also be released from nitric acid plants and other types of industrial processes involving the generation and/or use of nitric acid (HNO_3).

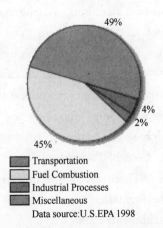

Figure 18-6　Major Sources of Nitrogen Oxides (1997)

Nitric oxide is a colorless, odorless gas. It is essentially insoluble in aqueous liquids. Nitrogen dioxide is moderately soluble in aqueous liquids. At low temperatures such as those often present in ambient air, nitrogen dioxide can form a dimer compound (N_2O_4) that has a distinctly reddish-brown color. This compound contributes to the brown haze that is often associated with photochemical smog incidents. Nitrogen dioxide has a pungent acid odor.

Both nitrogen oxide compounds are toxic; however, the ambient levels of nitrogen oxides are usually well below the concentrations believed to contribute to adverse health effects. The low ambient concentrations are due primarily to the relatively rapid reactions that occur when NO and NO_2 are emitted into the atmosphere.

The main reason for regulating nitrogen oxide emissions is the suppression of these atmospheric reactions, which create ozone and other reaction products that are associated with adverse health effects. Nitrogen oxides are one of the most important reactants in ozone formation. In fact, nitrogen dioxide is the main gas phase species responsible for the absorption of light in the photochemical reactions that cause smog formation.

2. Formation Mechanisms

Nitrogen oxides are formed during the combustion of fuel in the presence of air. At elevated temperatures, nitrogen from the air and the fuel reacts with oxygen to form nitrogen oxides. Approximately 90% to 95% of the nitrogen oxides generated in combustion processes are in the form of nitric oxide (NO). Once in the atmosphere, the NO experiences a variety of photochemical and thermal reactions to form NO_2. Accordingly, the total mass emissions of nitrogen oxides from the unit are usually expressed in the form of NO_2. Nitrogen oxides are formed from two different sources:

- Atmospheric nitrogen (N_2) entering the combustion zone as part of the combustion air.
- Nitrogen present in the fuel or waste.

The nitrogen oxides generated from atmospheric nitrogen are often termed thermal NO_x because they are formed in the high temperature areas around burner flames in combustion chambers. Nitrogen oxides generated from the fuel or wastes are termed fuel NO_x. Not all of the fuel nitrogen compounds are released during combustion. Unlike sulfur, a significant fraction of the fuel nitrogen remains in the bottom ash or in the fly ash.

The complex sets of reactions responsible for nitrogen oxides generation are very sensitive to high oxygen concentrations and high gas temperatures in the combustion zone. Nitrogen oxides emissions are highest during high boiler or incinerator loads because the highest gas temperatures occur under these conditions.

3. Control Techniques

There are two primary control techniques: (1) combustion modifications to suppress the formation of nitrogen oxides, and (2) add-on controls to reduce nitrogen oxides to molecular nitrogen.

Combustion Modifications

The purpose of combustion modifications is altering the conditions that contribute to the formation of both thermal and fuel NO_x. Most of these techniques involve a reduction in the peak gas temperatures, a reduction in the oxygen concentrations in the high temperature areas of the burner flames, and/or a reduction in the residence time of combustion products in the high temperature areas of the burner flame. A partial list of the combustion modifications that have been used to reduce NO_x formation is provided below:

- Low excess air operation.
- Off-stoichiometric combustion.
- Flue gas recirculation.
- Fuel reburning.

Low excess air operation simply involves a reduction in the total quantity of air used in the combustion process. All combustion systems use slightly more air than theoretically needed to ensure complete combustion of the fuel. By reducing the excess air levels down to the lowest possible level, the oxygen concentrations in the high temperature zone of the combustion process can be minimized, thereby reducing NO_x formation.

Off-stoichiometric combustion involves the mixing of the fuel and air in a way that reduces the peak gas temperatures and peak oxygen concentrations. Usually, a portion of the combustion flame is operated with very low oxygen levels (fuel rich) to allow a major portion of the fuel oxidation to occur under conditions where NO_x formation is suppressed. Combustion is completed in the remaining portion of the flame and/or combustion chamber by providing the remainder of the oxygen needed for complete fuel oxidation. There are a variety of different approaches for achieving off-stoichiometric firing conditions. These methods include low NO_x burners, overfire air (OFA), and burners-out-of service (BOOS). Low NO_x burners control the mixing of fuel and air in a pattern that keeps the flame temperature low and dissipates the heat quickly. OFA refers to operating the lower burners as fuel rich and placing air injection nozzles above the burners to complete the combustion process. BOOS is performed by operating alternate burners in the combustion zone as fuel rich, air rich, or air only.

Flue gas recirculation involves the return of combustion gases to the burner area of the boiler. The slightly cooled combustion gas from the boiler exit is mixed back with the burner flame to reduce the peak flame temperatures, thereby suppressing NO_x formation. This approach requires a separate recirculation fan and duct system.

Fuel reburning involves the operation of the main burners in a boiler at very low excess air (fuel rich conditions). Between 10% to 20% of the total fuel is injected into the boiler through a series of ports. This creates fuel rich conditions across the entire combustion chamber. The partially oxidized compounds formed in the burner and reburn fuel injection area, which is located in the middle region of the boiler, are then oxidized completely in the upper region of the boiler. A series of overfire air ports are used in this upper region to provide all of the air needed for complete combustion.

These combustion modifications are usually capable of reducing nitrogen oxides levels

30% to 50% from the levels that would exist in less sophisticated combustion system designs. There are several practical limits to the combustion modifications. If the combustion conditions are altered too much, some partially oxidized organic compounds and carbon monoxide can form due to impaired oxidation conditions. Flame instability can occur from minimizing oxygen and temperature levels. Also, by operating at fuel-rich conditions, the combustion processes can become vulnerable to operating problems.

Add-On Controls

Due to the limitations of combustion modifications, add-on control systems are being developed to decrease nitrogen oxide emissions below the levels possible by means of combustion modifications alone. There are two categories of add-on control systems that are applicable to boilers and other combustion processes.

- Selective Non-Catalytic Reduction (SNCR).
- Selective Catalytic Reduction (SCR).

Both types of systems inject ammonia or urea into the gas stream to reduce nitrogen oxides to molecular nitrogen and water.

Selective Non-Catalytic Reduction

In selective non-catalytic reduction systems, the ammonia (NH_3) or urea is injected into a very hot gas zone where thermal reactions leading to the chemical reduction of nitrogen oxides can occur.

These reactions are completed within the boiler, and no waste products are generated. However, if the ammonia or urea is injected into an area that is too cold, the reduction efficiency of nitrogen oxides is low, and some of the reducing gas (NH_3) can be emitted to the atmosphere. SNCR systems are capable of reducing nitrogen oxides from 20% to 60%.

Selective Catalytic Reduction

In selective catalytic reduction, beds containing catalysts reduce nitrogen oxides to molecular nitrogen and water. The catalysts are usually composed of tungsten and vanadium deposited through a substrate that is extruded into a honeycomb arrangement. The gas stream passes through the channels in the honeycomb. There are usually two or three separate catalyst beds in series. NO_x reduction efficiencies ranging from 75% to 90% are possible when the following conditions are met:

- The amount of catalyst is sufficient.
- The catalyst is in good condition.
- The ammonia reagent flow is sufficient.
- The ammonia is adequately distributed across the gas stream.

SCR systems are now being used for numerous gas turbines and a growing number of coal- and oil-fired boilers.

Unit 19　Summary of Environmental Impact Assessment

1. Definitions

An action is used in this sense of any engineering of industrial project, legislative proposal, policy, program or operational procedure with environmental implications. An environmental impact assessment (EIA) is an activity designed to identify and predict the impact of an action on the biogeophysical environment and on man's health and well-being, and to interpret and communicate information about the impacts.

2. Operational Procedures

(1) Environmental impact assessments should be an integral part of all planning for major actions, and should be carried out at the same time as engineering, economic, and sociopolitical assessments.

(2) In order to provide guidelines for environment impact assessments, national goals and policies should be established which take environmental considerations into account; these goals and policies should be widely promulgated.

(3) The institutional arrangements for the process of environmental impact assessment should be determined and made public. Here it is essential that the roles of the various participants (decision-maker, assessor, proponent, reviewer, other expert advisors, the public and inter-national bodies) be designated. It is also important that timetables for the impact assessment process be established, so that proposed actions are not held up unduly and the assessor and the reviewer are not so pressed that they undertake only superficial analyses.

(4) An environmental impact assessment should contain the following:

① a description of the proposed action and of alternatives;

② a prediction of the nature and magnitude of environmental effects (both positive and negative);

③ an identification of human concerns;

④ a listing of impact indicators as well as the methods used to determine their scales of magnitude and relative weights;

⑤ a prediction of the magnitudes of the impact indicators and of the total impact, for the project and for alternatives;

⑥ recommendation for acceptance, remedial action, acceptance of one or more of the alternatives, of rejection;

⑦ recommendation for inspection procedures.

(5) Environment impact assessments should include study of all relevant physical, biological, economic, and social factors.

(6) At a very early stage in the process of environment impact assessment, inventories should be prepared of relevant sources of data and of technical expertise.

(7) Environment impact assessments should include study of alternatives, including that of no action.

(8) Environment impact assessments should include a spatial frame of reference much larger than the area encompassed by the action, e. g. larger than the "factory fence" in the case of an engineering project.

(9) Environment impact assessments should include both mid-term and long-term predictions of impacts. In the case of engineering projects, for example, the following time frames should be covered:

① during construction;

② immediately after completion of the development;

③ two to three decades later.

(10) Environmental impact should be assessed as the difference between the future state of the environment if the action took place and the state if no action occurred.

(11) Estimates of both the magnitude and the importance of environmental impacts should be obtained. (Some large effects may not be very important to society, and vice versa.)

(12) Methodologies for impact assessment should be selected which are appropriate to the nature of the action, the data base, and the geographic setting. Approaches which are too complicated or too simple should both be avoided.

(13) The affected parties should be clearly identified, together with the major impacts for each party.

Research should be encouraged in the following areas:

(1) Post-audit reviews environmental impact assessment for accuracy and completeness in order that knowledge of assessment methods may be improved. (No systematic post-audit program has as yet been initiated in any country with experience in impact assessment.)

(2) Study of methods suitable for assessing the environmental effects of social and institutional programmes, and of other activities of the non-construction type.

(3) Study of criteria for environmental.

(4) Study of quantifying value judgments on the relative worth of various components of environmental quality.

(5) Development of modeling techniques for impact assessments, with special emphasis on combined physical, biological, socio-economic systems.

(6) Study of sociological effects and impacts.

(7) Study of methods for communicating the results of highly technical assessments to the non-specialist.

Vocabulary and Phrases

well-being　幸福
legislative　*adj.* 立法的，立法机关的
integral　*adj.* 完整的；整体的；构成整体所需要的
sociopolitical　*adj.* 社会政治的
biogeoophysical　*adj.* 生物的，地球的；物理的
guideline　*n.* 准则，指导路线
promulgate　*v.* 颁布，公布

institutional　*adj.* 惯例的，制度的
inventory　*n.* 目录，报表
participant　*n.* 参加者，参与者
proponent　*n.* 建议者，提议者
unduly　*ad.* 过度地，过分地
post-audit　*n.* 后检查
superficial　*adj.* 肤浅的，表面的
criterion　*n.* 标准，规范
indicator　*n.* 指示物

Notes

in the sense of　有……的意义
environmental impact assessment　环境影响评价
Here it is essential that…　it 为形式主语，代表 that 从句，意为"……是基本的"
hold up　阻挡，使停
in the process of　在……的过程中
in the case of　就……来说，至于
vice versa　反之亦然，反过来也一样

Question and Exercises

1. Put the following into Chinese

Action　biogeoophysical environment　superficial analysis　remedial action　an identification of human concerns　relevant physical and biological factors　a spatial frame of reference

2. Put the following into English

环境影响评价　地理环境　替代方案的研究　基础数据　环境质量标准　有关环境质量的各组分相对值　模型技术

3. Translate the following paragraphs into Chinese

(1) An environmental impact assessment (EIA) is an activity designed to identify and predict the impact of an action on the biogeophysical environment and on man's health and well-being, and to interpret and communicate information about the impacts.

(2) The institutional arrangements for the process of environmental impact assessment should be determined and made public. Here it is essential that the roles of the various participants (decision-maker, assessor, proponent, reviewer, other expert advisors, the public and inerbodies) be designated. It is also important that timetables for the impact assessment process be established, so that proposed actions are not held up unduly and the assessor and the reviewer are not so pressed that they undertake only superficial analyses.

Further Reading (1)

Environmental Assessment as a Project Management Tool

One of the main strengths of environmental assessment (EA) is its flexibility. All projects have a planning process in which EA can be integrated. Given its sensitivity to the social and economical as well as environmental impacts of projects, the EA process can be used in a project to accomplish many different objectives.

Shortcomings can be avoided in the project planning process. For example, a project, which failed to adequately consult the community at the outset can take advantage of the EA to involve the community in a necessary exchange of ideas and views. The EA can help establish and strengthen decision-making and communication mechanisms within a project. It can also pave the way for introducing innovations.

An EA may reveal sound environmental, social or economic reasons for shifting a project's direction. In view of the primacy accorded the opinions and aspirations of local people, the EA process may also function as a project control mechanism. While the EA should not be expected to correct all the weakness of a flawed planning process, when properly designed and executed, it can be a valuable tool for project implementation.

When the role of the EA is more restricted, the situation can work in reverse. Other project planning activities can be used to gather necessary information for the EA and to create support for the EA process. Each project manager must decide how much importance to accord each planning activity.

The Benefits of Environmental Assessment

Most governments and donor agencies acknowledge the contribution of EA to improved project design. The weakness of EA in the past has been largely due to poor techniques and the failure to pay attention to findings at the implementation stage. A review of current environmental practices found the major benefits of the EA process for project sponsors to be:

(1) Reduced cost and time of project implementation.

(2) Cost-saving modifications in project design.

(3) Increased project acceptance.

(4) Avoided impacts and violations of laws and regulations.

(5) Improved project performance.

(6) Avoided treatment/clean up costs.

The benefits to local communities from taking part in environmental include:

(1) A healthier local environment (forests, water sources, agricultural potential, recreational potential, aesthetic values, and clean living in urban areas).

(2) Improved human health.

(3) Maintenance of biodiversity.

(4) Decreased resource use.

(5) Fewer conflicts over natural resource use. Increased community skills, knowledge and pride.

This is a general overview of the many benefits offered by effective EA.

The Cost of Environmental Assessment

Given the dearth of research in the field, it is not surprising that there is little information on the cost of carrying out EAs on community development projects. However, we can look to the experience with large projects for some indication of the costs involved. According to the World Bank, the cost of an EA rarely exceeds one percent of the total projects cost. Mitigation measures usually account for three to five percent of total project cost. These figures do not include the cost of environmental damage caused by a project, which has not undergone an EA.

In large projects, the availability of related data and studies can help lower the cost of EA as a proportion of total cost. However, this is not as applicable to small community development projects, since so little data exists in this area. Also, it is much easier to keep environmental assessment costs down to one percent on a project whose budget is $ 20000000.

Given the modest budgets of most community development projects, it is imperative to find ways to limit costs. Over time, many believe that the costs of assessing small projects will eventually become proportionate to those of larger ones.

Here are some ideas for cutting costs:

(1) Incorporate the EA into other project planning activities such as feasibility studies.

(2) Seek the technical and financial assistance of government departments and other partners.

(3) Avoid the high associated with hiring technical specialists and building material by promoting community involvement.

(4) Costs usually diminish with experience and with the appropriate EA support mechanisms.

Further Reading (2)

Environmental Impact Assessment of Air Quality

A comprehensive assessment of a proposed project's air quality impacts involves the following steps:

(1) Establish background air quality levels.

(2) Identify applicable air quality criteria and standards.

(3) Forecast future air pollutant emissions with and without the project.

(4) Forecast future ambient air pollutant concentration with and without the project.

(5) Compare predicted air quality with applicable standards.

(6) Modify plans, if necessary, to deal with potential air quality problems.

The first step, establishing background levels of air quality is carried out only for air quality indicators likely to be influenced by the proposal. For example, if the proposed project were a high way, the main indicators of interest would be hydrocarbons, nitrogen oxides, carbon monoxide, and photochemical oxidants. In the United States, data on background air quality levels are available from local, regional, and state air quality management agencies, various state implementation plans proposed in response to federal air quality laws, and EPA's computerized data retrieval systems. The EPA data include measurements from the federal Continues Air Monitoring Program (CAMP) stations established in the early 1960s.

Applicable air quality criteria and standards are determined in the second step of the assessment process. The EPA has issued criteria relating levels of air pollutants to human health and welfare. In the United States, standards used in impact assessments include the national ambient air quality standards and pertinent state or local standards.

The NAAQS, as illustrated by carbon monoxide standards in 1980, stipulate both concentration and time of exposure. The CO standards require that the average concentration for any 1-hr period be less than 9 parts of CO per million parts of ambient air. These CO measurements are said to have an "averaging time" of 1-hr. The NAAQS also require average CO concentration during any 8-hr period to be less than 35 parts per million (ppm). Limits on CO are not be exceeded more than once per year. As of 1980, the NAAQS applied to particulates, SO_X, CO, NO_2, hydrocarbons, ozone, and lead. The national standards are revised periodically and up-to-date versions are obtainable from EPA.

In the third step of the assessment process, the proposed project's emissions are estimated in units of weight (or mass) per time period. Procedures for estimating emissions are reviewed in the next section.

The fourth step, predicting changes in the ambient concentrations of air quality indicators due to a new discharge, is often complex. In fact, sometimes it is not carried out and the assessment considers only the increased emissions form proposed project. Reasons for not estimating concentration include (1) inadequate understanding of the underlying physical and chemical processes and (2) unwillingness to commit the time and money needed to utilize existing forecasting procedures.

The final steps in a comprehensive impact assessment are to compare forecasted concentrations with applicable standards and to modify the proposed project if expected air quality degradation is unacceptable. Air-borne residuals are commonly reduced by changing combustion processes and using emission control devices such as scrubbers and filters. However, there are many options for mitigating adverse air quality effects that do net involve control devices. Examples include reducing the scale of a facility or changing the locations of discharges.

Forecasts of air pollutant emissions and concentrations are carried out at various levels of sophistication. Some impact assessments are limited to quick and simple estimates of increase in emissions. Others use elaborate computer-based mathematical models to translate increase in emissions into changes in concentrations at various times and places.

Part D Urban Solid Waste and Sound Environment

Unit 20 Solid Waste Characteristics

The sight of a dustbin overflowing and the stench rising from it, the all too familiar sights and smells of a crowded city, you look away from it and hold your nose as you cross it. Have you ever thought that you also have a role to play in the creation of this stench? That you can also play a role in the lessening of this smell and making this waste bin look a little more attractive if you follow proper methods of disposal of the waste generated in the house?

Since the beginning, humankind has been generating waste, be it the bones and other parts of animals they slaughter for their food or the wood they cut to make their carts. With the progress of civilization, the waste generated became of a more complex nature. At the end of the 19th century the industrial revolution saw the rise of the world of consumers. Not only did the air get more and more polluted but the earth itself became more polluted with the generation of nonbiodegradable solid waste. The increase in population and urbanization was also largely responsible for the increase in solid waste.

1. What is Solid Waste?

In natural systems, there is no such thing as waste. Everything flows in a natural cycle of use and reuse. Living organisms consume materials and eventually return them to the environment, usually in a different form, for reuse. Solid waste (or trash) is a human concept. It refers to a variety of discarded materials, not liquid or gas, these are deemed useless or worthless. However, what is worthless to one person may be of value to someone else, and solid wastes can be considered to be misplaced resources.

The statutory definition of a solid waste is not based on the physical form of the material, (i. e. , whether or not it is a solid as opposed to a liquid or gas), but on the fact that the material is a waste. Environmental Protection Agency (EPA) defines solid waste as any garbage or refuse, sludge from a wastewater treatment plant, water supply treatment plant, or air pollution control facility and other discarded material, including solid, liquid, semi-solid, or contained gaseous material resulting from industrial, commercial, mining, and agricultural operations, and from community activities. Nearly everything we do leaves behind some kind of waste.

The definition specifically excludes certain materials from the definition of solid waste, such as domestic sewage and special nuclear material covered by the Atomic Energy

Act are not solid wastes. Other materials that would normally be classified as solid wastes may qualify for exclusions from regulation if a generator petitions for a variance from classification as a solid waste.

In fact, in 2006, U. S. residents, businesses, and institutions produced more than 251 million tons of municipal solid waste, which is approximately 4. 6 pounds of waste per person per day. In addition, American industrial facilities generate and dispose of approximately 7. 6 billion tons of industrial solid waste each year.

2. Sources of Solid Waste

There are two basic sources of solid wastes: non-municipal and municipal. Non-municipal solid waste is the discarded solid material from industry, agriculture, mining, and oil and gas production. It makes up almost 99 percent of all the waste in the United States. Some common items that are classified as non-municipal waste are: construction materials (roofing shingles, electrical fixtures, bricks); waste-water sludge; incinerator residues; ash; scrubber sludge; oil/gas/mining waste; railroad ties, and pesticide containers. Sources and typical waste generators of solid wastes are presented in Table 20-1.

Sources and Typical generator of Solid Wastes Table 20-1

Source	Typical waste generators	Types of solid wastes
Residential	Single and multifamily dwellings.	Food wastes, paper, cardboard, plastics, textiles, leather, yard wastes, wood, glass, metals, ashes, special wastes (e. g. , bulky items, consumer electronics, white goods, batteries, oil, tires, and household hazardous wastes.).
Industrial	Light and heavy manufacturing, fabrication, construction sites, power and chemical plants.	Housekeeping wastes, packaging, food wastes, construction and demolition materials, hazardous wastes, ashes, special wastes.
Commercial	Stores, hotels, restaurants, markets, office buildings, etc.	Paper, cardboard, plastics, wood, food wastes, glass, metals, special wastes, hazardous wastes.
Institutional	Schools, hospitals, prisons, government centers.	Same as commercial.
Construction and demolition	New construction sites, road repair, renovation sites, demolition of buildings	Wood, steel, concrete, dirt, etc.
Municipal services	Street cleaning, landscaping, parks, beaches, other recreational areas, water and wastewater treatment plants.	Street sweepings; landscape and tree trimmings; general wastes from parks, beaches, and other recreational areas; sludge.
Process (manufacturing, etc.)	Heavy and light manufacturing, refineries, chemical plants, power plants, mineral extraction and processing.	Industrial process wastes, scrap materials, off-specification products, slay, tailings.
Agriculture	Crops, orchards, vineyards, dairies, feedlots, farms.	Spoiled food wastes, agricultural wastes, hazardous wastes (e. g. , pesticides).

3. Types of Solid Waste

Solid waste can be classified into different types depending on their source:

a) Household waste is generally classified as municipal waste,

b) Industrial waste as hazardous waste, and

c) Biomedical waste or hospital waste as infectious waste.

Municipal Solid Waste

Municipal solid waste is made up of discarded solid materials from residences, businesses, and city buildings. It makes up a small percentage of waste in the United States, only a little more than one percent of the total. Municipal solid waste consists of materials from plastics to food scraps. The most common waste product is paper (about 40 percent of the total).

Other common components are: yard waste (green waste), plastics, metals, wood, glass and food waste. The composition of the municipal wastes can vary from region to region and from season to season. Food waste, which includes animal and vegetable wastes resulting from the preparation and consumption of food, is commonly known as garbage.

Hazardous Waste

Some solid wastes are detrimental to the health and well-being of humans. These materials are classified as hazardous wastes. Hazardous wastes are defined as materials which are toxic, carcinogenic (cause cancer), mutagenic (cause DNA mutations), teratogenic (cause birth defects), highly flammable, corrosive or explosive. Although hazardous wastes in the United States are supposedly regulated, some obviously hazardous solid wastes are excluded from strict regulation; these include: mining, hazardous household and small business wastes.

Hospital Waste

Hospital waste is generated during the diagnosis, treatment, or immunization of human beings or animals or in research activities in these fields or in the production or testing of biologicals. It may include wastes like sharps, soiled waste, disposables, anatomical waste, cultures, discarded medicines, chemical wastes, etc. These are in the form of disposable syringes, swabs, bandages, body fluids, human excreta, etc. This waste is highly infectious and can be a serious threat to human health if not managed in a scientific and discriminate manner. It has been roughly estimated that of the 4kg of waste generated in a hospital at least 1kg would be infected.

4. Waste Disposal Methods

Most solid waste is either sent to landfills (dumped) or to incinerators (burned). Ocean dumping has also been a popular way for coastal communities to dispose of their solid wastes. In this method, large barges carry waste out to sea and dump it into the ocean. That practice is now banned in the United States due to pollution problems it created. Most municipal and non-municipal waste (about 60%) is sent to landfills. Landfills are popular because they are relatively easy to operate and can handle of a lot of waste material. There are two types of landfills: sanitary landfills and secure landfills.

In a sanitary landfill solid wastes are spread out and compacted in a hole, canyon area or a giant mound. Modern sanitary landfills are lined with layers of clay, sand and plastic. Each day after garbage is dumped in the landfill, it is covered with clay or plastic to prevent redistribution by animals or the wind.

Rainwater that percolates through a sanitary landfill is collected in the bottom liner. This liquid leachate may contain toxic chemicals such as dioxin, mercury, and pesticides. Therefore, it is removed to prevent contamination of local aquifers. The groundwater near the landfill is closely monitored for signs of contamination from the leachate.

As the buried wastes are decomposed by bacteria, gases such as methane and carbon dioxide are produced. Because methane gas is very flammable, it is usually collected with other gases by a system of pipes, separated and then either burned off or used as a source of energy (e.g., home heating and cooking, generating electricity). Other gases such as ammonia and hydrogen sulfide may also be released by the landfill, contributing to air pollution. These gases are also monitored and, if necessary, collected for disposal. Finally, when the landfill reaches its capacity, it is sealed with more layers of clay and sand. Gas and water monitoring activities, though, must continue past the useful life of the landfill.

About 15 percent of the municipal solid waste in the United States is incinerated. Incineration is the burning of solid wastes at high temperatures ($>1000\,^\circ\text{C}$). Though particulate matter, such as ash, remains after the incineration, the sheer volume of the waste is reduced by about 85 percent. Ash is much more compact than unburned solid waste. In addition to the volume reduction of the waste, the heat from the trash that is incinerated in large-scale facilities can be used to produce electric power. This process is called waste-to-energy. There are two kinds of waste-to-energy systems: mass burn incinerators and refuse-derived incinerators.

In mass burn incinerators all of the solid waste is incinerated. The heat from the incineration process is used to produce steam. This steam is used to drive electric power generators. Acid gases from the burning are removed by chemical scrubbers.

Any particulates in the combustion gases are removed by electrostatic precipitators. The cleaned gases are then released into the atmosphere through a tall stack. The ashes from the combustion are sent to a landfill for disposal.

It is best if only combustible items (paper, wood products, and plastics) are burned. In a refuse-derived incinerator, non-combustible materials are separated from the waste. Items such as glass and metals may be recycled. The combustible wastes are then formed into fuel pellets which can be burned in standard steam boilers. This system has the advantage of removing potentially harmful materials from waste before it is burned. It also provides for some recycling of materials. As with any combustion process, the main environmental concern is air quality. Incineration releases various air pollutants (particulates, sulfur dioxide, nitrogen oxides, and methane) into the atmosphere. Heavy metals (e. g. , lead, mercury) and other chemical toxins (e. g. , dioxins) can also be released. Many communities do not want incinerators within their city limits. Incinerators are also costly to build and to maintain when compared to landfills.

Vocabulary and Phrases

stench n. 恶臭
lessening n. 减少，变小
cart n. 大车，手推车
urbaniztion n. 城市化
trash n. 垃圾，废物
garbage n. 垃圾，厨余物
refuse n. 垃圾，弃物；v. 拒绝
sludge n. 污泥，泥浆，沉淀物
petition v. 请愿，祈求；n. 申请
railroad tie n. 铁道枕木
pesticide n. 杀虫剂，农药
tire n. 轮胎，橡胶
demolish v. 拆除，推翻
scrap n. 碎片，废料，食物残渣
detrimental adj. 有害的，伤害的，不利的
cancinogenic adj. [生] 致癌的
mutagenic adj. [生] 诱变的
teratogenic adj. [生] 致畸的
corrosive adj. 腐蚀的，破坏的；n. 腐蚀剂
immunization n. 有免疫力，免疫作用，免疫预防针
anatomical adj. 解剖的，解剖学的，组织的
syringe n. 注射器
swab n. 拖把

bandage n. 绷带
excreta n. 排泄物，粪便
landfill n. 垃圾填埋场，垃圾填埋
incinerator n. 焚化炉
barge n. 驳船
sanitary landfill 卫生填埋
secure landfill 安全填埋
canyon n. 峡谷
leachate n. 渗滤液
dioxin n. [化] 二氧(杂)芑
aquifer n. 含水层
methane n. 甲烷
flammable adj. 易燃的，可燃的
ammonia n. 氨，氨水
hydrogen sulfide n. 硫化氢
incineration n. 焚化，焚烧
sheer adj. 绝对的，完全的
mass burn incinerators 全燃烧焚化炉
scrubber n. 洗涤器
electrostatic adj. 静电的，静电学的
precipitator n. 沉淀器，[静电] 滤尘器
refuse-derived incinerator 垃圾焚烧炉
pellet n. 小球，小弹丸
sulfur dioxide 二氧化硫
nitrogen oxides 氮氧化物

Notes

Environmental Protection Agency (EPA) （美国）环境保护署

Atom Energe Act　原子能法案

waste-to-energy　废弃物能源

Problem and Exercises

1. What kind of solid wastes do you often see? Where are they from?
2. Translate the fourth and seventh paragraphs of the text into Chinese.
3. List the sources and types of solid wastes.

Unit 21　Municipal Solid Waste Processing

Municipal solid waste (MSW), also called urban solid waste, is a waste type that includes predominantly household waste (domestic waste) with sometimes the addition of commercial wastes collected by a municipality within a given area. They are in either solid or semisolid form and generally exclude industrial hazardous wastes.

1. What Is Included in Municipal Solid Waste?

MSW—otherwise known as trash or garbage—consists of everyday items such as productpackaging, grass clippings, furniture, clothing, bottles, food scraps, newspapers, appliances, and batteries. Not included are materials that also may be disposed in landfills but are not generally considered MSW, such as construction and demolition materials, municipal wastewater treatment sludges, and non-hazardous industrial wastes.

The U. S. Environmental Protection Agency (EPA) uses two methods to characterize the 254.1 million tons of MSW generated in 2007. The first is by material (paper and paperboard, yard trimmings, food scraps, plastics, metals, glass, wood, rubber, leather and textiles, and other); the second is by several major product categories. The product-based categories are containers and packaging; nondurable goods (e. g. , newspapers); durable goods (e. g. , appliances); food scraps; and other materials. See Table 21-1 for product category definitions.

Municipal Solid Waste Sources and Examples　　　　Table 21-1

Sources and Examples	Example Products
Residential (single-and multi-family homes)	Newspapers, clothing, disposable tableware, food packaging, cans and bottles, food craps, yard trimmings
Commercial (office buildings, retail and wholesale establishments, restaurants)	Corrugated boxes, food scraps, office papers, disposable tableware, paper napkins, yard trimmings
Institutional (schools, libraries, hospitals, prisons)	Cafeteria and restroom trash can wastes, office papers, classroom wastes, yard trimmings
Industrial (packaging and administrative; not process wastes)	Corrugated boxes, plastic film, wood pallets, lunchroom wastes, office papers

2. Functional elements of Municipal Solid waste

The functional elements of Municipal Solid waste are as follows:

(1) Waste generation

Waste generation encompasses activities in which materials are identified as no longer being of value and are either thrown away or gathered together for disposal.

(2) Waste handling and separation, storage and processing at the source

Waste handling and separation involves the activities associated with management of waste until they are placed in storage container for collection. Handling also encompasses the movement of loaded containers to the point of collection. Separation of waste components is an important step in the handling and storage of solid waste at the source.

(3) Collection

The functional element of collection includes not only the gathering of solid waste and recyclable materials, but also the transport of these materials, after collection, to the location where the collection vehicle is emptied. This location may be a materials processing facility, a transfer station or a landfill disposal site.

(4) Separation and processing and transformation of solid wastes

The types of means and facilities that are now used for the recovery of waste materials that have been separated at the source include curbside collection, drop off and buy back centers. The separation and processing of wastes that have been separated at the source and the separation of commingled wastes usually occur at a materials recovery facility, transfer stations, combustion facilities and disposal sites.

(5) Transfer and transport

This element involves two steps: i) the transfer of wastes from the smaller collection vehicle to the larger transport equipment. ii) the subsequent transport of the wastes, usually over long distances, to a processing or disposal site.

(6) Disposal

Today the disposal of wastes by landfilling or landspreading is the ultimate fate of all solid wastes, whether they are residential wastes collected and transported directly to a landfill site, residual materials from Materials Recovery Facilities (MRFs), residue from the combustion of solid waste, compost or other substances from various solid waste processing facilities. A modern sanitary land is not a dump; it is an engineered facility used for disposing of solid wastes on land without creating nuisances or hazards to public health or safety, such as the breeding of rats and insects and the contamination of ground water.

3. Management of Municipal Solid Waste

Management of Municipal Solid Waste (MSW) continues to be a high priority for communities in the 21st century. The concept of integrated solid waste management-source reduction of wastes before they enter the waste stream, recovery of generated wastes for recycling (including composting), and environmental management through combustion with energy recovery and landfilling that meet current standards—is being used by communities as they plan for the future.

EPA's 1989 Agenda for Action endorsed the concept of integrated waste management, by which municipal solid waste is reduced or managed through several different practices, which can be tailored to fit a particular community's needs. The components of the hierarchy are:

(1) Source reduction (or waste prevention), including reuse of products and on (or backyard) composting of yard trimmings.

(2) Recycling, including off-site (or community) composting.

(3) Combustion with energy recovery.

(4) Disposal through landfilling.

Source Reduction

Source reduction is gaining more attention as an important solid waste management option. Source reduction, often called "waste prevention", is defined by EPA as "any change the design, manufacturing, purchase, or use of materials or products (including packaging) to reduce their amount or toxicity before they become municipal solid waste. Prevention also refers to the reuse of products or materials". Thus, source reduction activities affect the waste stream before the point of generation.

Source reduction encompasses a very broad range of activities by private citizens, communities, commercial establishments, institutional agencies, and manufacturers and distributors. Examples of source reduction actions include:

(1) Redesigning products or packages so as to reduce the quantity of materials or the toxicity of the materials used, by substituting lighter materials for heavier ones and lengthening the life of products to postpone disposal.

(2) Using packaging that reduces the amount of damage or spoilage to the product.

(3) Reducing amounts of products or packages used through modification of current practices by processors and consumers.

(4) Reusing products or packages already manufactured.

(5) Managing non-product organic wastes (food scraps, yard trimmings) through backyard composting or other on-site alternatives to disposal.

Recycling (Including Composting)

The second component of our waste management hierarchy is recycling, including off-site (or community) composting. Residential and commercial recycling turns materials and products that would otherwise become waste into valuable resources. Materials like glass, metal, plastics, paper, and yard trimmings are collected, separated, and sent to facilities that can process them into new materials or products.

Before recyclable materials can be processed and recycled into new products, they must be collected. Most residential recycling involves curbside recyclables collection, drop-off programs, buy-back operations, and/or container deposit systems. Collection of recyclables from commercial establishments is usually separate from residential recyclables collection programs.

Processing recyclable materials is performed at MRFs, mixed waste processing facilities, and mixed waste composting facilities. Some materials are sorted at the curb and require less attention. Other materials are sorted into categories at the curb, such as a paper category and a container category, with additional sorting at a facility (MRF). There is a more recent trend towards MRFs that can sort recyclable materials that are picked up unsorted (single-stream recycling). Mixed waste can also be processed to pull out recyclable and compostable materials.

Mixed waste processing facilities are less common than conventional MRFs, but there are several facilities in operation in the United States. Mixed waste processing facilities receive mixed solid waste (including recyclable and non-recyclable materials), which is then loaded on conveyors. Using both mechanical and manual (high and low technology) sorting, recyclable materials are removed for further processing.

Mixed waste composting starts with unsorted MSW. Large items are removed, as well as ferrous and other metals, depending on the type of operation. Mixed waste composting takes advantage of the high percentage of organic components of MSW, such as paper, food scraps and yard trimmings, wood, and other materials. Yard trimmings composting is much more prevalent than mixed waste composting. On-site management of yard trimmings (back yard composting) is discussed earlier in this chapter, and is classified as source reduction, not recycling.

Combustion

MSW combustion with energy recovery increased substantially between 1980 and 1990 (from 2.7 million tons in 1980 to 29.7 million tons in 1990). Since 1990, the quantity of MSW combusted with energy recovery has only increased slightly.

Most of the municipal solid waste combustion currently practiced in this country incorporates recovery of an energy product (generally steam or electricity). The resulting energy reduces the amount needed from other sources, and the sale of the energy helps to offset the cost of operating the facility. In past years, it was common to burn municipal solid waste in incinerators solely as a volume reduction practice; energy recovery became more prevalent in the 1980s.

Total U.S. MSW combustion with energy recovery, referred to as waste-to-energy (WTE) combustion, had a 2007 design capacity of 94721 tons per day. There were 87 WTE facilities in 2007, down from 102 in 2000. In tons of capacity per million persons, the Northeast region had the most MSW combustion capacity in 2007.

In addition to facilities combusting mixed MSW (processed or unprocessed), there is a small but growing amount of combustion of source-separated MSW. In particular, rubber tires have been used as fuel in cement kilns, utility boilers, pulp and paper mills, industrial boilers, and dedicated scrap tire-to-energy facilities. In addition, there is combustion of wood wastes and some paper and plastic wastes, usually in boilers that already burn some other type of solid fuel.

Landfills

A landfill, also known as a dump (and historically as a midden), is a site for the disposal of waste materials by burial and is the oldest form of waste treatment. Historically, landfills have been the most common methods of organized waste disposal and remain so in many places around the world. Landfills may include internal waste disposal sites (where a producer of waste carries out their own waste disposal at the place of production) as well as sites used by many producers. Many landfills are also used for other waste management purposes, such as the temporary storage, consolidation and transfer, or processing of waste material (sorting, treatment, or recycling).

Modern landfills are well-engineered facilities that are located, designed, operated, and monitored to ensure compliance with federal regulations. Solid waste landfills must be designed to protect the environment from contaminants which may be present in the solid waste stream. The landfill siting plan—which prevents the siting of landfills in environmentally-sensitive areas—as well as on-site environmental monitoring systems—which monitor for any sign of groundwater contamination and for landfill gas—provide additional safeguards. In addition, many new landfills collect potentially harmful landfill gas emissions and convert the gas into energy.

Municipal Solid Waste Landfills (MSWLFs) receive household waste. MSWLFs can also receive non—hazardous sludge, industrial solid waste, and construction and demolition debris. Federal MSWLFs standards include:

Location restrictions—ensure that landfills are built in suitable geological areas away from faults, wetlands, flood plains, or other restricted areas.

Composite liners requirements—include a flexible membrane (geomembrane) overlaying two feet of compacted clay soil lining the bottom and sides of the landfill, protect groundwater and the underlying soil from leachate releases.

Leachate collection and removal systems—sit on top of the composite liner and removes leachate from the landfill for treatment and disposal.

Operating practices—include compacting and covering waste frequently with several inches of soil help reduce odor; control litter, insects, and rodents; and protect public health.

Groundwater monitoring requirements—requires testing groundwater wells to determine whether waste materials have escaped from the landfill.

Closure and postclosure care requirements—include covering landfills and providing long—term care of closed landfills.

Corrective action provisions—control and clean up landfill releases and achieves roundwater protection standards.

Financial assurance—provides funding for environmental protection during and after landfill closure (i. e., closure and postclosure care).

During 2007, about 54 percent of MSW was landfilled in the United State, down somewhat from 55.1 percent in 2006. The number of MSW landfills decreased substantially over the past 18 years, from nearly 8000 in 1988 to 1754 in 2006—while average landfill size increased. At the national level, capacity does not appear to be a problem, although regional dislocations sometimes occur.

Vocabulary and Phrases

Municipal solid waste n. 城市固体废物
productpackaging n. 产品包装
appliance n. 器具，器械，装置
tableware n. 餐具
napkin n. 餐巾，餐纸，毛巾，布
restroom n. ［美］公用厕所；盥洗室；卫生间
lunchroom n. 便餐馆，餐厅，小吃馆
curbside n. 街头，路边
commingled v. 混合，参合，合并
compost n. 混合废料，堆肥
nuisance n. 讨厌的人或物，麻烦事
endorse v. 认可，赞同
tailored adj. 定制的，定做的
hierarchy n. 等级制度，统治集团，特权阶级
off-site n. ［化］工地外；厂区外

conveyor n. 运送者；传送者；（财产）转让人，传送带；传送装置
unsorted adj. 未分类的，未整理的，为分选的
ferrous adj. 铁的，含铁的，［化］亚铁的，二价铁的
cement kiln 水泥窑
utility boiler 电站锅炉，动力锅炉
midden n. 粪堆，垃圾堆
consolidation n. 巩固，加强，固化
fault n. ［地］断层，过失，错误
flood plain 河漫滩，河滩
composite liner 复合防渗层
membrane n. 膜，薄膜，隔膜
geomembrane 土工膜，地质处理膜
litter n. 废弃物，垃圾
rodent n. 啮齿类动物

Notes

corrugated box 瓦楞纸箱，是一种经过模切、压痕、钉箱或粘箱制成的应用最广的包装制品，用量一直是各种包装制品之首。

buy back centers 回收中心

Materials Recovery Facilities (MRFs) 美国废料回收设施

Municipal Solid Waste Landfills (MSWLFs) 城市固体废弃物填埋

Problems and Exercises

1. What is the process of landfill?
2. What is your present concept of MSW disposal?
3. Put the following sentence into English:
   a) 大部分的城市垃圾焚烧都伴随着能源的利用。
   b) 填埋是最早的垃圾处理方法。
4. Write an abstract of this article.

Unit 22　Treatment, Storage, and Disposal of Hazardous Waste Management

Hazardous waste is any solid, liquid, or gaseous waste material that may pose substantial hazards to human health and the environment if improperly treated, stored, transported, disposed of, or otherwise managed. Every industrial country has had problems with managing hazardous wastes. Improper waste management has necessitated expensive cleanup operations in many instances. Efforts are under way internationally to remedy past problems caused by hazardous waste and to prevent future problems through source reduction (eliminating hazardous wastes at the source), recycling, treatment, and proper disposal of hazardous wastes.

Modern hazardous waste regulations in the U. S. began with the Resource Conservation and Recovery Act (RCRA) which was enacted in 1976. The primary contribution of RCRA was to create a "cradle to grave" system of record keeping for hazardous wastes. Hazardous wastes must be tracked from the time they are generated until their final disposition.

A U. S. facility that treats, stores or disposes of hazardous waste must obtain a permit for doing so under the RCRA. Generators of and transporters of hazardous waste must meet specific requirements for handling, managing, and tracking waste. Through the RCRA, Congress directed the United States Environmental Protection Agency (EPA) to create regulations to manage hazardous waste. Under this mandate, the EPA developed strict requirements for all aspects of hazardous waste management including the treatment, storage, and disposal of hazardous waste. In addition to these federal requirements, states may develop more stringent requirements or requirements that are broader in scope than the federal regulations.

1. Hazardous Waste Generators

Many types of businesses generate hazardous waste. Some are small areas that may be located in a community. For example, dry cleaners, automobile repair shops, hospitals, exterminators, and photo processing centers all generate hazardous waste. Some hazardous waste generators are larger companies like chemical manufacturers, electroplating companies, and oil refineries. A hazardous waste generator is any person or site whose processes and actions create hazardous waste.

Generators are divided into three categories based upon the quantity of waste they produce:

Large Quantity Generators (LQGs) generate 1000 kilograms per month or more of hazardous waste, more than 1 kilogram per month of acutely hazardous waste, or more than 100 kilograms per month of acute spill residue or soil.

Small Quantity Generators (SQGs) generate more than 100 kilograms, but less than 1000 kilograms, of hazardous waste per month.

Conditionally Exempt Small Quantity Generators (CESQGs) generate 100 kilograms or less per month of hazardous waste, or 1 kilogram or less per month of acutely hazardous waste, or less than 100 kilograms per month of acute spill residue or soil.

Each class of generator must comply with its own set of requirements. All generators are also required to:

(1) Obtain an EPA Identification number (available from State environmental offices).

(2) Comply with the manifest system.

(3) Handle wastes properly before shipment (packaging, labeling, marking, placarding, accumulation time, etc.).

(4) Comply with record keeping and reporting requirements.

(5) Comply with any additional State requirements for generators (contact your State environmental office for more information).

2. Hazardous Waste Transporters

All aspects of hazardous waste transportation are closely regulated by EPA. Transporters are individuals or entities that move hazardous waste from one site to another by highway, rail, water, or air. This includes transporting hazardous waste from a generator's site to a facility that can recycle, treat, store, or dispose of the waste. It can also include transporting treated hazardous waste to a site for further treatment or disposal.

3. Hazardous Waste Storage

Through RCRA, Congress directed EPA to regulate all aspects of hazardous waste. As a result, EPA developed strict regulations for the treatment, storage, and disposal of hazardous waste. States may implement stricter requirements than the Federal regulations as needed.

Storage is the holding of waste for a temporary period of time prior to the waste being treated, disposed, or stored elsewhere. Hazardous waste is commonly stored prior to treatment or disposal, and must be stored in containers, tanks, containment buildings, drip pads, waste piles, or surface impoundments that comply with the RCRA regulations.

Containers—A hazardous waste container is any portable device in which a hazardous waste is stored, transported, treated, disposed, or otherwise handled. The most common hazardous waste container is the 55-gallon drum. Other examples of containers are tanker trucks, railroad cars, buckets, bags, and even test tubes.

Tanks—Tanks are stationary devices constructed of non-earthen materials used to store or treat hazardous waste. Tanks can be open-topped or completely enclosed and are constructed of a wide variety of materials including steel, plastic, fiberglass, and concrete.

Drip Pads—A drip pad is a wood drying structure used by the pressure-treated wood industry to collect excess wood preservative drippage. Drip pads are constructed of non-earthen materials with a curbed, free-draining base that is designed to convey wood preservative drippage to a collection system for proper management.

Containment Buildings——Containment buildings are completely enclosed, self-supporting structures (i. e. , they have four walls, a roof, and a floor) used to store or treat non-containerized hazardous waste.

Waste Piles—A waste pile is an open, uncontained pile used for treating or storing waste. Hazardous waste waste piles must be placed on top of a double liner system to ensure leachate from the waste does not contaminate surface or ground water supplies.

Surface Impoundments—A surface impoundment is a natural topographical depression, man-made excavation, or diked area such as a holding pond, storage pit, or settling lagoon. Surface impoundments are formed primarily of earthen materials and are lined with synthetic plastic liners to prevent liquids from escaping.

4. Hazardous Waste Treatment

Treatment is any process that changes the physical, chemical, or biological character of a waste to make it less of an environmental threat. Treatment can neutralize the waste; recover energy or material resources from a waste; render the waste less hazardous; or make the waste safer to transport, store, or dispose. Hazardous wastes undergo different treatments in order to stabilize and dispose of them.

Recycling

Many HWs can be recycled into new products. Examples might include lead-acid batteries or electronic circuit boards where the heavy metals can be recovered and used in new products. Another example is the ash generated by coal-fired power plants; these plants produced two types of these residues: fly and bottom ash. Fly ash particles have a low density, are very fine, and are removed by air pollution control devices. On the other side, bottom ash is a dense, dark, gravely substance that remains on the bottom of combustion chambers. After these types of ashes go though the proper treatment, they could bind to other pollutants and convert them into easier-to-dispose solids, or they could be used as pavement filling. Such treatments reduce the level of threat of harmful chemicals, like fly and bottom ash, while also recycling the safe product and helping the environment.

Portland Cement

Another commonly used treatment is cement based solidification and stabilization. Cement is used because it can treat a range of hazardous wastes by improving physical characteristics and decreasing the toxicity and transmission of contaminants. The cement produced is categorized into 5 different divisions, depending on its strength and components. This process of converting sludge into cement might include the addition of pH adjustment agents, phosphates, or sulfur reagents to reduce the settling or curing time, increase the compressive strength, or reduce the leach ability of contaminants.

Neutralization

Some HW can be processed so that the hazardous component of the waste is eliminated making it a non-hazardous waste. An example of this might include a corrosive acid that is neutralized with a basic substance so that it is no-longer corrosive. Another mean to neutralize some of the waste is pH adjustment. pH is an important factor on the leaching activity of the hazardous waste. By adjusting the pH of some toxic materials, we are reducing the leaching ability of the waste.

Incineration, Destruction and Waste-to-energy

A hazardous waste may be "destroyed" for example by incinerating it at a high temperature. Flammable wastes can sometimes be burned as energy sources. For example many cement kilns burn hazardous wastes like used oils or solvents. Today incineration treatments not only reduce the amount of hazardous waste. They also generate energy throughout the gases released in the process. It is known that this particular waste treatment releases toxic gases produced by the combustion of by-product or other materials and this can affect the environment. However, current technology has developed more efficient incinerator units that control these emissions to a point that this treatment is considered a more beneficial option. There are different types of incinerators and they vary depending on the characteristics of the waste. Starved air incineration is another method used to treat hazardous wastes. Just like in common incineration, burning occurs, however controlling the amount of oxygen allowed proves to be significant to reduce the amount of harmful by-products produced. Starved Air Incineration is an improvement of the traditional incinerators in terms of air pollution. Using this technology it is possible to control the combustion rate of the waste and therefore reduce the air pollutants produce in the process.

Hazardous Waste Landfill (sequestering, isolation, etc.)

A hazardous waste may be sequestered in a hazardous waste landfill or permanent disposal facility. "In terms of hazardous waste, a landfill is defined as a disposal facility or part of a facility where hazardous waste is placed in or on land and which is not a pile, a land treatment facility, a surface impoundment, an underground injection well, a salt dome formation, a salt bed formation, an underground mine, a cave, or a corrective action management unit."

Pyrolysis

Some hazardous waste types may be eliminated using pyrolisis in an ultra high temperature electrical arc, in inert conditions to avoid combustion. This treatment method may be preferable to high temperature incineration in some circumstances such as in the destruction of concentrated organic waste types, including PCBs, pesticides and other POPs.

Disposal

Disposal is the placement of waste into or on the land. Disposal facilities are usually designed to permanently contain the waste and prevent the release of harmful pollutants to the environment. The most common hazardous waste disposal practice is placement in a land disposal unit such as a landfill, surface impoundment, waste pile, land treatment unit, or injection well. Land disposal is subject to requirements under EPA's Land Disposal Restrictions Program.

Underground injection wells are the most commonly used disposal method for liquid hazardous waste. Because of their potential impact upon drinking water resources, injection wells are also regulated under the Safe Drinking Water Act (SDWA) and by the Underground Injection Control (UIC) Program.

Vocabulary and Phrases

substantial *adj.* 相当的，实际的，重要的
track *v.* 跟踪
mandate *n.* 命令，指示
stringent *adj.* 严格的，严厉的，迫切的
exterminator *n.* 灭鼠药，杀虫剂
electroplating *n.* 电镀，电镀术
acute *adj.* 严重的，剧烈的，敏感的，[医] 急性的
spill *n.* 溢出，溢出物
residue *n.* 残余物，残渣
manifest *n.* 清单，名单
label *v.* 贴标签于，用签条标明，把……归类为；*n.* 标签，标记，符号
placard *n.* 海报，布告；*v.* 贴布告于，在……上张贴海报
entities *n.* 实体
pad *n.* 垫子，缓冲器
waste pile 废石堆
impoundment *n.* 蓄水，积水，搁置
bucket *n.* 水桶，吊桶
earthen *adj.* 土制的，陶制的

fiberglass *n.* [美] 玻璃纤维，玻璃棉
drippage *n.* 滴落，滴下的水
topographical *adj.* 地形的，地形学的
depression *n.* 洼地，凹地，盆地
dike *n.* 堤，坝，障碍物；*v.* 筑堤防护
lagoon *n.* 污水池，咸水池，[地] 泄湖
neutralize *v.* 抵消，中和
fly ash 飞灰
bottom ash 底灰
Portland cement 硅酸盐水泥
phosphate *n.* 磷酸盐
sulfur reagent 硫试剂
curing *n.* 固化，养护
compressive strength 抗压强度
cement kiln 水泥窑
Starved air incineration 贫气焚化炉，两级焚化炉
sequester *v.* 使隔离
pyrolisis *n.* [化] 热解(作用)；高温分解
concentrated *adj.* 浓缩的，集中的

Notes

1. Resource Conservation and Recovery Act (RCRA) 资源回收保护法，1976 年美国

国会通过的一部基于国家规划，以保护环境和人民健康，防止固体废弃物不正确处理，并鼓励保护自然资源的法律。

2. Environmental Protection Agency (EPA) （美国）环境保护署

3. PCBs—polychlorinated biphenyl 多氯联苯

4. POPs—persistent organic pollutants 持久性有机污染物

5. Safe Drinking Water Act (SDWA) （美国）安全饮用水法案，1974年由美国国会通过，其目的是通过对美国公共饮用水供水系统的规范管理，以确保公众的健康。该法律于1986年和1996年进行修改，要求采取很多行动来保护饮用水及其水源——河流、湖泊、水库、泉水和地下水水源。

6. Underground Injection Control (UIC) 地下注水控制

Problems and Exercises

1. What is hazadous waste?
2. What specific requirements must Generators of hazardous waste meet?
3. The incineration process is useful for volume reduction, but may lead to secondary pollution problems, is it? Give some examples.
4. What kind of environmental damage does the improperly handling of hazardous wastes cause?

Further Reading

Current Status and Future Focus of Hazardous Waste Management in China

In China, hazardous wastes are defined as any poisonous, flammable, explosive, caustic, reactive, or infective waste material which is listed in the National Catalog of Hazardous Wastes, or any waste materials which hold dangerous characteristics according to the national standards and methods of identification. The proper management, treatment, and disposal of hazardous wastes is extremely important, otherwise they will become harmful to ecology and the environment, as well as to public health. Since the time taken to generate, transport, treat, and dispose of hazard wastes is very variable, it is difficult to monitor and manage hazardous wastes efficiently. It is now necessary to identify advanced technologies, raise generous amounts of funding, and establish appropriate treatment/disposal sites. At this time, the improvement of hazardous waste monitoring techniques is one of the most important environmental problems throughout the world.

1. Current Status and Characteristics of Hazardous Waste Pollution in China

The amount of hazardous waste generated in China is enormous. It is estimated that the total amount of industrial solid wastes generated in 1999 was about 800 million tons. Industrial hazardous solid waste reached 10.15 million tons, accounting for 3% of all industrial wastes. There are 47 types of hazard wastes listed in the National Catalog of Hazardous Wastes. However, the hazardous waste catalog and the waste quantities generated differ widely among the various districts and even countries. General speaking, hazardous waste produced in China consists of industrial waste and municipal waste (social waste), both of which have different characteristics.

Distribution Characteristics of Industrial Hazardous Waste

Generally, waste producers are distributed extensively throughout China, and the wastes produced are relatively concentrated in certain areas.

Looking at the industrial distribution, the amount of hazardous waste produced in nonmetal mineral manufacturing, chemical materials and chemical products manufacturing, metal manufacturing, machine manufacturing, sanitation, the weaving industry, and the paper and papermaking industry combined account for nearly 50% of the total.

As for the geographical distribution, hazardous wastes are mainly produced in the provinces of ZheJiang, Henan, According to the Solid Law, the Environmental Protection Department should be in charge of managing hazardous wastes in China, as well as the establishment of related policies and their implementation. Local environmental protection departments and the Solid Waste Management Center should be in charge of managing hazardous wastes at the local level. The main systems and measures for waste management include a system for declaring and registering hazardous wastes, a system for making producers responsible for disposal, obligatory disposal, and the centralized disposal of hazardous waste, a system of charging fees for hazardous waste, and a system of permissions for the collection, storage, and disposal of hazardous waste.

The Main problems in the Management of Hazardous Waste

There are no related executive regulations at the State Council level. Laws and obligations relating to waste issues have not yet been specified. The implementation of policies for waste issues has not been made a priority, so environmental standards and technical policies have not been perfected, and are currently inadequate. There are only about 20 management centers for handling solid wastes in China, most of which are still under construction or not yet in full working order. Many other problems have arisen with regard to the management of hazardous waste. (1) Management methods have not yet been adapted to suit the increasing development of the market economy in China, and management staff lack knowledge about proper handling, etc., of hazardous waste. (2) Although industrial wastes are now being collected and recovered by producers and users, there are still many problems with regard to the proper collection of public types of hazardous waste. (3) The exchange net of hazardous waste and recovery technologies is not yet very sophisticated, and secondary pollution is severe. (4) Facilities for treatment and disposal have not been perfected.

2. Principles for the Management of Hazardous Waste in China

According to the current status of hazardous waste treatment and disposal, and with reference to experiences with the management of hazardous waste in developed countries, the principles for waste management processes in China should be as described below.

The Minimum Generation Principle

Enterprises are encouraged to use more advanced, cleaner technologies to avoid or decrease the amount of hazardous waste output. For the waste that is produced, the recovery rate should be maximized. For those wastes that cannot be recovered or treated, it is necessary to ensure their safe storage. The discharge of any hazardous waste directly into the environment must be prohibited.

The Pivot Management Principle

Owing to the vast amount and many types of hazardous waste produced in China, it is not possible to treat all waste using one basic standard. The best solution is to prioritize, and first to control those hazardous wastes that occur in the largest quantities.

Principles of Pollution Control and Centralized Treatment and Disposal

Owing to the various characteristics and quantities of hazardous wastes, it is necessary to determine whether they should be disposed of on site or in a centralized location. Regional treatment/disposal sites for hazardous waste should have large capacities, and should be able to serve a large area. It is very important to adopt the most advanced technologies available in order to decrease treatment/disposal costs and pollution risks.

Principle of Treatment/Disposal Cost to be Paid by Waste Producers

Environmental law states that the cost of the management and disposal of hazardous waste must be covered by the waste producers.

Principle of Public Participation

It is necessary to raise public environmental awareness and enhance public participation in the process of waste management, treatment, and disposal. Waste can be most efficiently managed by the government, or by private enterprises that operate under public supervision.

Principle of Industrialization and Marketing of Waste Treatment Facilities

In the long term, waste treatment/disposal facilities should be managed by private enterprises. Hazardous waste treatment/disposal facilities could be built by private funding, and then managed and operated by other independent corporations after being established. The operation of the whole hazardous waste management process will also be gradually taken over by the marketing mechanism.

3. Objectives and Strategies for Hazardous Waste Management in China

Objectives of hazardous waste management According to the results of the study "Action Plan for Hazardous Waste Management in China," the objectives of hazardous waste management are detailed in the Technological Policy for Pollution Prevention of Hazardous Wastes, published by China's state EPA, State ETC, and Science and Technology Ministry.

(1) Short-term objective. By 2005, all hazardous waste in the focus areas and in cities should be safely stored away and disposed of to the extent that this is possible. Clinical wastes in the main areas and in cities should also be safely stored and disposed of.

(2) Medium-term objective. By 2010, hazardous waste in the focus areas and in cities

should be suitably treated and disposed of in a way that is not harmful to the environmental.

(3) Long-term objective. By 2015, hazardous waste in most cities should be treated and disposed of in way that does not harm the environment.

Strategies for the Management of Hazardous Waste Management

To achieve a program of environmentally sound hazardous waste management, it is necessary to design an action plan that takes the whole hazardous waste management process into account. An overall policy system and a national framework of law has been established in China in order to complete the regulation system for waste management, and to enhance its enforcement. Technological and economic policies, as well as standard and technical guidelines, are not well developed, so it is necessary to complete the creation of a system of technical and economic policies and related regulations. These mainly include technological policies for pollution control of hazardous wastes, an environmental tax mechanism on products containing hazardous substances, a recovery system by producers, and the establishment of policies which will eventually eliminate harmful products altogether.

The systems and standards which need to be carried out immediately include a license management system, standards for pollution control at hazardous waste landfills, standards for pollution control for the storage of hazardous wastes, and technical assessment guidelines for the capacity of hazardous waste management organizations.

In order to strengthen the administrative management of hazardous waste, cooperation between governmental departments must be strong. It is also necessary to strengthen the functioning of government enforcement of waste management policies, to increase their number, to improve the quality of management centers for solid wastes, and to establish effective monitoring mechanisms in order to ensure efficient supervision of producers and their behavior (i. e., whether or not their actions take the environment into consideration).

Strengthening the whole hazardous waste management process will require improving the management of transportation, treatment, and recovery of hazardous waste, developing data registration, improving data management system, establishing information systems for the decision-making process, enhancing registration veracity through penalties, training staff at local environmental bureaus, and developing an information system for the decision-making process for any issues dealing with waste management.

To increase hazardous waste recovery rates and improve disposal technologies, several new activities must be set in place. A local exchange net for hazardous waste must be established. Wastes must be used as secondary sources. Local centers for spent oil recovery must be established. Recovery management of spent lead-acid batteries must be regulated. The recovery of spent nickel-cadmium batteries must gradually be increased.

In order to improve training and international cooperation, training centers should be built to improve technological training and raise social awareness about environmental issues. The environmental management staff should first be well trained. It is also important to strengthen international cooperation in order to obtain economic, and technical investments from developed countries, and to receive favorable technological transfers. Cooperation with UN agencies, and other international organizations should be established in order to fulfill the obligations, and rights stated in the Basel Convention, as well as to strengthen cooperation between people, public organizations, learning institutes, funding organizations, and other interested people so that there can be fruitful exchanges and communication on scientific techniques and construction projects where the pollution control of hazardous wastes has been successful.

Unit 23 Adverse Health Effects of Noise

1. Introduction

The perception of sounds in day-to-day life is of major importance for human well-being. Communication through speech, sounds from playing children, music, natural sounds in parklands, parks and gardens are all examples of sounds essential for satisfaction in every day life. Conversely, this document is related to the adverse effects of sound (noise). According to the International Program on Chemical Safety (WHO 1994), an adverse effect of noise is defined as a change in the morphology and physiology of an organism that results in impairment of functional capacity, or an impairment of capacity to compensate for additional stress, or increases the susceptibility of an organism to the harmful effects of other environmental influences. This definition includes any temporary or long-term lowering of the physical, psychological or social functioning of humans or human organs. The health significance of noise pollution is given in this chapter under separate headings, according to the specific effects: noise-induced hearing impairment; interference with speech communication; disturbance of rest and sleep; psychophysiological, mental-health and performance effects; effects on residential behavior and annoyance; as well as interference with intended activities. This chapter also considers vulnerable groups and the combined effects of sounds from different sources. Conclusions based on the details given as they relate to guideline values.

2. Effects of Noise on Residential Behavior and Annoyance

Noise annoyance is a global phenomenon. A definition of annoyance is "a feeling of displeasure associated with any agent or condition, known or believed by an individual or group to adversely affect them". However, apart from "annoyance", people may feel a variety of negative emotions when exposed to community noise, and may report anger, disappointment, dissatisfaction, withdrawal, helplessness, depression, anxiety, distraction, agitation, or exhaustion. Thus, although the term annoyance does not cover all the negative reactions, it is used for convenience in this document.

Noise can produce a number of social and behavioral effects in residents, besides annoyance. The social and behavioral effects are often complex, subtle and indirect. Many of the effects are assumed to be the result of interactions with a number of non-auditory variables. Social and behavioral effects include changes in overt everyday behavior patterns (e.g. closing windows, not using balconies, turning TV and radio to louder levels, writing petitions, complaining to authorities); adverse changes in social behavior (e.g. aggression, unfriendliness, disengagement, non-participation); adverse changes in social indicators (e. g. residential mobility, hospital admissions, drug consumption, accident rates); and changes in mood (e. g. less happy, more depressed).

Although changes in social behavior, such as a reduction in helpfulness and increased aggressiveness, are associated with noise exposure, noise exposure alone is not believed to be sufficient to produce aggression. However, in combination with provocation or pre-existing anger or hostility, it may trigger aggression. It has also been suspected that people are less willing to help, both during exposure and for a period after exposure. Fairly consistent evidence shows that noise above 80dB (A) is associated with reduced helping behavior and increased aggressive behavior. Particularly, there is concern that high-level continuous noise exposures may contribute to the susceptibility of schoolchildren to feelings of helplessness.

The effects of community noise can be evaluated by assessing the extent of annoyance (low, moderate, high) among exposed individuals; or by assessing the disturbance of specific activities, such as reading, watching television and communication. The relationship between annoyance and activity disturbances is not necessarily direct and there are examples of situations where the extent of annoyance is low, despite a high level of activity disturbance. For aircraft noise, the most important effects are interference with rest, recreation and watching television. This is in contrast to road traffic noise, where sleep disturbance is the predominant effect.

A number of studies have shown that equal levels of traffic and industrial noises result in different magnitudes of annoyance. This has led to criticism of averaged dose-response curves determined by meta-analysis, which assumed that all traffic noises are the same have synthesized curves of annoyance associated with three types of traffic noise (road, air, railway). In these curves, the percentage of people highly or moderately annoyed was related to the day and night continuous equivalent sound level, L_{dn}. For each of the three types of traffic noise, the percentage of highly annoyed persons in a population started to increase at an L_{dn} value of 42dB (A), and the percentage of moderately annoyed persons at an L_{dn} value of 37dB (A). Aircraft noise produced a stronger annoyance response than road traffic, for the same L_{dn} exposure, consistent with earlier analyses (Kryter 1994; Bradley 1994a). However, caution should be exercised when interpreting synthesized data from different studies, since five major parameters should be randomly distributed for the analyses to be valid: personal, demographic, and lifestyle factors, as well as the duration of noise exposure and the population experience with noise.

Annoyance in populations exposed to environmental noise varies not only with the acoustical characteristics of the noise (source, exposure), but also with many non-acoustical factors of social, psychological, or economic nature. These factors include fear associated with the noise source, conviction that the noise could be reduced by third parties, individual noise sensitivity, the degree to which an individual feels able to control the noise (coping strategies), and whether the noise originates from an important economic activity. Demographic variables such as age, sex and socioeconomic status, are less strongly associated with annoyance. The correlation between noise exposure and general annoyance is much higher at the group level than at the individual level, as might be expected. Data from 42 surveys showed that at the group level about 70% of the variance in annoyance is explained by noise exposure characteristics, whereas at the individual level it is typically about 20%.

When the type and amount of noise exposure is kept constant in the meta-analyses, differences between communities, regions and countries still exist. This is well demonstrated by a comparison of the dose-response curve determined for road-traffic noise and that obtained in a survey along the North-South transportation route through the Austrian Alps. The differences may be explained in terms of the influence of topography and meteorological factors on acoustical measures, as well as the low background noise level on the mountain slopes.

Stronger reactions have been observed when noise is accompanied by vibrations and contains low frequency components, or when the noise contains impulses, such as shooting noise. Stronger, but temporary, reactions also occur when noise exposure is increased over time, in comparison to situations with constant noise exposure. Conversely, for road traffic noise, the introduction of noise protection barriers in residential areas resulted in smaller reductions in annoyance than expected for a stationary situation.

To obtain an indicator for annoyance, other methods of combining parameters of noise exposure have been extensively tested, in addition to metrics such as L_{Aeq}, 24h and L_{dn}. When used for a set of community noises, these indicators correlate well both among themselves and with L_{Aeq}, 24h or L_{dn} values. Although L_{Aeq}, 24h and L_{dn} are in most cases acceptable approximations, there is a growing concern that all the component parameters of the noise should be individually assessed in noise exposure investigations, at least in the complex cases.

3. Recommendations

The potential health effects of community noise include hearing impairment; startle and defense reactions; aural pain; ear discomfort speech interference; sleep disturbance; cardiovascular effects; performance reduction; and annoyance responses. These health effects, in turn, can lead to social handicap; reduced productivity; decreased performance in learning; absenteeism in the workplace and school; increased drug use; and accidents. In addition to health effects of community noise, other impacts are important such as loss of property value. In these guidelines the international literature on the health effects of community noise was reviewed and used to derive guideline values for community noise. Besides the health effects of noise, the issues of noise assessment and noise management were also addressed. Other issues considered were priority setting in noise management; quality assurance plans; and the cost-efficiency of control actions.

The following recommendations were considered appropriate:

(1) Governments should consider the protection of populations from community noise

as an integral part of their policy for environmental protection.

(2) Governments should consider implementing action plans with short-term, medium-term and long-term objectives for reducing noise levels.

(3) Governments should adopt the health guidelines for community noise as targets to be achieved in the long-term.

(4) Governments should include noise as an important issue when assessing public health matters and support more research related to the health effects of noise exposure.

(5) Legislation should be enacted to reduce sound pressure levels, and existing laws should be enforced.

(6) Municipalities should develop low-noise implementation plans.

(7) Cost-effectiveness and cost-benefit analyses should be considered as potential instruments when making management decisions.

(8) Governments should support more policy-relevant research into noise pollution.

Vocabulary and Phrases

noise-induced   *adj.* 噪声引起的
psychophysiological   *adj.* 精神生理学的，心理生理学的
mental-health   *n.* 心理健康
vulnerable   *adj.* 易受攻击的，易受……的攻击
auditory   *adj.* 耳的，听觉的
overt   *adj.* 明显的，公然的
provocation   *n.* 激怒，刺激
aircraft   *n.* 飞行器、航空器
demographic   *adj.* 人口统计学的
acoustical   *adj.* 听觉的，声学的
socioeconomic   *adj.* 社会经济学的
vibration   *n.* 振动
impulse   *n.* 脉冲
cardiovascular   *adj.* 心脏血管的

Notes

the adverse effects of sound (noise)   声音的负面效应
be defined as   被定义为
result in impairment of   导致……损伤
temporary or long-term lowering of   暂时或长期降低……
noise pollution   噪声污染
community noise   城市噪声
high-level continuous noise exposures   高水平连续噪声暴露
be evaluated by   通过……评价
road traffic noise   交通噪声
day and night continuous equivalent sound level   昼夜连续等效声级
noise source   噪声源
background noise level   噪声背景水平
low frequency component   低频成分

Question and Exercises

1. What are the main adverse effects of sound?
2. What's the noise-induced annoyance?
3. Under what conditions, noise has stronger reactions to humans?
4. What kinds of the potential health effects does community noise include?

Further Reading

Harmful Effects of Noise

Commonly thought of as unwanted sound, noise might describe one of the new rock bands for you but not so for the multitude of teenagers who listen to it from clamped headphones. Unfortunately the damage caused by listening to overly loud music results in a loss of hearing, even at a young age. Not only teenagers, every single person is exposed to potentially damaging noise everyday, from car horns, barking dogs, gunshots, screeching tyres and of course, loud music. The blaring noises in most high population cities must in no way be underestimated. Little do people know how sounds of such high intensity can cause damage to the ear.

Hearing loss may also occur as a result of diseases such as diabetes, infections or drugs. It can be inherited or be a result of physical damage to the ears or serious injuries to the head. The most common causes of hearing impairment are noise and ageing. Even though people of all ages can possess a hearing loss, younger people usually hear better at lower volumes than older people do. As they grow up, their power of hearing diminishes. That probably explains why parents do not dare move a muscle after they have spent a strenuous hour making the baby go to sleep. There's no telling what the baby, with its acute listening powers, might hear and get disturbed by so much for the sixth sense.

According to the World Health Organization's Guidelines for Community Noise, noise is an increasing public health problem. Noise can have the following adverse health effects: hearing loss, sleep disturbances, cardiovascular and psychophysiologic problems, performance reduction, annoyance responses and adverse social behavior.

1. Noise-induced Hearing Loss

NIHL (noise-induced hearing loss) can be caused by a one-time exposure to loud sound as well as by repeated exposure to sounds at various loudness levels over an extended period of time. Hearing loss is now a common ailment all over the world. Considering the statistics, the situation is pretty bad. In developing countries the burden of hearing impairment is estimated to be twice as large as in developed countries, probably because of a lot of untreated ear infections.

How noise affects hearing?

Sounds are transmitted as vibrations from the outer ear to the 30000 delicate hair cells of the inner ear. Exposure to harmful sounds causes damage to these cells as well as the hearing nerve. These structures can be injured by two kinds of noise: loud impulse noise, such as an explosion, or loud continuous noise, such as that generated in a woodworking shop.

Exposure to impulse and continuous noise may cause only a temporary hearing loss, in which the hair cells vibrate vigorously and sometimes violently. This often results in tinnitus (a ringing, buzzing or roaring in the ears or head) and temporary hearing loss. If the hearing recovers, the temporary hearing loss is called a temporary threshold shift, which largely disappears 16 to 48 hours after exposure to loud noise.

With continued exposure to loud noise, the temporary hearing loss can become permanent by a destruction of the hair cells. Hearing in the higher pitch range is usually affected first. This often results in a reduction in the clarity of speech, especially in a background of noise. It is easier to hear men, whose voices are deeper, than women and children.

What noise is too loud?

The loudness of sound is measured in units called decibels. Regular exposure to noise above a decibel level of 85dB (A) can permanently damage your hearing. As opposed to a muscle which can be strengthened with over stimulation, your ears do not grow stronger but will sustain greater hearing loss with increased exposure to noise. Even some commonly heard sounds can be potentially damaging. Remember, the louder the sound, the shorter the allowable exposure time. The Occupational Safety and Health Administration (OSHA) allows exposure to a 90dB (A) sound for eight hours, 95dB (A) for four hours, or 100dB (A) for two hours, etc. Sounds of less than 80 decibels, even after long exposure, are unlikely to cause hearing loss.

Where are we at risk?

Listening to very loud music can start to permanently damaged hearing. Rock concerts pose risky here. Many rock musicians start to go deaf from their own music-which forces them out of their business. People who sit in front of the speakers during a concert often suffer loss of hearing for days afterwards. For some, the damage and injury to the hearing can become permanent.

Using earphones, if the volume is cranked up too high, can also be dangerous. Since the earphones are directly placed against the ears, there is no buffering and high intensity sound waves blast directly on the eardrum. When listening in a noisy environment, like in an airplane, the person will increase the volume to a dangerous level. There are noise reduction earphones that damper the outside noise, so that the volume does not have to be so high. Inexpensive earphones often do not filter out spikes of sound that are of a very high volume and possibly harmful to hearing, but they happen so quickly that you don't even notice them.

Symptoms of NIHL?

The symptoms of NIHL increase gradually over a period of continuous exposure.

Sounds may become distorted or muffled, and it may be difficult for the person to understand speech. The individual may not be aware of the loss, but it can be detected with a hearing test.

Minimize your potential for hearing loss?

If you notice ringing in your ears shortly after exposure to a loud sound, your ears are crying "ouch". You should avoid that sound or wear hearing protectors when you are exposed to that sound. No matter what type of hearing protectors you choose, be sure to use them correctly. If they are not inserted properly, you may get a false sense of security in believing that you are protecting your hearing when in reality you are not.

Here are some helpful ear-saving tips:

(1) Wear hearing protection at home, especially when working with power tools, chainsaws, and woodworking equipment.

(2) Be careful of firearm noise.

(3) Be careful of recreational noise, particularly motorcycles and speedboats.

(4) Avoid loud music or unnecessary exposure to noise at work.

(5) Escape. Let your ears rest in a quiet environment.

(6) Protect children who are too young to protect themselves.

(7) Make family, friends, and colleagues aware of the hazards of noise.

(8) Have a medical examination by an otolaryngologist, (a physician who specializes in diseases of the ears, nose, throat, head, and neck), and a hearing test by an audiologist, (a health professional trained to identify and measure hearing loss and to rehabilitate persons with hearing impairments).

2. Hearing Aids

It is a commonly voiced fact that ear pain is the most difficult one to bear, except for toothache. Also considering the enormous inconvenience that is caused due to defective hearing, techniques for the cure of ear ailments and hearing loss must be swiftly administered to all kinds of people. More than half of the people who use hearing aids have reported that in the initial stages, their hearing improved considerably. People with hearing problems, who did not use aids, suffered even more as a result; their hearing power worsened with time.

Hearing aids are better than ever. Many hearing-impaired people would benefit from using them, but only one out of five, who need hearing aids, actually have one. There are many types and degrees of hearing loss and so are there many types of hearing aids with a wide range of functions and features to address individual needs.

Today, hearing aids can be programmed to automatically respond to minute changes in sounds, etc. However, many people still think that hearing aids do not function well, are expensive, unsightly and uncomfortable to wear. On the contrary, most hearing aids are small, discrete and well-designed. As for being uncomfortable, this is not really true. When you have become used to wearing them, your overall comfort is improved.

Unit 24　Engineering Noise Control

As with any occupational hazard, control technology should aim at reducing noise to acceptable levels by action on the work environment. Such action involves the implementation of any measure that will reduce noise being generated, and or will reduce the noise transmission through the air or through the structure of the workplace. Such measures include modifications of the machinery, the workplace operations, and the layout of the workroom. In fact, the best approach for noise hazard control in the work environment is to eliminate or reduce the hazard at its source of generation, either by direct action on the source or by its confinement.

Practical considerations must not be overlooked; it is often unfeasible to implement a global control program all at once. The most urgent problems have to be solved first; priorities have to be set up. In certain cases, the solution may be found in a combination of measures which by themselves would not be enough; for example, to achieve part of the required reduction through environmental measures and to complement them with personal measures (e. g. wearing hearing protection for only 2～3 hours), bearing in mind that it is extremely difficult to make sure that hearing protection is properly fitted and properly worn.

1. Noise Control Strategies

Prior to the selection and design of control measures, noise sources must be identified and the noise produced must be carefully evaluated. Procedures for taking noise measurements in the course of a noise survey are discussed above. To adequately define the noise problem and set a good basis for the control strategy, the following factors should be considered:

(1) type of noise,
(2) noise levels and temporal pattern,
(3) frequency distribution,
(4) noise sources (location, power, directivity),
(5) noise propagation pathways, through air or through structure,
(6) room acoustics (reverberation).

In addition, other factors have to be considered; for example, number of exposed workers, type of work, etc. If only one or two workers are exposed, expensive engineering measures may not be the most adequate solution and other control options should be considered; for example, a combination of personal protection and limitation of exposure.

The need for control or otherwise in a particular situation is determined by evaluating noise levels at noisy locations in a facility where personnel spend time. If the amount of time spent in noisy locations by individual workers is only a fraction of their working day, then local regulations may allow slightly higher noise levels to exist. Where possible, noise levels should be evaluated at locations occupied by workers' ears.

Normally the noise control program will be started using as a basis A-weighted immission or noise exposure levels for which the standard ISO 11690-1 recommends target values and the principles of noise control planning. A more precise way is to use immission and emission values in frequency bands as follows.

The desired (least annoying) octave band frequency spectrum for which to aim at the location of the exposed worker is level of 90dB (A) for an overall. If the desired level after control is 85dB (A), then the entire curve should be displaced downwards by 5dB (A). The curve is used by determining the spectrum levels (see chapter 1) in octave bands and plotting the results on the graph to determine the required decibel reductions for each octave band. Clearly it will often be difficult to achieve the desired noise spectrum, but at least it provides a goal for which to aim.

It should be noted that because of the way individual octave band levels are added logarithmically, an excess level in one octave band will not be compensated by a similar decrease in another band. The overall A-weighted sound level due to the combined contributions in each octave band is obtained by using the decibel addition procedure.

Any noise problem may be described in terms of a source, a transmission path and a receiver (in this context, a worker) and noise control may take the form of altering any one or all of these elements. The noise source is where the vibratory mechanical energy originates, as a result of a physical phenomenon, such as mechanical shock, impacts, friction or turbulent airflow. With regard to the noise produced by a particular machine or process, experience strongly suggests that when control takes the form of understanding the noise-producing mechanism and changing it to produce a quieter process, as opposed to the use of a barrier for control of the transmission path, the unit cost per decibel reduction is of the order of one tenth of the latter cost. Clearly, the best controls are those implemented in the original design. It has also been found that when noise control is considered in the initial design of a new machine, advantages manifest themselves resulting in a better machine overall. These unexpected advantages then provide the economic incentive for implementation, and noise control becomes an incidental benefit. Unfortunately, in most industries, occupational hygienists are seldom in the position of being able to make fundamental design changes to noisy equipment. They must often make do with what they are supplied, and learn to use effective "add-on" noise control technology, which generally involves either modification of the transmission path or the receiver, and sometimes the source.

If noise cannot be controlled to an acceptable level at the source, attempts should then be made to control it at some point during its propagation path; that is, the path along which the sound energy from the source travels. In fact, there may be a multiplicity of paths, both in air and in solid structures. The total path, which contains all possible avenues along which noise may reach the ear, has to be considered.

As a last resort, or as a complement to the environmental measures, the noise control problem may be approached at the level of the receiver, in the context of this document, the exposed worker (s).

In existing facilities, controls may be required in response to specific complaints from within the workplace, and excessive noise levels may be quantified by suitable measurements as described previously. In proposed new installations, possible complaints must be anticipated, and expected excessive noise levels must be estimated by some procedure. As it is not possible to entirely eliminate unwanted noise, minimum acceptable levels of noise must be formulated and these levels constitute the criteria for acceptability which are generally established with reference to appropriate regulations in the workplace.

In both existing and proposed new installations an important part of the process is to identify noise sources and to rank order them in terms of contributions to excessive noise. When the requirements for noise control have been quantified, and sources identified and ranked, it is possible to consider various options for control and finally to determine the cost effectiveness of the various options. As was mentioned earlier, the cost of enclosing a noise source is generally much greater than modifying the source or process producing the noise. Thus an argument, based upon cost effectiveness, is provided for extending the process of source identification to specific sources on a particular item of equipment and rank ordering these contributions to the limits of practicality.

2. Control of Noise at the Source

To fully understand noise control, fundamental knowledge of acoustics is required. Although well covered in the specialized literature, some fundamental concepts are necessary, and some additional concepts relevant to noise control are hereby reviewed.

To control noise at the source, it is first necessary to determine the cause of the noise and secondly to decide on what can be done to reduce it. Modification of the energy source to reduce the noise generated often provides the best means of noise control. For example, where impacts are involved, as in punch presses, any reduction of the peak impact force (even at the expense of a longer time period over which the force acts) will dramatically reduce the noise generated.

Generally, when a choice of mechanical processes is possible to accomplish a given task, the best choice, from the point of view of minimum noise, will be the process which minimizes the time rate of change of force or jerk (time rate of change of acceleration). Alternatively, when the process is aerodynamic a similar principle applies; that is, the process which minimizes pressure gradients will produce minimum noise. In general, whether a process is mechanical or fluid mechanical, minimum rate of change of force is associated with minimum noise.

3. Control of Noise Propagation

With regard to control of the noise during its propagation from the source to the receiver (generally the worker) some or all of the following actions need to be considered.

(1) Use of barriers (single walls), partial enclosures or full enclosure of the entire item of equipment.

(2) Use of local enclosures for noisy components on a machine.

(3) Use of reactive or dissipative mufflers; the former for low frequency noise or small exhausts, the latter for high frequencies or large diameter exhaust outlets.

(4) Use of lined ducts or lined plenum chambers for air handling systems.

(5) Reverberation control-the addition of sound absorbing material to reverberant spaces to reduce reflected noise fields. Note that care should be taken when deciding upon this form of noise control, as direct sound arriving at the receiver will not be affected. Experience shows that it is extremely unusual to achieve noise reductions in excess of 3 or 4dB (A) using this form of control which can be exorbitantly expensive when large spaces or factories are involved. In flat rooms the spatial sound distribution is of interest.

(6) Active noise control, which involves suppression, reflection or absorption of the noise radiated by an existing sound source by use of one or more secondary or control sources.

To understand how best to design the propagation path controls mentioned above, the following concepts will be discussed: the determination of whether the problem arises from airborne or structure-borne transmission of the energy; noise absorption, reflection and reverberation; and transmission loss and isolation. This will be followed by a discussion of some of the controls which can be designed and implemented.

4. Receiver Control

Receiver control in an industrial situation is generally restricted to providing headsets and/or ear plugs for the exposed workers. It must be emphasized that this is a last resort treatment and requires close supervision to ensure long term protection of workers' hearing. The main problems lie in ensuring that the devices fit adequately to provide the rated sound attenuation and that the devices are properly worn. Extensive education programs are needed in this regard. Hearing protection is also uncomfortable for a large proportion of the workforce; it can lead to headaches, fungus infections in the ear canal, a higher rate of absenteeism and reduced work efficiency. It is worth remembering that the most protection that a properly fitted headset/earplug combination will provide is 30dB (A), due to conduction through the bone structure of the head. In most cases, the noise reduction obtained is much less than this.

Another option which is sometimes practical for receiver control is to enclose personnel in a sound reducing enclosure. This is often the preferred option in facilities where there are many noisy machines, many of which can be operated remotely. Guidelines which should be followed during design and construction are: doors, windows and wall panels should be well sealed at edges; interior surfaces of enclosure should be covered with sound absorptive material; all ventilation openings should provided with acoustic attenuators.

Vocabulary and Phrases

occupational   *adj.* 职业的，
transmission   *n.* 播送，发射，传动，传送，传输
workplace   *n.* 工作场所，车间
directivity   *n.* 方向性，指向性
acoustics   *n.* 声学
reverberation   *n.* 反响，回响，反射
A-weighted   *adj.* A 计权的
decibel   *n.* 分贝

logarithmically   *adv.* 用对数，对数地
vibratory   *adj.* 振动的，振动性的
hygienist   *n.* 卫生学者
aerodynamic   *adj.* 空气动力学的
barrier   *n.* （阻碍通道的）障碍物，栅栏，屏障
headset   *n.* 戴在头上的耳机或听筒
attenuator   *n.* 衰减器

Notes

frequency distribution   （噪声）频率分布
noise propagation pathway   噪声传播途径
frequency band   频程
octave band   倍频带
frequency spectrum   频谱
punch press   冲床，压力机
partial enclosures or full enclosure   局部或全部隔声间
dissipative muffler   消声器
the spatial sound distribution   声音空间分布
ear plug   耳塞
wall panel   墙板

Question and Exercises

1. How to describe a noise problem?
2. To control noise at the source, which method is the most effective?
3. Please describe the function of reactive or dissipative mufflers
4. What are included in the measurements of active noise control?

Further Reading

Control of Noise Propagation

With regard to control of the noise during its propagation from the source to the receiver (generally the worker) some or all of the following methods need to be considered.

1. Airborne vs Structure-borne Noise

Very often in existing installations it is relatively straightforward to track down the original source (s) of the noise, but it can sometimes be difficult to determine how the noise propagates from its source to a receiver. A classic example of this type of problem is associated with noise on board ships. When excessive noise (usually associated with the ship's engines) is experienced in a cabin close to the engine room (or in some cases far from the engine room), or on the deck above the engine room, it is necessary to determine how the noise propagates from the engine. If the problem is due to airborne noise passing through the deck or bulkheads, then a solution may include one or more of the following: enclosing the engine, adding sound absorbing material to the engine room, increasing the sound transmission loss of the deck or bulkhead by using double wall constructions or replacing the engine exhaust muffler.

On the other hand, if the noise problem is caused by the engine exciting the hull into vibration through its mounts or through other rigid connections between the engine and the hull (for example, bolting the muffler to the engine and hull), then an entirely different approach would be required. In this latter case it would be the mechanically excited deck, hull and bulkhead vibrations which would be responsible for the unwanted noise. The solution would be to vibration isolate the engine (perhaps through a well constructed floating platform) or any items such as mufflers from the surrounding structure. In some cases, standard engine vibration isolation mounts designed especially for a marine environment can be used.

As both types of control are expensive, it is important to determine conclusively and in advance the sound transmission path. The simplest way to do this is to measure the noise levels in octave frequency bands at a number of locations in the engine room with the engine running, and also at locations in the ship where the noise is excessive. Then the engine should be shut down and a loudspeaker placed in the engine room and driven so that it produces noise levels in the engine room sufficiently high that they are readily detected at the locations where noise reduction is required.

Usually an octave band filter is used with the speaker so that only noise in the octave band of interest at any one time is generated. This aids both in generating sufficient level and in detection. The noise level data measured throughout the ship with just the loudspeaker operating should be increased by the difference between the engine room levels with the speaker as source and with the engine as source, to give corrected levels for comparison with levels measured with the engine running. The most suitable noise input to the speaker is a recording of the engine noise, but in some cases a white noise generator may be acceptable. If the corrected noise levels in the spaces of concern with the speaker excited are substantially less than those with the engine running, then it is clear that engine isolation is the first noise control which should be implemented. In this case, the best control that could be expected from engine isolation would be the difference in noise levels in the space of concern with the speaker excited and with the engine running.

If the corrected noise levels in the spaces of concern with the speaker excited are similar to those measured with the engine running, then acoustic noise transmission is the likely path, although structure-borne noise may also be important but at a slightly lower level. In this case, the treatment to minimise airborne noise should be undertaken and after treatment, the speaker test should be repeated to determine if the treatment has been effective and to determine if structure-borne noise has subsequently become the problem.

Another example of the importance of determining the noise transmission path is demonstrated in the solution of an intense tonal noise in the cockpit of a fighter airplane which was thought to be due to a pump, as the frequency of the tone corresponded to a multiple of the pump rotational speed. Much fruitless effort was expended to determine the sound transmission path until it was shown that the source was the broadband aerodynamic noise at the air conditioning outlet into the cockpit and the reason for the tonal quality was because the cockpit responded modally. The frequency of strong cockpit resonance coincided with the multiple of the rotational speed of the pump but was unrelated. In this case, the obvious lack of any reasonable transmission path led to an alternative hypothesis and a solution.

2. Isolation of Noise and Transmission Loss

The noise generated by a source can be prevented from reaching a worker by means of an obstacle to its propagation, conveniently located between the source and worker. This is the concept of sound isolation. Although one would ideally like the obstacle to isolate the noise completely, in practice, some of the noise always passes through it and the amount by which the noise is reduced by the obstacle, in dB (A), is dependent on the noise reducing properties of the material (its "transmission loss") and the acoustic properties of the room into which the noise is being transmitted.

Transmission loss through a partition depends on the type of material of which it is made and it varies as a function of frequency. For usual industrial noise, the transmission loss through a partition increases by about 6dB (A) for each doubling of its weight per unit of surface area. Therefore, the best sound isolating materials are those which are compact, dense, and heavy.

3. Enclosures

The first task to consider in the design of an acoustic enclosure of a noise source is to determine the transmission paths from the source to the receiver and order them in relative importance. For example, on close inspection it may transpire that, although the source of noise is readily identified, the important acoustic radiation originates elsewhere, from structures mechanically connected to the source. In this case structure-borne sound is more important than the airborne component. In considering enclosures for noise control one must always guard against such a possibility; if structure-borne sound is the problem, an enclosure to contain airborne sound can be completely useless.

The wall of an enclosure may consist of several elements, each of which may be characterised by a different transmission loss. For example, the wall may be constructed of panels of different materials, it may include permanent openings for passing materials or cooling air in and out of the enclosure, and it may include windows for inspection and doors for access.

4. Acoustic Barriers

Since detailed information on the calculation of the Insertion Loss of a single barrier (indoors and outdoors) is available in the literature this clause is concerned with the basic rules for the use of indoor barriers.

Barriers are placed between a noise source and a receiver as a means of reducing the direct sound observed by the receiver. In rooms, barriers suitably treated with sound-absorbing material may also slightly attenuate reverberant sound field levels by increasing the overall room absorption.

Barriers are a form of partial enclosure usually intended to reduce the direct sound field radiated in one direction only. For non-porous barriers having sufficient surface density, the sound reaching the receiver will be entirely due to diffraction around the barrier boundaries.

Now we will consider the effect of placing a barrier in a room where the reverberant sound field and reflections from other surfaces cannot be ignored.

In estimating the Insertion Loss of a barrier installed in a large room the following assumptions are implicit:

(1) The transmission loss of the barrier material is sufficiently large that transmission through the barrier can be ignored. A transmission loss of 20dB (A) is recommended.

(2) The sound power radiated by the source is not affected by insertion of the barrier.

(3) The receiver is in the shadow zone of the barrier; that is, there is no direct line of sight between source and receiver.

(4) Interference effects between waves diffracted around the side of the barrier, waves diffracted over the top of the barrier and reflected waves are negligible. This implies octave band analysis.

5. Mufflers and Lined Ducts

Muffling devices are commonly used to reduce noise associated with internal combustion engine exhausts, high pressure gas or steam vents, compressors and fans. These examples lead to the conclusion that a muffling device allows the passage of fluid while at the same time restricting the free passage of sound. Muffling devices might also be used where direct access to the interior of a noise containing enclosure is required, but through which no steady flow of gas is necessarily to be maintained. For example, an acoustically treated entry way between a noisy and a quiet area in a building or factory might be considered as a muffling device.

Muffling devices may function in any one or any combination of three ways: they may suppress the generation of noise; they may attenuate noise already generated; and they may carry or redirect noise away from sensitive areas. Careful use of all three methods for achieving adequate noise reduction can be very important in the design of muffling devices, for example, for large volume exhausts.

Two terms, insertion loss, IL, and transmission loss, TL, are commonly used to describe the effectiveness of a muffling system. The insertion loss of a muffler is defined as the reduction (in decibels) in sound power transmitted through a duct compared to that transmitted with no muffler in place. Provided that the duct outlet remains at a fixed point in space, the insertion loss will be equal to the noise reduction which would be expected at a reference point external to the duct outlet as a result of installing the muffler. The transmission loss of a muffler, on the other hand, is defined as the difference (in decibels) between the sound power incident at the entry to the muffler to that transmitted by the muffler.

Muffling devices make use of one or the other or a combination of the two effects in their design. Either sound propagation may be prevented (or strongly reduced) by reflection (generally as the result of using orifices and expansion chambers), or sound may be dissipated, generally by the use of sound absorbing material. Muffling devices based upon reflection are called reactive devices and those based upon dissipation are called dissipative devices. A duct lined with sound absorbing material on its walls is one form of dissipative muffler.

6. Sound Absorption and Reflection

Sound absorption is the phenomenon by which sound is absorbed by transformation of acoustic energy into ultimately thermal energy (heat). Although some absorption always happens when a sound wave encounters an obstacle, it happens in an appreciable manner when the sound wave is incident on a sound absorbing material.

Sound absorbing materials are fibrous, lightweight and porous, possessing a cellular structure of intercommunicating spaces. It is within these interconnected open cells that acoustic energy is converted into thermal energy. Thus the sound-absorbing material is a dissipative structure which acts as a transducer to convert acoustic energy into thermal energy. The actual loss mechanisms in the energy transfer are viscous flow losses caused by wave propagation in the material and internal frictional losses caused by motion of the material's fibres. The absorption characteristics of a material are dependent upon its thickness, density, porosity, flow resistance, fibre orientation, and the like.

Common porous absorption materials are made from vegetable, mineral or ceramic fibres (the latter for high temperature applications) and elastomeric foams, and come in various forms. The materials may be prefabricated units, such as glass blankets, fibreboards, or lay-in.

7. Reverberation

When sound reflects within boundaries, it "accumulates" as a result of the addition of the reflected sound to the original sound. Sound may continue even after the original source stops this is called "reverberation".

Thus a reverberant field is one which is characterized by sound which has been reflected from at least one surface in a particular room or enclosure. When the enclosure boundaries are hard and reflective, the reverberant field can easily dominate the sound arriving directly (without reflection) from a particular sound source and this will become increasingly likely as the distance from the sound source is increased.

8. Active Noise Control

Active control of noise is the process of reducing existing noise by the introduction of additional noise by means of one or more secondary (or control) noise sources. The introduced noise may achieve the required noise reduction by way of any one or combination of three different physical mechanisms.

One mechanism which is often used to describe the active control of noise in the popular press is that of sound field cancellation; that is, the introduced control sound is anti-phase to the original sound and cancellation results. This mechanism characterizes cases where noise reduction is achieved in small local areas surrounding a control source; however, local areas of cancellation are always balanced by other areas of reinforcement where the sound level is increased. This type of control mechanism, which may be called "local cancellation", characterizes the process involved in the control of noise around a passenger's head in an aircraft or motor vehicle using a loudspeaker embedded in the head rest of the seat or by use of a headset or earmuff containing a loudspeaker.

A second mechanism, which will be called suppression of sound generation, is possible and may be understood on the basis of the following considerations. If it were possible to make the entire control sound field (or almost all of it) 180 degree out of phase with the original (primary) field, then the sound radiated by the primary source would be effectively "cancelled" leaving one to wonder where all the energy had gone. The answer is that in this case, the control mechanism is not really cancellation; the sound field generated by the control sources has effectively "unloaded" the primary source, changing its radiation impedance so that it radiates much less sound (even though the motion of the physical source such as a vibrating surface may remain unchanged). In this case, the control sources act to suppress the sound power radiated by the primary source by making its radiation impedance reactive with only a negligible real part.

To achieve effective suppression of the primary source output by presenting a purely reactive impedance to it, the control sources must be large enough and located such that they are capable of presenting the required impedance to the primary source. In one dimensional wave guides, such as air conditioning ducts, these constraints are relatively easy to satisfy and the distance between the control and primary sources is not too important. However, in 3-D space, the control source in general will need to be close to the primary source to affect its radiation impedance significantly. It will also need to be of similar size with a similar volume velocity output.

A third mechanism of active noise control is that of absorption by the control sources. In this case, the primary sound field energy is used to assist in driving the control source (for example the speaker cone if the control source is a loudspeaker). However, the acoustical efficiency of loudspeakers and other artificial noise generators is so poor, that electrical energy is still needed to drive the source with sufficient amplitude and at the correct phase to enable it to absorb energy from the sound field. Except for plane wave sound propagation in ducts, this mechanism is likely to result only in areas of reduced noise close to the control source.

9. Separation of Source and Receiver

Another type of noise propagation control is the separation, which can be by distance or in time. As the direct field radiated by a source generally decreases by 6dB for each doubling of the distance from it (after the initial 1 metre), separating the source and receiver by distance is beneficial.

Noisy operations can also be separated in time; that is, they are performed out of the usual shift.

Part E Urban Eco-Environment

Unit 25 Benefits of Trees In Urban Areas

Trees are major capital assets in cities across the United States. Just as streets, sidewalks, public buildings and recreational facilities are a part of a community's infrastructure, so are publicly owned trees. Trees and, collectively, the urban forest—are important assets that require care and maintenance the same as other public property. Trees are on the job 24 hours every day working for all of us to improve our environment and quality of life.

Urban forest provides many environmental benefits to our community. Aside from the obvious aesthetic benefits, trees within our urban forest improve our air, protect our water, save energy, and improve economic sustainability.

1. Urban Forests Improve Our Air

Carbon Sequestration:

Heat from earth is trapped in the atmosphere due to high levels of carbon dioxide (CO_2) and other heat-trapping gases that prohibit it from releasing heat into space—creating a phenomenon known as the "greenhouse effect". Trees remove (sequester) CO_2 from the atmosphere during photosynthesis to form carbohydrates that are used in plant structure/function and return oxygen back to the atmosphere as a byproduct. About half of the greenhouse effect is caused by CO_2. Trees therefore act as a carbon sink by removing the carbon and storing it as cellulose in their trunk, branches, leaves and roots while releasing oxygen back into the air.

Trees also reduce the greenhouse effect by shading our homes and office buildings. This reduces air conditioning needs up to 30%, thereby reducing the amount of fossil fuels burned to produce electricity. This combination of CO_2 removal from the atmosphere, carbon storage in wood, and the cooling effect makes trees a very efficient tool in fighting the greenhouse effect. One tree that shades your home in the city will also save fossil fuel, cutting CO_2 buildup as much as 15 forest trees.

Each person in the U.S. generates approximately 2.3 tons of CO_2 each year. A single mature tree can absorb carbon dioxide at a rate of 48 lbs./year and release enough oxygen back into the atmosphere to support 2 human beings. A healthy tree stores about 13 pounds of carbon annually or 2.6 tons per acre each year. An acre of trees absorbs enough CO_2 over one year to equal the amount produced by driving a car 26000 miles. An estimate of carbon emitted per vehicle mile is between 0.88~1.06 lbs. CO_2/mi. Thus, a car driven 26000 miles will emit between 22880 lbs CO_2 and 27647 lbs. CO_2. Thus, one acre of tree cover in Brooklyn can compensate for automobile fuel use equivalent to driving a car between 7200 and 8700 miles.

Planting trees remains one of the cheapest, most effective means of drawing excess CO_2 from the atmosphere. If every American family planted just one tree, the amount of CO_2 in the atmosphere would be reduced by one billion lbs annually. This is almost 5% of the amount that human activity pumps into the atmosphere each year. The U.S. Forest Service estimates that all the forests in the United States combined sequestered a net of approximately 309 million tons of carbon per year from 1952 to 1992, offsetting approximately 25% of U.S. human-caused emissions of carbon during that period. Over a 50-year lifetime, a tree generates $31250 worth of oxygen, provides $62000 worth of air pollution control, recycles $37500 worth of water, and controls $31250 worth of soil erosion. Now, approximately 800 million tons of carbon are stored in U.S. urban forests with a $22 billion equivalent in control costs.

Reduction of Other Air Pollutants:

Trees also remove other gaseous pollutants such as Sulfur Dioxide (SO_2), Ozone (O_3), Nitrogen oxides and small (<10 microns) particles by absorbing them with normal air components through the stomates in the leaf surface. In some region, there is up to a 60% reduction in street level particulates with trees.

In one urban park (212 ha.) tree cover was found to remove daily 48 lbs. particulates, 9 lbs nitrogen dioxide, 6 lbs sulfur dioxide, and 2 lbs. carbon monoxide ($136/day value based upon pollution control technology) and 100 lbs of carbon. One sugar maple (12″ DBH) along a roadway removes in one growing season 60mg cadmium, 140mg chromium, 820mg nickel, and 5200mg lead from the environment. Planting trees and expanding parklands improves the air quality of Los Angeles county. A total of 300 trees can counter balance the amount of pollution one person produces in a lifetime.

2. Urban Forests Protect Our Water

Trees reduce topsoil erosion, prevent harmful land pollutants contained in the soil from getting into our waterways, slow down water run-off, and ensure that our groundwater supplies are continually being replenished. For every 5% of tree cover added to a community, stormwater runoff is reduced by approximately 2%. Research by the USFS shows that in a 1 inch rainstorm over 12 hours, the interception of rain by the canopy of the urban forest in Salt Lake City reduces surface runoff by about 11.3 million gallons, or 17%. These values would increase as the canopy increases. Along with breaking the fall of rainwater, tree roots remove nutrients harmful to water ecology and quality.

Trees act as natural pollution filters. Their canopies, trunks, roots, and associated soil and other natural elements of the landscape filter polluted particulate matter out of the flow toward the storm sewers. Reducing the flow of stormwater reduces the amount of pollution that is washed into a drainage area. Trees use nutrients like nitrogen, phosphorus, and potassium—byproducts of urban living—which can pollute streams.

3. Urban Forests Save Energy

Homeowners that properly place trees in their landscape can realize savings up to 58% on daytime air conditioning and as high as 65% for mobile homes. If applied nationwide to buildings not now benefiting from trees, the shade could reduce our nation's consumption of oil by 500000 barrels of oil/day. The maximum potential annual savings from energy conserving landscapes around a typical residence ranged from 13% in Madison up to 38% in Miami. Projections suggest that 100 million additional mature trees in US cities (3 trees for every unshaded single family home) could save over $2 billion in energy costs per year.

Trees lower local air temperatures by transpiring water and shading surfaces. Because they lower air temperatures, shade buildings in the summer, and block winter winds, they can reduce building energy use and cooling costs. Help to cool cities by reducing heat sinks. Heat sinks are 6~19°F warmer than their surroundings. A tree can be a natural air conditioner. The evaporation from a single large tree can produce the cooling effect of 10 room size air conditioners operating 24 hours/day.

USFS estimates the annual effect of well-positioned trees on energy use in conventional houses at savings between 20%~25% when compared to a house in a wide-open area.

4. Urban Forests Can Extend the Life of Paved Surfaces

Because the asphalt paving on streets contain stone aggregate in an oil binder. Without tree shade, the oil heats up and volatizes, leaving the aggregate unprotected. Vehicles then loosen the aggregate and much like sandpaper, the loose aggregate grinds down the pavement. Streets should be overlaid or slurry sealed every 7~10 years over a 30~40 year period, after which reconstruction is required.

A slurry seal costs approximately $0.27/sq. ft. or $50000/linear mile. Because the oil does not dry out as fast on a shaded street as it does on a street with no shade trees, this street maintenance can be deferred. The slurry seal can be deferred from every 10 years to every 20~25 years for older streets with extensive tree canopy cover.

5. Urban Forests Can Increase Traffic Safety

Trees can also enhance traffic calming measures, such as narrower streets, extended curbs, roundabouts, etc. Tall trees give the perception of making a street feel narrower, slowing people down. Closely spaced trees give the perception of speed (they go by very quickly) slowing people down. A treeless street enhances the perception of a street being wide and free of hazard, thereby increasing speeds. Increased speed leads to more accidents. Trees can serve as a buffer between moving vehicles and pedestrians. Street trees also forewarn drivers of upcoming curves. If the driver sees tree trunks curving ahead before seeing the road curve, they will slow down and be more cautious when approaching curves.

6. Urban Forests Can Improve Economic Sustainability

The scope and condition of a community's trees and, collectively, its urban forest, is usually the first impression a community projects to its visitors. A community's urban forest is an extension of its pride and community spirit. Studies have shown that: (1) trees enhance community economic stability by attracting businesses and tourists; (2) people linger and shop longer along tree-lined streets; (3) apartments and offices in wooded areas rent more quickly and have higher occupancy rates; (4) businesses leasing office spaces in developments with trees find their workers are more productive and absenteeism is reduced.

7. Urban Forests Can Increase Real Estate Values

Property values increase 5%~15% when compared to properties without trees (depends on species, maturity, quantity and location). A 1976 study that evaluated the effects of several different variables on homes in Manchester, Connecticut, found that street trees added about $2686 or 6% to the sale price of a home. A more recent study indicated that trees added $9500, or more than 18 percent, to the average sale price of a residence in a suburb of Rochester, New York.

8. Urban Forests Can Increase Sociological Benefits

Two University of Illinois researchers studied how well residents of the Chicago Robert Taylor Housing Project (the largest public housing development in the world) were doing in their daily lives based upon the amount of contact they had with trees, and came to the following conclusions:

(1) Trees have the potential to reduce social service budgets, decrease police calls for domestic violence, strengthen urban communities, and decrease the incidence of child abuse according to the study. Chicago officials heard that message last year. The city government spent $10 million to plant 20000 trees, a decision influenced by Kuo's and Sullivan's research, according to the Chicago Tribune.

(2) Residents who live near trees have significantly better relations with and stronger ties to their neighbors.

(3) Researchers found fewer reports of physical violence in homes that had trees outside the buildings. Of the residents interviewed, 14% of residents living in barren conditions have threatened to use a knife or gun against their children versus 3% for the residents living in green conditions.

(4) Studies have shown that hospital patients with a view of trees out their windows recover much faster and with fewer complications than similar patients without such views.

(5) A Texas A & M study indicates that trees help create relaxation and well being. A U. S. Department of Energy study reports that trees reduce noise pollution by acting as a buffer and absorbing 50% of urban noise.

Unlike urban areas in the eastern U. S. , canopy cover in some region decreases along an urban to rural gradient. In other words, since most trees have been planted much of the tree cover is in urban areas as opposed to "natural lands". Therefore, estimated pollutant uptake rates are higher for residential compared to natural or unmanaged lands. Possible management implications of these estimates are that air pollutant uptake benefits from tree planting may be optimized by planting in areas where air pollutant concentrations are elevated and where relatively high planting densities can be achieved thereby enhancing the health of urban dwellers.

Vocabulary and Phrases

aesthetic   *adj.* 美学的，审美的
asphalt   *n.* 沥青
binder   *n.* 胶粘剂
cadmium   *n.* [化]镉，元素符号 Cd
carbohydrate   *n.* [化]碳水化合物，糖类
carbon monoxide   一氧化碳
carbon sink   碳汇
cellulose   *n.* [化]纤维素
chromium   *n.* [化]铬，元素符号 Cr
forewarn   *v.* 警告，有言在先
gallon   加仑(体积单位)
grind   *n.* 磨碎，小块；*v.* 研磨
infrastructure   *n.* 基础设施，建筑物
lbs   磅(重量单位)
linger   *v.* 徘徊，盘旋
nationwide   *adj.* 全国的

nickel   *n.* [化]镍，元素符号 Ni
pedestrian   *n.* 行人，*adj.* 徒步的
phosphorus   *n.* [化]磷，元素符号 P
photosynthesis   *n.* [生]光合作用
potassium   *n.* [化]钾，元素符号 K
recreation   *n.* 娱乐，休闲
replenish   *v.* 补充，添加
roundabout   *n.* 弯路；*adj.* 迂回的
sandpaper   *n.* 砂纸
sequester   *v.* 固定
slurry   *n.* 泥浆
stomata   *n.* [生]气孔
sugar maple   糖槭（树木）
transpire   *v.* 蒸发
volatize   *v.* 挥发

Note

gallon   是英美常用的体积单位，1 加仑＝4.55 升(英制)，1 加仑＝3.785 升(美制)。
lbs   是英美常用的质量单位，1 磅换算成千克就是 0.45359 千克。
carbon sequester   指空气中的 CO_2 被植物或土壤贮存的过程。

Question and Exercises

1. Do you think which benefits of trees is the most critical?
2. Please give an example to indicate sociological benefits of urban trees.
3. Could you give a simple method to calculate the tree benefits in urban area?

Unit 26 Urban Woodlands Reducing the Particulate Pollution

It has quite recently been realised that particulate pollution is a very real and serious problem, with the potential to cause the most severe and damaging health effects of any pollutant. It is also universally accepted that trees improve the quality of urban life. The literatures illustrates that trees can be effective in reducing the impacts of damaging forms of particulate pollution such as PM_{10}.

1. Effects of Particulate Pollution to Trees

The effects that many pollutant particles have been shown to have on some tree species are summarized from various literature sources. From these literatures it can be seen that particles can produce a wide variety of effects on the physiology of trees. Heavy metals and other toxic particles have been shown to accumulate, causing damage and death of some species. This damage has mainly been reported to result from the phytotoxicity of these particles. However, that has been stated that a significant source of damage can be the abrasive action of their turbulent deposition, which they showed increased callus tissue formation on leaf surfaces. With further regard to the phytotoxicity of heavy metals, although many concentrations encountered in the literature are sufficient to cause ill-health in humans, they are often enough to cause only minimal toxic effects in trees. Heavy loads of atmospheric particles, such as those which can occur close to unpaved roads and opencast quarries, also result in the occlusion of stomata, decreasing the efficiency of gaseous exchange. The resultant "crust" of particles that can form on leaf and bark surfaces disrupts other physiological processes, such as bud break, pollination and light absorption/reflectance. Some authors also reported a number of indirect effects such as the predisposition of plants to infection by pathogens and the long-term alteration of genetic structure. Conversely, particulates have been shown to produce positive growth responses in a few.

2. Trees as Particle Sources and Sinks

Naturally released VOCs (Volatile Organic compounds) can condense and agglomerate with other atmospheric particles, producing summertime hazes over some forest canopies, VOCs, both natural and man-made, are also important precursors in the formation of O_3, leading to the occasional and often misleading claims that trees can actually increase air-pollution episodes. The production of pollen by trees is also a source of particles which can have quite severe health effects in the form of hay-fever. It is interesting to note that a very high pollen count of 1000 grains m^{-3}, each of approximately $30 \mu m$ diameter, produces a particle mass of $14 \mu gm^{-3}$, much less than the daily average AQS of $50 \mu gm^{-3}$ set by EPAQS.

Trees and other vegetation are effective at trapping and absorbing many pollutant particles. These particles are produced from the expansion and escape of gases during combustion and are markers of anthropogenic sources. Some researchers stated that forest canopies are more effective at capturing particles than any other vegetation type, due to their much greater surface roughness. This increases turbulent deposition and impaction processes by causing localized increases in wind speed. The relationship between turbulence and wind speed and increased particle deposition has been shown to be highly significant in forest ecosystems. The same relationship has also been illustrated in the urban environment, whereby the complex surface structure facilitates an intricate pattern of turbulent air flow. The importance of these examples is that both ecotypes provide a rough surface which results in turbulent atmospheric mixing, the removal of a significant surface boundary layer resistance and thus the efficient deposition of pollutants. In addition to the effects pollutant particles have on trees, aspects of the effectiveness with which they are captured are also summarised from the literature. In general, the effectiveness of particle uptake by trees is increased if their leaf and bark surfaces are rough or sticky. Variations in the structure and microroughness of a leaf also affect patterns of particle deposition. As was shown by researcher who found that in exposures of conifers to very fine particles ($\sim 0.5 \mu m$) in a wind tunnel, deposition was much greater in the stomatal regions of needles. For coarser particles it appears that increased stickiness of the surface facilitates greater particulate capture, whilst for finer particles it is the roughness of the surface that has the greater influence on uptake.

3. The Role of Urban Woodlands

Within the past decade there have been many reports on the need for increased tree planting in the urban environment. Since then the community forests have been established close to large urban settlements, where the emphasis is placed upon leisure and amenity. Also, the development of Local Agenda 21 (a spin-off from UNCED) recognised the importance of urban trees in improving the health of inhabitants and the need for local authorities to develop long-term strategies for urban tree management.

About 10% of the UK land area is in urban use, within which there are many parks, gardens (public and private), and open spaces. However, unfortunately, some of these areas are derelict or mismanaged. The need to redevelop them is recognised by government policy, and there are numerous grants available. It is also recognised that the establishment of trees is a suitable use for such land. This could be in a permanent or temporary form, since the time-lag between securing and developing an urban site can take between 2 to 20 years, a fact that Bradshaw et al. (1995) and Baines (1996) have realised, leading to their suggestions of establishing short rotation woody crops as an interim land use.

There is currently a view that too little use is being made of urban trees (Arboricultural Association, 1996). However, the UK government's response to the damning "Park Life" report shows that it is beginning to take the matter more seriously, although to what extent is disputed, especially when research is highlighting the environmental and economic benefits which come from the establishment of more trees in our towns and cities. There are now many urban forestry strategies beginning to be implemented, with the primary aim of improving the quality of urban life, such as the Black Country Urban Forest, which mirror some of the North American undertakings of over a decade ago.

The complexity of the urban environment greatly influences the establishment and management of trees which are planted and grow there. Federer (1971) described three broad classes of urban ecosystem in which trees are grown. These were:

(1) areas with high evaporative or transpirative surfaces; e.g. parks and wide streets-hot during the day and cold at night;

(2) areas open to the sky but dry; e.g. large squares and car-parks—similar microclimate to (1), but drier; and

(3) areas sheltered by buildings; e.g. narrow streets and courtyards-cool during the day and warm at night (i.e. small diurnal temperature fluctuations).

In addition to, and in some cases because of, these ecosystem types, the micrometeorology and sources of urban air pollutants cause "hotspots" and different ranges of pollution concentrations, a factor compounded by the frequent temperature inversions which trap pollutants in built-up areas, particularly during the winter and especially in London. This has led to the conviction of many authors that a great deal of emphasis should be placed on the selection of appropriate tree species for their particular urban settings.

Broadleaves are generally considered to be more hardy of pollution than many coniferous species. But since pollution concentrations have been sufficiently reduced from the Clean Air Acts, the presence of conifers in urban areas may well, and arguably should, increase, an initiative supported by Peter Thoday, a past president of the Institute of Horticulture. Conifers have also been shown to be more efficient scavengers of particulate lead than broadleaves. The appeal of broadleaf planting in high-pollution areas results mainly from the fact that most broadleaves renew their filtering apparatus, i.e. their leaves, each year, therefore decreasing their accumulated annual load of toxic particles. However, as the leaves senesce these can then accumulate in the soil, causing other physiological damage, particularly to the root system and particularly if the particles are not readily leached or mobilised from the soil. Obversely, since the vast majority of conifers keep their foliage throughout the winter and continue to transpire, they continue to accumulate atmospheric toxins. This has two main effects: firstly, the load of particles on conifers is generally higher than for broadleaves, thereby making conifers more efficient at improving urban air quality; and secondly, as a result of the first effect, the higher load of toxins on conifers can result in prolonged and more severe physiological damage. The value of conifers at absorbing pollution comes from a variety of factors further to their evergreen habit, most notably their speed of establishment, very high surface areas, and their particular effectiveness at absorbing particles.

4. The Improvement of Urban Air Quality by Woodland

　　For urban trees, critical loads can be regarded as the accumulated amount of a pollutant which will result in physical damage (i. e. a threshold for injury). Farmer (1995) discussed the critical loads of plants for a variety of pollutants including particulates, and Caborn (1965) identified the mechanisms by which some tree species are able to avoid damage specifically from particles. These include altering the timing of bud break or leaf fall, and the ability to produce new shoots when injured. Due to these and other physiological mechanisms, some trees are better able to survive smoky and polluted conditions. Broadmeadow and Freer-Smith (DOE, 1996) have suggested that pollutant-tolerant trees which exhibit high stomatal conductances should be planted in high-pollution"hot-spots" in order to absorb contaminants and, therefore, improve air quality, an approach also recommended by Good (1990). The concordant increase in transpiration that is often present in species exhibiting high stomatal conductances (obviously dependent on the specific stomatal size and density) can improve the efficiency with which particles are captured by leaf surfaces. Tong (1991) states that this mechanism operates by the capture of particles on the film of moisture produced by transpiration. This would be particularly true for water-soluble sulphate and nitrate particles. Very fine particles ($<1\mu m$) can behave like gaseous pollutants and diffuse across concentration gradients due to the action of Brownian motion. By this mechanism, such particles can be attracted onto the surfaces of the substomatal cavity, as was observed by Thompson et al. (1984).

　　High rates of transpiration, in addition to shading and pollutant uptake effects, are also a factor in the reduction of localised particulate concentrations by lowering urban air temperatures. Moll (1996) states that 12% of air-pollution problems in cities are attributable to urban heat-island effects. This is due to the temperature-dependent formation of many pollutants, such as VOCs and O_3 and the dynamics of particulate dispersal. With regard to the latter, Fritschen and Edmonds (1976) found that the temperature inversions created by woodland canopies effectively alklows particles to permeate for miles within the woodland, escaping in plumes from canopy gaps and boundaries with water or bare ground.

　　A study by Impens and Delcarte (1979) showed that the interception of particles by vegetation was massively greater for street trees, due to their proximity to high intensities of road traffic. Their study recognized the importance of urban-tree establishment to create dust falters in towns and cities. They also realised that the areas of highest pollution concentrations, usually in central locations where trees could be most effectively used, were those most lacking in urban greenery. The reasons for this were partly explained by the hostility of these environments to trees, with factors such as low soil fertility, poor drainage, compaction, and vandalism, in addition to poor air quality, affecting survivability. Madders and Lawrence (1981) state that these types of pressures can compound the influence of air pollution on urban trees, and so describe the necessity of appropriate design, species choice, and location of pollution control tree plantings. They mention that the most effective use of trees as particulate falters is achieved when planting occurs as close as possible to a source, forming a buffer around it. Research by Spitsyna and Skripal'shchikova (1991) revealed that more than 50% of the large amount of dust produced from an open-cast coal mine in the Kansk steppe ($252 kgha^{-1}$ in one summer at the leading edge of the surrounding woodland) was intercepted by just a 15-m-wide stand of birch trees.

Quantification of the benefits of urban trees in removing particulate pollution have been calculated by McPherson et. al. (1994) who estimated that the trees of Chicago removed approximately 234 tons of PM_{10} in 1991, improving average hourly air quality by 0.4% (2.1% in heavily wooded areas). The opportunity cost of this service was estimated at \$9.2 million. Similarly, Nowak et al. (1997) calculated that the trees of Philadelphia improved air quality by 0.72% in the process of PM_{10} reduction. The opportunity cost of which was calculated at \$1.9 million. Freer-Smith and Broadmeadow (1996) also attempted quantification of the air improving properties of trees by the application of a mathematical model based on physiological and meteorological data. This model did not quantify particulate uptake but did show that urban trees could absorb significant amounts of both sub-stomatal cavity, and SO_2 (up to 21 and 20% of exposure concentrations during episodes of O_3 and SO_2 respectively) both of which are associated with particulate forms of pollution; i.e. VOCs and sulphate.

Trees capture particles through a number of simple physical processes. As such, their effectiveness in particle uptake results mainly from the properties and area of their surfaces. In capturing pollutant particles, trees expose themselves to a variety of toxic effects. The susceptibility to damage varies greatly between species, as can the efficiency of pollutant uptake. This has important implications for the planting and management of appropriate species in urban areas. The information reviewed here suggests that future urban planting should focus on the increased use of conifers.

The pattern of particle deposition in urban areas is much more complicated than is indicated by a small number of spot measurements. To buffer these high and transitory concentrations of particles, tree planting should be optimised in local pollution 'hotspots'. On a wider scale, this also highlights the importance of suburban woodlands in reducing "background" concentrations of fine particles.

It is widely recognised that the fraction of particles most damaging to human health are those derived from anthropogenic sources, and while there is still an obvious and vital need to reduce pollution emissions at source, there is growing, and historical evidence that trees can be used to improve air quality. In spite of this, the use of trees and woodlands, generally and specifically, for the removal of particulates in urban spaces has been far from optimal. As a result, there is room for improvement in our understanding of how trees can be most effectively utilised in the urban environment. We conclude that further research is required, in particular, to quantify accurately particulate uptake by urban plantings, to establish why some species are more effective at removal of particulates, and to identify the practical measures by which particle removal by trees can be maximised.

Vocabulary and Phrases

abrasive   *n.* 磨料，*adj.* 有研磨作用的
agglomerate   *n.* 凝聚，*adj.* 凝聚的，结块的，*v.* 凝聚
amenity   *n.* 舒适宜人
appeal   *v.* 呼吁，申诉
arboriculture   *n.* [林] 树木栽培(学)
birch   *n.* [植] 桦树
broadleaf   *n.* [植] 阔叶(树种)
bud   *n.* [植] 芽，蓓蕾
buffer   *n.* 缓冲
callus   *n.* [植] 愈伤组织
canopy   *n.* [植] 冠层
combustion   *n.* 燃烧，(生物体内的)氧化生热过程
condense   *v.* 冷凝，使凝结
coniferous   *n.* [植] 针叶(树木)
critical loads   临界负荷
derelict   *adj.* 被抛弃的，被废弃的
drainage   *n.* [地] 排水
ecotype   *n.* [生] 生态型
falter   *n.* 摇晃，*v.* 摇晃，晃动
foliage   *n.* [植] 树叶
haze   *n.* 烟雾
heat-island effects   热导效应
interim   *n.* 间歇，*adj.* 间歇的
intricate   *adj.* 错综复杂的
micrometeorology   *n.* [气] 微气候，微气象
mobilize   *v.* 动员，使流通

needle   *n.* [植] 针叶
nitrate   *n.* [化] 硝酸盐
occlusion   *n.* 闭塞，阻塞
opencast quarry   露天采石场
optimize   *v.* 优化
pathogen   *n.* [生] 病原体
physiological processes   生理过程
physiology   *n.* [生] 生理学
phytotoxicity   *n.* [植] 植物毒害
pollen   *n.* [植] 花粉
pollination   *n.* [植] 授粉
precursor   *n.* 先锋，前体，母体
predisposition   *n.* 预处理
proximity   *n.* 临近，接近
resultant   *adj.* 组合的，合成的，作为结果而发生的
rotation   *n.* 旋转，转动，[农] 轮作
scavenger   *n.* 清道夫；[化] 清除剂，净化剂
stomatal conductance   气孔导度
sulphate   *n.* [化] 硫酸盐
survivability   *n.* [植] 存活能力
toxin   *n.* [生] 毒素
transitory   *adj.* 暂时的，瞬时的
turbulence   *n.* 紊流，[气] 湍流
vandalism   *n.* 破坏
wind tunnel   风洞

Note

  anthropogenic source   人为源，由于认为活动而导致的各种污染源
  EPAQS 为 Expert Panel on Air Quality Standards 的缩写

VOCs (Volatile Organic compounds)　挥发性有机物
pollutant-tolerant tree　耐污染树木
substomatal cavity　气孔腔，叶片表层气孔凹下的部分

Question and Exercises

1. Please summarize the roles of urban woodlands.
2. How to improve urban air quality by woodland?
3. How to consider the roles of forestry in urban planning?

Unit 27 Ecological Structure, Functions and Services of Green Roofs

Buildings change the flow of energy and matter through urban ecosystems, often causing environmental problems. These problems can be partially mitigated by altering the buildings' surficial properties. Roofs can represent up to 32% of the horizontal surface of built-up areas and are important determinants of energy flux and of buildings' water relations. The addition of vegetation and soil to roof surfaces can lessen several negative effects of buildings on local ecosystems and can reduce buildings' energy consumption. Living, or green, roofs have been shown to increase sound insulation, fire resistance, and the longevity of the roof membrane. They can reduce the energy required for the maintenance of interior climates, because vegetation and growing plant media intercept and dissipate solar radiation. Green roofs can also mitigate storm-water runoff from building surfaces by collecting and retaining precipitation, thereby reducing the volume of flow into storm-water infrastructure and urban waterways. Other potential benefits include green-space amenity, habitat for wildlife, air-quality improvement, and reduction of the urban heat-island effect. Architects have applied green-roof technology worldwide, and policymakers and the public are becoming more aware of green-roof benefits. Although green roofs are initially more expensive to construct than conventional roofs, they can be more economical over the life span of the roof because of the energy saved and the longevity of roof membranes.

Although green roofs represent a distinct type of urban habitat, they have been treated largely as an engineering or horticultural challenge, rather than as ecological systems. The environmental benefits provided by green roofs derive from their functioning as ecosystems. The first goal of this article is to describe the history and components of living-roof ecosystems; the second is to review the ways in which the structure of a green roof—including vegetation, growing medium, and roof membrane—determines its functions.

1. History of Green Roofs

Roof gardens, the precursors of contemporary green roofs, have ancient roots. The earliest documented roof gardens were the hanging gardens of Semiramis in what is now Syria, considered one of the seven wonders of the ancient world. Today, similarly elaborate roof-garden projects are designed for high-profile international hotels, business centers, and private homes. These green roofs, known for their deep substrates and variety of plantings as "intensive" green roofs, have the appearance of conventional ground-level gardens, and they can augment living and recreation space in densely populated urban areas. Intensive green roofs typically require substantial investments in plant care. Furthermore, they emphasize the active use of space and carry higher aesthetic expectations than "extensive" green roofs, which generally have shallower soil and low-growing ground cover.

Extensive green roofs are a modern modification of the roof-garden concept. They typically have shallower substrates, require less maintenance, and are more strictly functional in purpose than intensive living roofs or roof gardens. In their simplest design, extensive green roofs consist of an insulation layer, a waterproofing membrane, a layer of growing medium, and a vegetation layer. This basic green-roof design has been implemented and studied in diverse regions and climates worldwide.

The modern green roof originated at the turn of the 20th century in Germany, where vegetation was installed on roofs to mitigate the damaging physical effects of solar radiation on the roof structure. Early green roofs were also employed as fire retardant structures. There are now several competing types of extensive green-roof systems, which provide similar functions but are composed of different materials and require different implementation protocols.

In the 1970s, growing environmental concern, especially in urban areas, created opportunities to introduce progressive environmental thought, policy, and technology in Germany. Green-roof technology was quickly embraced because of its broad-ranging environmental benefits, and interdisciplinary research led to technical guidelines, the first volume of which was published in 1982 by the Landscape, Research, Development and Construction Society. Many German cities have since introduced incentive programs to promote green-roof technology and improve environmental standards. Building law now requires the construction of green roofs in many urban centers. Such legal underpinnings of green-roof construction have had a major effect on the widespread implementation and success of green-roof technology throughout Germany. Green-roof coverage in Germany alone now increases by approximately 13.5 million square meters (m^2) per year. Haemmerle (2002) calculates that approximately 14% of all new flat roofs in Germany will be green roofs; the total area covered by green roofs is unknown. The market for sloped green roofs is also developing rapidly, and accessible green roofs have become a driving force in neighborhood revitalization.

2. Green-roof Vegetation

Rooftop conditions are challenging for plant survival and growth. Moisture stress and severe drought, extreme (usually elevated) temperatures, high light intensities, and high wind speeds increase the risk of desiccation and physical damage to vegetation and substrate. Plants suitable for extensive green roofs share adaptations that enable them to survive in harsh conditions. These plants have stress-tolerant characteristics, including low, mat-forming or compact growth; evergreen foliage or tough, twiggy growth; and other drought-tolerance or avoidance strategies, such as succulent leaves, water storage capacity, or CAM (crassulacean acid metabolism) physiology. However, frequent drought-related disturbance to green-roof vegetation also favors some ruderal species that can rapidly occupy gaps. Green-roof communities are dynamic, and with time, vegetation is likely to change from the original composition.

Since the 1980s, researchers have tested many herbaceous and woody taxa in different rooftop conditions. Heinze (1985) compared combinations of various *Sedum* species, grasses, and herbaceous perennials, planted at two substrate depths in simulated roof platforms. *Sedum* species outperformed the other taxa, except in consistently moist substrate deeper than 10 centimeters (cm). In these conditions, a taller grass and herbaceous canopy layer created shaded conditions that proved unfavorable to the *Sedum* species. Other studies support the suitability of low-growing *Sedum* species for use in green roofs because of their superior survival in substrate layers as thin as 2 to 3cm. Physical rooftop conditions, suitability for plant growth, and the cost of various substrates have also been examined.

The composition and character of green-roof vegetation depend on many factors. To a large extent, substrate depth dictates vegetation diversity and the range of possible species. Shallow substrate depths between 2 and 5cm have more rapid rates of desiccation and are more subject to fluctuations in temperature, but can support simple *Sedum*-moss communities. Substrate depths of 7 to 15cm can support more diverse mixtures of grasses, geophytes, alpines, and drought-tolerant herbaceous perennials, but are also more hospitable for undesirable weeds.

Green-roof substrates tend to be highly mineral based, with small amounts of organic matter (approximately 10% by weight). The mineral component may come from a variety of sources, and can be of varying weight depending on the load capacity of the roof. Light expanded clay granules and crushed brick are two common materials. There is increasing interest in the use of locally derived lightweight granular waste materials as sustainable sources for green-roof substrates.

Climatic conditions, especially rainfall and extreme temperatures, may restrict the use of certain species or dictate the use of irrigation. Native plants are generally considered ideal choices for landscapes because of their adaptations to local climates, and the native stress-tolerant floras (particularly dry grassland, coastal, and alpine floras) of many regions offer opportunities for trial and experiment. Furthermore, policies for biodiversity and nature conservation may favor the establishment of locally distinctive and representative plant communities. Unfortunately, many native plants appear to be unsuitable for conventional extensive green-roof systems because of the roofs' harsh environmental conditions and typically shallow substrate depths. In a study at Michigan State University, only 4 of the original 18 native prairie perennial species growing in 10cm of substrate persisted after three years. In comparison, all 9 nonnative species of *Sedum* used in the study thrived.

In theory, almost any plant taxon could be used for green-roof applications, assuming it is suited to the climatic region, grown in appropriate substrate at an adequate depth, and given adequate irrigation. Wind stress resulting from building height and form may affect the selection of plants, and visibility and accessibility are other selection criteria. Although

Sedum remains the most commonly used genus for green roofs, the scope for green-roof vegetation is wide, and many possibilities have yet to be realized.

3. Ecosystem Services Provided by Green Roofs

The green-roof benefits investigated to date fall into three main categories: storm-water management, energy conservation, and urban habitat provision. These ecosystem services derive from three main components of the living roof system: vegetation, substrate (growing medium), and membranes. Plants shade the roof surface and transpire water, cooling and transporting water back into the atmosphere. The growing medium is essential for plant growth but also contributes to the retention of storm water. The membranes are responsible for waterproofing the roof and preventing roof penetration by roots.

Storm-water Management

Urban areas are dominated by hard, nonporous surfaces that contribute to heavy runoff, which can overburden existing storm-water management facilities and cause combined sewage overflow into lakes and rivers. In addition to exacerbating flooding, erosion, and sedimentation, urban runoff is also high in pollutants such as pesticides and petroleum residues, which harm wildlife habitats and contaminate drinking supplies.

Conventional storm-water management techniques include storage reservoirs and ponds, constructed wetlands, and sand filters; however, these surface-area intensive technologies may be difficult to implement in dense urban centers. Green roofs are ideal for urban storm-water management because they make use of existing roof space and prevent runoff before it leaves the lot. Green roofs store water during rainfall events, delaying runoff until after peak rainfall and returning precipitation to the atmosphere through evapotranspiration. The depth of substrate, the slope of the roof, the type of plant community, and rainfall patterns affect the rate of runoff. Studies in Portland, Oregon, and East Lansing, Michigan, showed that rainfall retention from specific green roofs was 66% to 69% for roofs with more than 10cm of substrate. Rainfall retention varied from 25% to 100% for shallower substrates in other studies. Green roofs can reduce annual total building runoff by as much as 60% to 79%, and estimates based on 10% green-roof coverage suggest that they can reduce overall regional runoff by about 2.7%. In general, total runoff is greater with shallower substrate and steeper slopes. Although green roofs can reduce runoff, they do not solve the problem of reduced recharge of groundwater in urban areas.

Improving Roof Membrane Longevity

Waterproofing membranes on conventional dark roofs deteriorate rapidly in ultraviolet (UV) light, which causes the membranes to become brittle. Such membranes are consequently more easily damaged by the expansion and contraction caused by widely fluctuating roof temperatures. By physically protecting against UV light and reducing temperature fluctuations, green roofs extend the life span of the roof's waterproofing membrane and improve building energy conservation. Temperature stabilization of the waterproofing membranes by green-roof coverage may extend their useful life by more than 20 years; some green roofs in Berlin have lasted 90 years without needing major repairs. In Ottawa, Canada, Liu (2004) found that an unvegetated reference roof reached temperatures higher than 70℃ in summer, while the surface temperature of the green roof only reached 30℃. The membrane on the reference roof reached 30℃ on 342 of the 660 days of the study, whereas the membrane underneath the green roof only reached that temperature on 18 days.

Summer Cooling

During warm weather, green roofs reduce the amount of heat transferred through the roof, thereby lowering the energy demands of the building's cooling system. Wong and colleagues (2003) found that the heat transfer through a green roof in Singapore over a typical day was less than 10% of that of a reference roof. Research in Japan found reductions in heat flux on the order of 50% per year, and work in Ottawa found a 95% reduction in annual heat gain. A study in Madrid showed that a green roof reduced the cooling load on an eight-story residential building by 6% during the summer. In a peak demand simulation, the cooling load was reduced by 10% for the entire building and by 25%, 9%, 2%, and 1% for the four floors immediately below the green roof. For a typical residential house in Toronto, the cooling load for the month of July was reduced by 25% for the building and by 60% for the floor below the green roof. Green roofs will have the greatest effect on energy consumption for buildings with relatively high roof-to-wall area ratios.

In the summer, green roofs reduce heat flux through the roof by promoting evapotranspiration, physically shading the roof, and increasing the insulation and thermal mass. Gaffin and colleagues applied energy-balance models to determine how effectively green roofs evaporate and transpire water vapor compared with other vegetated surfaces. During the summer of 2002, experimental green roofs at Pennsylvania State University performed equivalently to irrigated or wet habitats, indicating that evapotranspiration may be the most important contributor toward reducing summer building energy consumption under green roofs. Of course, green roofs are not the only technology that can provide summer cooling; enhanced insulation may be able to provide equivalent energy savings and can be combined with green roofs to further advantage. Evaporative roofs are another example of such a technology; water is sprayed on the roof surface to induce evaporative cooling. Rigorous comparisons of multiple roofing systems are necessary to evaluate prospects for optimal building energy savings.

Urban Heat Island

In urban environments, vegetation has largely been replaced by dark and impervious surfaces (e.g., asphalt roads and roofs). These conditions contribute to an urban heat island, wherein urban regions are significantly warmer than surrounding suburban and rural areas, especially at night. This effect can be reduced by increasing albedo (the reflection of incoming radiation away from a surface) or by increasing vegetation cover with sufficient soil moisture for evapotranspiration. A regional simulation model using 50% green-roof coverage distributed evenly throughout Toronto showed temperature reductions as great as 2℃ in some areas.

Urban Habitat Values

Green-roof habitats show promise for contributing to local habitat conservation. Studies have documented invertebrate and avian communities on a variety of living-roof types in several countries. Green roofs are commonly inhabited by various insects, including beetles, ants, bugs, flies, bees, spiders, and leafhoppers. Rare and uncommon species of beetles and spiders have also been recorded on green roofs. Species richness in spider and beetle populations on green roofs is positively correlated with plant species richness and topographic variability. Green roofs have also been used by nesting birds and native avian communities. Rare plants and lichens often establish spontaneously on older roofs as well. These findings have mobilized local and national conservation organizations to promote green-roof habitat, particularly in Switzerland and the United Kingdom. Furthermore, these results have encouraged discussion of green-roof design strategies to maximize biodiversity.

Living roofs also provide aesthetic and psychological benefits for people in urban areas. Even when green roofs are only accessible as visual relief, the benefits may include relaxation and restoration, which can improve human health. Other uses for green roofs include urban agriculture: food production can provide economic and educational benefits to urban dwellers. Living roofs also reduce sound pollution by absorbing sound waves outside buildings and preventing inward transmission.

Community and Landscape Properties

How important is the living portion of green roofs to their functioning? Although plants are an important component of green roofs, recent work shows that the growing medium alone can greatly reduce runoff from a green roof. The medium alone reduced runoff by approximately 50% in comparison with a conventional gravel roof; adding vegetation to the medium resulted in negligible further reductions. Other research shows that the depth of the growing medium is the main determinant of runoff retention. However, water availability and season affect the ability of the growing medium to retain water. When water is readily available, evapotranspiration rates are much greater on vegetated roofs than on roofs with growing medium alone, especially in the summer. Complicating our understanding of green-roof functions is the shading of the roof surface by vegetation, which may reduce evaporation from the soil surface.

With respect to thermal benefits, simulation models show that taller vegetation leads to greater thermal benefits in tropical environments, but these models do not separate the additive effects of soil and vegetation. Experiments on green roofs suggest that most of the summer cooling benefits from green roofs are attributable to evapotranspiration, but the relative contributions of vegetation and substrates cannot be separated out by these analyses. A study using small-scale constructed models showed that reductions to heat flux through the roof at peak daily temperatures were greater in vegetated soil roofs than in soil roofs alone, with 70% of the maximum reduction attributable to the soil and the remainder to the vegetation. Therefore, transpiration from living plants is most likely responsible for a substantial proportion of the cooling benefits of green roofs, and that proportion could be boosted further by selecting species with high leaf conductivity or large surface areas.

Two properties of plant communities can influence green-roof performance: the ability to resist and recover from environmental fluctuations or disturbances, and the rate at which resources can be consumed. Using vegetation types that recover more rapidly from disturbance should increase the duration of functions made possible by living plants, such as transpiration. Greater resource use in green roofs should reduce runoff of water and nutrients. High species diversity is expected to encourage more complete resource use and greater biomass constancy within the growing season.

Very simple communities of low species diversity may be vulnerable to environmental fluctuations, but the notion that more species are inevitably better is not always tenable. Most of the functions of vegetation are dictated by the performance of dominant plant species, and these are likely to be relatively few in number. Although the promotion of native species and communities may be important for conservation, experimental evidence indicates that the functional, structural, and phenological properties of vegetation are more important than "nativeness" in promoting invertebrate biodiversity and other community attributes in level-ground urban gardens. In an experiment involving vegetation similar to that of extensive green roofs, there was no relationship between species diversity and water retention, but a diversity of functional types (e. g., rosette formers and grasses, as opposed to monocultures of either) was crucial to maximizing performance. Work by Kolb and Schwarz indicates that vegetation including diverse functional types has a greater positive influence on the thermal properties of green roofs than monocultural types of vegetation.

The limited size of green roofs as habitats has implications for the biodiversity and landscape properties of areas in which green roofs are installed. Little is known about the relationships between roof area, which may range from approximately 1 to 40000m^2 for an individual roof, and the habitat occupation rates of different taxa. At least two questions still need to be addressed: What are the relationships between other green-roof ecosystem services and roof area, and how do regional benefits relate to the landscape configuration of green-roof patches in urban areas? In summary, green-roof benefits are partially derived from the living components of the system, but more research is needed in determining the relationships between biotic community parameters and ecosystem functioning, with a view toward selecting biotic components that can improve green-roof performance.

Vocabulary and Phrases

albedo n. [物] 反照率
alpine n. [生] 高山植物
architect n. 建筑师
asphalt n. 沥青
avian n. 鸟，adj. 鸟的，来自于鸟的
avoidance n. 逃避
beetle n. [生] 甲虫
brittle adj. 易碎的，脆弱的
configuration n. 配置，格局
crassulacean acid 景天酸
desiccation n. 干燥
deteriorate v. 使恶化，使退化
dissipate v. 消散，分散
dictate v. 主导，支配
drought-tolerance 耐旱性
exacerbate v. 加剧
expectation n. 期望，愿望
evapotranspiration n. 蒸散
geophyte n. [生] 地下芽植物
granular n. 石块，砾石
harsh adj. 苛刻的，恶劣的
herbaceous n. 草本
horticulture n. 园艺

impervious adj. 不透水的
insulation n. 隔离，绝缘
invertebrate n. [生] 无脊椎动物
leafhopper n. [生] 蝉
lichen n. [植] 地衣
longevity n. 长寿
metabolism n. [生] 代谢
monoculture n. [农] 单作
perennial adj. 多年生的
phenological adj. 物候的
prairie n. [地] 干草原
psychological adj. 心里的，心理学的
revitalize v. 使更新，使恢复
sedum species 景天科植物
strategy n. 策略，对策
substrate n. 基质，介质
succulent adj. 肉质的，多汁的
surficial property 表面特征
taxon n. [生] 分类单元，复数为 taxa
ruderal n. 路旁杂草
underpin v. 加强....基础
waterproofing adj. 防水的
wonder n. 奇迹

Note

hanging gardens 空中花园，传说中在古巴比伦建成时最大、最美的园林景观。
moisture stress 水分胁迫，指水分不足的条件下，植物出现的受旱的现象。
stress-tolerant 胁迫耐性，只生物忍受胁迫的能力。生物对各种胁迫的适应有两种方式，一种是抗性，另一种是忍耐能力。

Question and Exercises

1. Which functions did the greening roofs provide?
2. What are the characteristics of green-roof vegetation?
3. If you plan to green roof, what do you consider?

Unit 28 The Ecological City: Metaphor versus Metabolism

1. Introduction

The concept of the ecological city has a long history within western theories of urbanism: from the notion of the perfect city as the embodiment of the perfect body (Renaissance) to the city as diseased body (19th Century) to urban sociology's empiricist turn to science to describe the effects of modernization on both urban inhabitants and urban form (20th Century). In all of these conceptual frameworks, natural processes were embraced as metaphors for the city rather than as its underlying metabolism. In the 20th century both urban sociologists and planners used scientific signifiers to describe urban processes in an attempt to bring the study of the city into a value system controlled by the assumed objectivity and rationality of science. Many of these values remain imbedded in contemporary urban planning models, even as modifications are made to incorporate notions of sustainability and to reincorporate the image of the historic city within the new city as a signifier for a sustainable city.

Within the last two decades this combination of modernization and historicism has had significant impact on the explosive growth of Chinese cities. Until the 1960s, when an increased awareness of the actual biological, geological, and hydrological systems that are at play in all cities began to influence urban thinking, all uses of the term "ecology" with reference to urban environments were metaphoric. The essays and case studies in The City, an important foundational text for urban sociology, built an argument for linking human behavior to an urban ecology that could be described through processes of extension and succession, concentration and decentralization. One can "think of urban growth as a resultant of organization and disorganization analogous to the anabolic and katabolic processes of metabolism in the body", Ernest Burgess, wrote in this influential volume. The actual metabolic processes of the city were distinct from the formal rationale of its design, a split that was heightened in the modern city as these metabolic functions were increasingly handled by technologically complex infrastructures. Over the past two to three decades, a way of thinking about the ecological processes of the city—as distinct from the city as an ecology—has been absorbed into a larger concept of the "sustainable" city. The idea of the sustainable city assigns ecology a role alongside other urban systems, be they economic, cultural, social, or political. Ideally, a sustainable city is one in which conservation of natural resources and environments are balanced with economic viability. However, like all open, evolving systems, the sustainable city is not without its conflicts. Of particular note are the conflicts that arise between economic goals and cultural traditions and developmental imperatives and existing ecologies. The recent pace and scale of development of Chinese cities, bring these conflicts to light; as a recent New York Times headline read: "Red China or Green?"

implying the extensive debate over pollution (both internal and exported) and sustainable development attendant to Chinese urbanization and modernization.

The explosive growth of Chinese mega-cities such as Beijing, Shanghai, Shenyang, and Wuhan has called attention to the problems of massive human migration, pollution, and the loss of arable land. These are noted daily in both the Western and Chinese press, with a significant amount of finger pointing on both sides. However, the effects of rapid urban development can be seen throughout China in its large cities, rural regions, and small, yet growing cities. For my case study I will take as my examples several recent projects in the city of Shunde in the Guangdong province. Shunde is one of a large network of cities that together interact to form the agricultural and industrial base of the Pearl River Delta, historically united by water, and now brought together via advanced technological connections, highways, and, in the future, light-rail. Shunde itself is not a megacity; currently it operates as a district of Foshan, (pop. 3.8 million, 2004) itself a satellite of Guangzhou. Nor, however, is it "rural" in the strict sense of the word. Its population, somewhere above one million and growing (not including its floating population), is comparable to the North American city of Phoenix (pop. 1.3 million, 2006). As such, it poses a set of questions that are more common to urban development around the country than the over-generalized analyses of mega-city growth or the critique of specific projects such as the Beijing Olympic Village or the Pudong in Shanghai, around which many Western discussions of Asian urbanism are built. Despite its massive industrial, agricultural, and horticultural output including furniture, refrigerators, and flowers, Shunde is relatively "unseen" by the global economy of which it is a part.

2. Shunde Urbanization

Shunde's growth can be understood in the context of its location and its history. Its origin dates to the 15th century; its economy has been based on agriculture and manufacture tied to its location in the water-based ecology and economy of the Delta. Shunde's vernacular architecture and urban patterns have two primary characteristics: a loose but dense network of fish and silk farms scattered throughout the water-based landscape and dense urban development following the contours of a mountainous topography further inland. Chosen as a "pilot" city in the 1980s, Shunde has rapidly modernized along these historic lines requiring new industrial zones, technical schools, residential areas, and administrative capacities. Further, as part of a large network of growing cities, Shunde's government and planners perceive the city's future to be directly tied to its ability to compete for both local and global recognition and sales of its products. As one city development brochure puts it: "creating the future with urbanization." This future, however, still has ties to the past; city leaders describe their methodology for growth as "being flexible and dexterous like the water," recognizing that the city's foundations sit within "the earliest artificial ecological system in human history." As the urban designer and academic Richard Marshall has noted with regard to Asian projects designed to allow cities to compete globally: "Among other things these projects provide two very important global advantages to their host locations. First they provide a particular type of urban environment where the work of globalization gets done and second, they provide a specific kind of global image that can be marketed in the global market place. " This is true even in the relatively small city of Shunde, which promotes itself both on the basis of its industrial strengths and the image of its green environment. The question: is ecology a component of the city's efforts toward sustainability or is it an artifact?

Shunde's "urbanization" is situated throughout its metropolitan region rather than being specifically based on expansion from its historic center. Indeed, rather than expanding the "Old City" Shunde has built the infrastructure for and begun to develop a "New City" closer to its river waterfronts and new industrial areas. Although there is a great deal of western-style development in its Old City - tall buildings, shopping malls, hotels, McDonalds, KFC-most development is occurring within the delta-landscape. This "desakota" region is "characterized by an intense mix of agricultural and non-agricultural activities that often stretch along corridors between large city cores. " This "patchwork" urbanization, as Marshall describes it, creates "a thick band of ambiguous fuzziness that denotes the transition from one to another-neither wholly urban nor wholly rural, but something new entirely. "

Marshall's description is apt, but it misses two aspects of the scenario as it pertains to places such as Shunde. First, there are many moments when the physical space created by development is anything but blurry. The superimposition of a 300m×300m urban grid over fish farms to create the New City, for instance, illustrates a jarring clash of old and new landscape. One can still be seen inside the other, and often, where one system ends, the other suddenly begins. Second, the desakota landscape may be an inherent developmental condition of the region created by the local ecology itself in which land and water, urban and rural have always been of the same piece. The desakota may not be solely a development of the late twentieth century. Although the phrase describes a condition in distinct contrast to western models of urban development such as that described by Burgess in his famous diagram of city-centered growth, the "village"/"town" dynamic seems inherent in the historic development of the region, in which land and water have been continually modified to promote agriculture, production and trade. However, now the scale, technology, and methodology of building have shifted dramatically, layering modern building and infrastructure on top of the region's ecological attributes through a massive reshaping of water and land. When looking at maps, planning documents, and drawings, what is always unclear is the accuracy of original conditions, what has been built versus merely planned, and what plan is currently in place.

3. Landscape Measure

In Shunde development is taking place on two tracks: the building of the pieces necessary for globalized competition and the promotion of local traditions. With globalization comes not just factories but the infrastructure to support them: roads, highways, water management, and the New City, too large to be accommodated within the old center and, therefore, necessitating the large-scale conversion of agricultural land (much of it water) to hardscape and modern buildings. Preservation is also a developmental imperative necessitating the survival of pieces of rural lifestyle and the building of new sites to celebrate history and tradition.

The Hong Kong-based urbanists Laurent Gutierrez and Valerie Portefaix aptly describe the geography of the Delta as the situation and characterization of its urban form: floating. "It is very natural that these small villages developed from an agricultural production to an industrialized economy. The multiplication of rural enterprises, small towns and transport infrastructure are the main determinants of today's organization. This transition not only involved farming activity but also the size of plots and accessibility to water. This also extends to the erasure of surrounding peaks to fill ponds and create artificial planes that accommodate a growing infrastructure and urban settlements."

The erasure of cultural traditions, for instance, family farming and manufacture (including fishing and silk), canal towns, family temples, and local Lingnan organization incorporating buildings and water, has been brought about through massive reorganization of the landscape, particularly water and land. Gutierrez and Portefaix's description of urbanization in the PRD explains the method by which the landscape of the water-based agricultural region is being remade into a plane for industrial growth (including industrialized agriculture). And planners in Shunde are extremely aware of the consequences as they describe the price of development to be the loss of local culture and extreme pollution. This "loss" informs both the form and style of contemporary developments and an increasing interest in ecological and heritage tourism. This turn toward tourism is understood as both a way to preserve traditional culture and promote the city for future economic development. In this dynamic, preservation is an after-effect and made possible by the speed of growth and expansion of wealth in the city.

4. Water

Water is at the center of all development in Shunde and the central feature of its ecology, so it is not surprising to find it at the center of questions of both sustainability and preservation. The villages of the region are organized around canals and centered on the relationship of fish farming for sustenance and mulberry bushes for silk production. The canals were the primary circulation method with alleys opening off them. They also served to connect the villages to larger towns and trading networks for their products. Although no longer significant to the region's industrial economy, preservation of the villages and restoration of their family temples have become an essential feature of heritage tourism. One piece of Shunde's promotional material reads: "Stepping into Fengjian village, you will see the setting sun turning the Mulberry fish ponds into red, the reticulated watercourses, the hit-and-miss cockleboats, the banana forest on the riverside, the blue stone plate roads leading to the back roads."

The region's canal-focused villages illustrate a form of family-based, communal living that remains extent in many new residential projects but without the integration of landscape and production. As family farming and industry have shifted to industrial-scaled agriculture and horticulture and the region's productive capabilities have moved into industrial and technology parks, the nuanced integration of living and working around water has disappeared. The future economy of the villages is tourism, particularly as younger inhabitants move to jobs and residential projects in and around the city.

However, water remains a centerpiece of residential developments, albeit stripped of its functions as resource and connector. The role of water as a symbol of the region's ecology is present in almost every new development, but particularly in residential projects where water courses through artificial streams, canals, ponds, fountains, and play areas. Here water is image rather than environment, a stand in for ecology. Separated from productive functions, its "artificiality" is of a different nature than the artificial ecology of the aquaculture built within rather than on top of the delta. This is not to say, however, that there is not an attentiveness to the sustainability of the water itself. But the style of the architecture and landscape is distinct from the functions of the water, a metaphor of ecology and an image of sustainability.

Water is also a strong element in the design of the New City, but largely as a formal element within the grid. Large canals organize two axes, one focused on a new city hall and civic plaza, the other a cross axis leading to the Gui Pan River, anticipating housing and commercial development at this edge. Water here serves to augment the monumentality of the district's open spaces, roads, and buildings, but does little to relieve the excessive scale of roadbeds designed for anticipated traffic in the final build-out.

5. Land

The regional ecology of the Pearl River Delta may have always been a patchwork, but today's patches are of a size and scale out of touch with traditional forms. New projects may incorporate water elements into their landscape themes, but typically to the detriment of the existing water ecology. It is here that the greatest effects of Western-style modernization can be seen in the wholesale erasure of traditional landscapes and agricultural methods. To create the new ground plane for industrial agriculture, industrial and technology parks, and entirely new cities, large amounts of the delta must be filled. The effect of this filling is evident, not only in the loss of fish farms, but also in the denuding and carving out of the area's hills and mountains. An overlay of new and planned projects on top of the pre-existing landscape around the Gui Pan River illustrates the extent of this process. Development rarely follows the contours of the hills or former canals and dikes. The new polytechnic, with its own dramatic and meandering water landscape runs up against the existing water systems without acknowledging them, probably in anticipation of further development. Each block of the New City swallows numerous fish ponds with little attention to existing land divisions or waterways. The inhabitants of old villages are relocated to new villages in leftover spaces between new developments. Existing hills are scoured to provide new fill. Green elements—trees, bushes, flowers, and grass—are abundant, lining every new road and highway requiring constant watering by water tankers throughout the cool night and tending by manual labor throughout the hot day. The process is a new form of tabula rasa development in which land is created to provide a flat, fresh plane for development and history has to be rebuilt. In contrast to the New City with its axial organization, large-scale grid, and neo-historicist and neo-modern architecture is the Shunfeng Mountain Park, equal if not larger in size than the New City. Lying between the New and Old Cities, the park, its landscape features, and architectural elements form a new history for the city out of fragments of the past. Here the land is excavated rather than filled (one can assume excavated material is used in nearby development) to create a lake that forms the centerpiece of a series of experiences that reference, and in some cases appropriate, historic garden motifs. One of the first pieces to open in the park, the Baolin Temple presents a case in point. It is reconstructed from pieces of a tenth century temple relocated from a nearby hill to serve as a ceremonial link between city and mountain. Passing through its rooms filled with Buddhas and souvenir shops, one arrives to the top to gain a distant view of the lake, bridges, pagodas, and an array of passive recreational spaces, but also a forest of electrical transmission lines and new developments in the vast region beyond.

Shunfeng Mountain Park and Baolin Temple present a number of questions to a Westerner. At first glance they are an amalgam of Chinese references including Behai Park and the Summer Palace in Beijing, of a completely different scale and design tradition than the Qinghui Garden (now also renovated and expanded), an important historic artifact in the Old City. Methodologically, they appear to be Western, aligned with Haussmann and Alphand's creation of the Parc des Buttes-Chaumont in nineteenth-century Paris, in which a quarry was converted into a park through the invention of a new landscape and history. Like the building of Central Park in New York, also in the Nineteenth-Century, old uses were moved out to create the new landscaped park. In these instances existing water patterns were occasionally maintained, although "renaturalized" through the addition of artificial elements that augmented their appearance. In what ways do these modern methodologies intersect with Chinese traditions? The landscape theorist Stanislaus Fung noting that "Chinese buildings continue to be caught in a perpetual cycle of building and rebuilding," suggests that the physical survival of buildings in China is not an integral part of urban "identity." Shunfeng Mountain Park creates a vast environmental panorama at the same time that it is an indicator of the city's need for attractive, ecologically mindful open space. Is this spectacular new park a means of extending the identity of the region or creating a new one through the appropriation of a past at a distance? Like the preservation of the canal towns, the Park and the New City use the past as a way toward the future but in a way that speaks to stylistic rather than cultural traditions, a form of cultural, social, and economic, if not ecological sustainability.

Sustainability—economic, cultural, social, political and ecological—must be a guiding theme of contemporary and future development of both historic places and vast new urban regions. Are eco-and heritage-tourism sufficient means to support cultural preservation? How do we understand ecology within the context of urban development in such a way as to truly preserve natural resources and landscape systems, agricultural and communal lifestyles, and housing traditions and culturally significant spaces while still addressing contemporary urban needs? In the unfolding urbanization of China, ecology and the larger issue of sustainability will have to be understood as more than "greening". This will necessitate a conceptual move from the metaphoric to the metabolic, a true integration of cultural tradition, regional ecological systems, and economic globalization.

The questions raised in this paper are meant to point to areas for further investigation not as indictments. The city and people of Shunde are living the experiment of "creating a future with urbanization" where "the future lies in the past", a difficult task that begins to take into account the need to balance economy, culture, and ecology. Stanislaus Fung in his reading of the traditional text on Chinese Gardens, Yuan ye, within both a traditional Chinese and a Western philosophical context offers one way of thinking through this dilemma: "The person who notices borrowing is not a universal subject; the moment when borrowing is noticed is not just happenstance or undetermined. Rather, the borrowing of views is discussed in Yuan ye as eventful encounter and depends on the notion of tradition, here conceived not as a tradition of stylized or designed objects but as embodied practices of daily living." Here Fung is speaking to the specific experiences within a Chinese garden, however his focus on "embodied practices of daily living" suggests a connection between the social and the ecological. His analysis also suggests a methodology for conceiving a sustainable, regional urbanism in the Pearl River Delta, building identity out of the land and its traditions.

Vocabulary and Phrases

albeit *conj.* 尽管，虽然，同though
amalgam *n.* 汞合金，汞剂
anabolic *adj.* [生] 合成代谢的
analogous *adj.* 类似的，相似的
aquaculture *n.* 水产养殖
artificiality *n.* 人工制造的物品

blurry *n.* 模糊，*adj.* 模糊的，隐晦的
brochure *n.* 小册子
Buddha *n.* 佛，菩萨
centerpiece *n.* 核心
civic plaza 文娱广场
clash *n.* 冲突，*v.* 抵触，磕碰

communal living 社区生活
critique n. 批判，批评
delta-landscape n. 河流三角洲景观
denude v. 剥蚀，剥夺
dexterous adj. 灵巧的，敏捷的
disorganization n. 解体
erasure v. 清除，抹掉
floating population 流动人口
historicism n. 历史循环论，传统崇拜，复古主义（尤指建筑风格）
incorporate v. 包括，纳入
indictment n. 起诉，诉状
influential adj. 有影响的，有权势的
jarring n. 冲突，不协调；adj. 不和谐的
katabolic adj. [生] 分解代谢的，异化的；同 catabolic
happenstance n. 偶然事件
imbodiment n. 体现，具体化
imbody v. 体现，具体化
meander n. 蜿蜒，v. 蜿蜒，扭曲，漫步
metaphor n. 比喻，隐喻
monumentality n. 纪念碑，纪念雕塑
necessitate v. 必要，必须
nuance v. 形成细微差别
pagoda n. 佛塔
patchwork n. 东拼西凑，拼凑物
panorama n. 全景
pilot city 试点城市
quarry n. 采石场
Renaissance n. 文艺复兴
satellite n. 卫星，卫星城
souvenir n. 纪念品
swallow v. 吞噬
tabula rasa 白板，空白状态
vernacular adj. 本地的，方言的
waterfront n. 靠水的地方，江边，海边

Note

Pearl River Delta，珠江三角洲。

Desakota，模式，1987年，McGee在研究亚洲发展中国家的城市化问题时，发现了一类分布在大城市之间的交通走廊地带，与城市相互作用强烈、劳动密集型的工业、服务业和其他非农产业增长迅速的原乡村地区——他借用印尼语将其称作为"Desakota"（desa即乡村，kota即城镇）。McGee认为，许多亚洲国家并未重复西方国家通过人口和经济社会活动向城市集中，城市和乡村之间存在显著差别，并以城市为基础的城市化过程（city-based urbanization）。而是通过乡村地区逐步向"Desakota"转化，非农人口和非农经济活动在"Desakota"集中，从而实现以区域为基础的城市化过程（regional basic urbanization）。在我国经常将"Desakota"译作城乡一体化区域。

Question and Exercises

1. Please give an example of eco-city.
2. For an eco-city, which factor is the most important?
3. Which is the Desakota model?

Further Reading

Zero-Carbon Cities

Once upon a time, when the world's population was a fraction of the 6.5 billion it is today, environmental issues were thought of as local problems. Writers, politicians, scientists, and activists have recorded the polluted, disease-producing conditions of urban centers for centuries. Benjamin Franklin petitioned the Pennsylvania Assembly in 1739 to stop dumping waste and remove tanneries from Philadelphia's commercial district, citing foul odors, lower property values, and disease. And yet, even the proto-environmentalist Franklin could not predict that centuries of local industrial recklessness would one day endanger the entire planet.

In 2007, environmental issues are literally global in that the carbon emissions of every industrialized country have accumulated to create the present damaging climate changes. From Al Gore's 2006 documentary An Inconvenient Truth to the release last month of "Climate Change 2007," a six-year study about global warming from the Intergovernmental Panel on Climate Change, there seems to be little doubt that the by-products of industrialization—greenhouse gases, particularly carbon dioxide (CO_2)—are responsible.

Even before the latest reports, a new model for confronting global warming has gained traction. The new model, variously referred to as zero-carbon, carbon-neutral, or fossil-free development, is a revolutionary attempt to stop and, where possible, reverse the damage. Whereas there are currently no such zero-carbon cities in the industrialized world, an experiment is unfolding that attempts to create a carbon-neutral city from scratch and provide a prototype for the future of all cities in one of the world's most environmentally distressed countries, China.

1. China Syndrome

Adjacent to booming Shanghai, plans are coming together for the prototypical city of Dongtan, which designers argue would be the world's first truly sustainable new urban development. London-based Arup and the Shanghai Industrial Investment Corporation (SIIC), the city's investment branch, have partnered to create a master plan for Dongtan, an area three quarters the size of Manhattan. The brief calls for integrated sustainable urban planning and design to create a city as close to carbon-neutral as possible within economic constraints. Located in sensitive wetlands on Chongming Island at the mouth of the Yangtze River just north of Shanghai, Dongtan's first phase, a marina village of 20000 inhabitants, will be unveiled at the 2010 World Expo in Shanghai. By 2020, nearly 80000 people are expected to inhabit the city's environmentally sustainable neighborhoods.

As a strategic partner, Arup is responsible for a range of services, including urban design, sustainable energy management, waste management, renewable energy process implementation, architecture, infrastructure, and even the planning of communities and social structures. Peter Head, director of Arup's sustainable urban design, leads the project for the firm from its London's office (during design, Arup is offsetting the emissions of its team's travel to and from the site in cooperation with emissions brokerage firm CO_2). "Renewable energy will be used to reduce particulate CO_2 emissions. Transport vehicles will run on batteries or hydrogen-fuel cells and not use any diesel or petrol, creating a relatively quiet city," he explains. Other priorities include recycling organic waste to reduce landfills and generate clean energy.

Head insists development won't affect the wetlands. "First of all, water usually discharged into the river will be collected, treated, and recycled within the city boundaries," he says. "There will be a 2-mile buffer zone of eco-farm between city development and the wetlands." While farming is water intensive, relatively small amounts of water reach the plants themselves. Head says Dongtan "will capture and recycle water in the city and use recycled water to grow green vegetables hydroponically. This makes the whole water cycle much more efficient."

There are, of course, questions about what kind of sustainable industries will provide jobs for Dongtan residents. City officials and their consultants anticipate jobs in education, including an Institute for Sustainable Cities. They expect to attract companies pursuing new technologies, food research and production, and health care. Of course, ecotourism will become a significant industry. The plans for the development have already garnered their fair share of press in the design world.

2. Dongtan's Ecological Footprint

The Stockholm Environment Institute (SEI) describes the global environmental imbalance succinctly: "Sustainability requires living within the regenerative capacity of the planet. Currently, human demand on the planet is exceeding the planet's regenerative capacity by about 20 percent. This is called 'overshoot.'"

Arup is developing an ideal ecological footprint for Dongtan to guide the master plan and prevent overshoot. The new city's ecological footprint will be determined by a modeling program called the Resources and Energy Analysis Program (REAP), developed by SEI and the Center for Urban and Regional Ecology at the University of Manchester. Unlike the traditional focus on air and water pollution, REAP concentrates on measuring the amount of resources consumed by the number of individuals occupying a defined area.

The best ecological footprint, of course, balances (nature's) supply and (human) demand, which is the goal at Dongtan. Head and his team are using REAP to determine the effectiveness of their planning decisions in achieving sustainability, as well as Arup's own sustainable design assessment tool, the Sustainable Project Appraisal Routine (SPeAR). The footprint will reveal how much productive land and sea is needed to provide the energy, food, and materials for daily consumption, and how much land is required to absorb human-generated waste. The program also calculates the emissions generated from the oil, coal, and gas burned, and determines how much land, air, and water is required to disperse these emissions.

Ecological footprinting is not without critics. Some experts argue that by applying generalizations and averages to per capita analysis, while not accounting for multiple usages of the same land, for instance, oversimplifies the conclusions. SEI claims that REAP is "the only software tool that can convert household (and commercial) expenditure at the national level into its associated environmental impact," rendering its forecasts more accurate at the product level with the inclusion of "bottom-up" data.

The U. K. has a short, but intense history of ecological footprinting. For years prior to Dongtan, British architect Bill Dunster and Arup had pursued the viability of harnessing renewable resources, achieving closed-loop material use, and creating site-resource autonomy. In 1999, the Peabody Trust, one of London's largest affordable housing providers, selected Arup, Bill Dunster Architects (since renamed BDa ZEDfactory), and the BioRegional Development Group to test their ideas creating the Beddington Zero (fossil) Energy Development (BedZED). The goal of the 83-home, mixed-income housing development in South London was to prove that carbon-neutral projects were cost-effective, practical, and therefore ready for the mainstream marketplace.

For the Dongtan team, BedZED's most important lesson was its holistic design approach, which eliminated obsolete systems from the beginning, rather than tacking on sustainability features in an effort to increase performance. Arup is applying total design integration to the project through a series of design specifications for architects. While construction on infrastructure is slated to begin this year, no architects have been engaged for the building program.

3. Think Globally, Act Locally

Innovation should not be conceded entirely to the U. K. or Europe. Progress can be seen domestically at the local and state levels, as federal guidelines are now being fortified with legislation. With much fanfare, Governor Arnold Schwarzenegger of California recently established a groundbreaking Low Carbon Fuel Standard (LCFS) for transportation fuels sold in the state. Elsewhere, cities are concentrating on buildings and land use, confronting the fact that U. S. buildings annually consume 43 percent of the nation's energy and generate 35 percent of its CO_2 emissions.

Nine years ago, Austin, Texas, recognized the impact that residential construction was having on the local environment and established a Green Building Program to provide financial incentives—not just voluntary guidelines—for all residential, commercial, and multifamily projects. Austin Energy, the city's community-owned utility, has developed renewable energy sources, including wind turbines, landfill methane gas recovery projects, and solar energy sites. Seattle's public electricity utility has adopted a policy of zero-net greenhouse gas emissions by selling its stake in a coal-fired steam plant and mitigating emissions from its remaining fossil-fuel plant. These aren't new strategies, but they are now being widely embraced by a design community that is more engaged in the bigger regional picture of each individual architectural project.

A zero-carbon nation will undoubtedly take more generations to achieve, but the political will to do so is gaining momentum and having Dongtan as a model will certainly help matters. This growing interest reveals a commitment to long-term goals, which has effectively been missing since Benjamin Franklin argued that the public has a right to live in environmentally healthy communities.

Appendix 1

The Science of Scientific Writing

George D. Gopen and Judith A. Swan
American Scientist, 1990, 78: 550-558

1. *If the reader is to grasp what the writer means, the writer must understand what the reader needs.*

Science is often hard to read. Most people assume that its difficulties are born out of necessity out of the extreme complexity of scientific concepts, data and analysis. We argue here that complexity of thought need not lead to impenetrability of expression; we demonstrate a number of rhetorical principles that can produce clarity in communication without oversimplifying scientific issues. The results are substantive, not merely cosmetic: Improving the quality of writing actually improves the quality of thought. The fundamental purpose of scientific discourse is not the mere presentation of information and thought, but rather its actual communication. It does not matter how pleased an author might be to have converted all the right data into sentences and paragraphs; it matters only whether a large majority of the reading audience accurately perceives what the author had in mind. Therefore, in order to understand how best to improve writing, we would do well to understand better how readers go about reading. Such an understanding has recently become available through work done in the fields of rhetoric, linguistics and cognitive psychology. It has helped to produce a methodology based on the concept of reader expectations.

2. *Writing with the Reader in Mind: Expectation and Context.*

Readers do not simply read; they interpret. Any piece of prose, no matter how short, may "mean" in 10 (or more) different ways to 10 different readers. This methodology of reader expectations is founded on the recognition that readers make many of their most important interpretive decisions about the substance of prose based on clues they receive from its structure.

This interplay between substance and structure can be demonstrated by something as basic as a simple table. Let us say that in tracking the temperature of a liquid over a period of time, an investigator takes measurements every three minutes and records a list of temperatures. Those data could be presented by a number of written structures. Here are two possibilities:

$T(time)=15'$, $T(temperature)=32°C$; $t=0'$, $T=25°C$; $t=6'$, $T=29°C$; $t=3'$, $T=27°C$; $t=12'$, $T=32°C$; $t=9'$, $T=31°C$

time (min)	Temperature (°C)
0	25
3	27
6	29
9	31
12	32
15	32

Precisely the same information appears in both formats, yet most readers find the second easier to interpret. It may be that the very familiarity of the tabular structure makes it easier to use. But, more significantly, the structure of the second table provides the reader with an easily perceived context (time) in which the significant piece of information (temperature) can be interpreted. The contextual material appears on the left in a pattern that produces an expectation of regularity; the interesting results appear on the right in a less obvious pattern, the discovery of which is the point of the table.

If the two sides of this simple table are reversed, it becomes much harder to read.

temperature (°C)	time (min)
25	0
27	3
29	6
31	9
32	12
32	15

Since we read from left to right, we prefer the context on the left, where it can more effectively familiarize the reader. We prefer the new, important information on the right, since its job is to intrigue the reader.

Information is interpreted more easily and more uniformly if it is placed where most readers expect to find it. These needs and expectations of readers affect the interpretation not only of tables and illustrations but also of prose itself. Readers have relatively fixed expectations about where in the structure of prose they will encounter particular items of its substance. If writers can become consciously aware of these locations, they can better control the degrees of recognition and emphasis a reader will give to the various pieces of information being presented. Good writers are intuitively aware of these expectations; that is why their prose has what we call "shape".

This underlying concept of reader expectation is perhaps most immediately evident at the level of the largest units of discourse. (A unit of discourse is defined as anything with a beginning and an end: a clause, a sentence, a section, an article, etc.) A research article, for example, is generally divided into recognizable sections, sometimes labeled Introduction, Experimental Methods, Results and Discussion. When the sections are confused—when too much experimental detail is found in the Results section, or when discussion and results intermingle—readers are often equally confused. In smaller units of discourse the functional divisions are not so explicitly labeled, but readers have definite expectations all the same, and they search for certain information in particular places. If these structural expectations are continually violated, readers are forced to divert energy from understanding the content of a passage to unraveling its structure. As the complexity of the content increases moderately, the possibility of misinterpretation or noninterpretation increases dramatically.

We present here some results of applying this methodology to research reports in the scientific literature. We have taken several passages from research articles (either published or accepted for publication) and have suggested ways of rewriting them by applying principles derived from the study of reader expectations. We have not sought to transform the passages into "plain English" for the use of the general public; we have neither decreased the jargon nor diluted the science. We have striven not for simplification but for clarification.

3. Reader Expectations for the Structure of Prose.

Here is our first example of scientific prose, in its original form:

The smallest of the URF's (URFA6L), a 207-nucleotide (nt) reading frame overlapping out of phase the NIH2terminal portion of the adenosinetriphosphatase (ATPase) subunit 6 gene has been identified as the animal equivalent of the recently discovered yeast H^+—ATPase subunit 8 gene. The functional significance of the other URF's has been, on the contrary, elusive. Recently, however, immunoprecipitation experiments with antibodies to purified, rotenone-sensitive NADH-ubiquinone oxido-reductase [hereafter referred to as respiratory chain NADH dehydrogenase or complex I] from bovine heart, as well as enzyme fractionation studies, have indicated that six human URF's (that is, URF1, URF2, URF3, URF4, URF4L, and URF5, hereafter referred to as ND1, ND2, ND3, ND4, ND4L, and ND5) encode subunits of complex I. This is a large complex that also contains many subunits synthesized in the cytoplasm.

Ask any ten people why this paragraph is hard to read, and nine are sure to mention the technical vocabulary; several will also suggest that it requires specialized background knowledge. Those problems turn out to be only a small part of the difficulty. Here is the passage again, with the difficult words temporarily lifted:

The smallest of the URF's, an [A], has been identified as a [B] subunit 8 gene. The functional significance of the other URF's has been, on the contrary, elusive. Recently, however, [C] experiments, as well as [D] I studies, have indicated that six human URF's [1-6] encode subunits of Complex I. This is a large complex that also contains many subunits synthesized in the cytoplasm.

It may now be easier to survive the journey through the prose, but the passage is still difficult. Any number of questions present themselves: What has the first sentence of the passage to do with the last sentence? Does the third sentence contradict what we have been told in the second sentence? Is the functional significance of URF's still "elusive"? Will this passage lead us to further discussion about URF's, or about Complex I, or both?

4. Information is interpreted more easily and more uniformly if it is placed where most readers expect to find it.

Knowing a little about the subject matter does not clear up all the confusion. The intended audience of this passage would probably possess at least two items of essential technical information: first, "URF" stands for "Uninterrupted Reading Frame," which describes a segment of DNA organized in such a way that it could encode a protein, although no such protein product has yet been identified; second, both ATPase and NADH oxido-reductase are enzyme complexes central to energy metabolism. Although this information may provide some sense of comfort, it does little to answer the interpretive questions that need answering. It seems the reader is hindered by more than just the scientific jargon.

To get at the problem, we need to articulate something about how readers go about reading. We proceed to the first of several reader expectations.

Subject-Verb Separation

Look again at the first sentence of the passage cited above. It is relatively long, 42 words; but that turns out not to be the main cause of its burdensome complexity. Long sentences need not be difficult to read; they are only difficult to write. We have seen sentences of over 100 words that flow easily and persuasively toward their clearly demarcated destination. Those well-wrought serpents all had something in common: Their structure presented information to readers in the order the readers needed and expected it.

The first sentence of our example passage does just the opposite: it burdens and obstructs the reader, because of all-too-common structural defect. Note that the grammatical subject ("the smallest") is separated from its verb ("has been identified") by 23 words, more than half the sentence. Readers expect a grammatical subject to be followed immediately by the verb. Anything of length that intervenes between subject and verb is read as an interruption, and therefore as something of lesser importance.

5. Beginning with the exciting material and ending with a lack of luster often leaves us disappointed and destroys our sense of momentum.

The reader's expectation stems from a pressing need for syntactic resolution, fulfilled only by the arrival of the verb. Without the verb, we do not know what the subject is doing, or what the sentence is all about. As a result, the reader focuses attention on the arrival of the verb and resists recognizing anything in the interrupting material as being of primary importance. The longer the interruption lasts, the more likely it becomes that the "interruptive" material actually contains important information; but its structural location will continue to brand it as merely interruptive. Unfortunately, the reader will not discover its true value until too late—until the sentence has ended without having produced anything of much value outside of that subject-verb interruption.

In this first sentence of the paragraph, the relative importance of the intervening material is difficult to evaluate. The material might conceivably be quite significant, in which case the writer should have positioned it to reveal that importance. Here is one way to incorporate it into the sentence structure:

The smallest of the URF's is URFA6L, a 207-nucleotide (nt) reading frame overlapping out of phase the NH2-terminal portion of the adenosinetriphosphatase (ATPase) subunit 6 gene; it has been identified as the animal equivalent of the recently discovered yeast H^+-ATPase subunit 8 gene.

On the other hand, the intervening material might be a mere aside that diverts attention from more important ideas; in that case the writer should have deleted it, allowing the prose to drive more directly toward its significant point:

The smallest of the URF's (URFA6L) has been identified as the animal equivalent of the recently discovered yeast H^+-ATPase subunit 8 gene.

Only the author could tell us which of these revisions more accurately reflects his intentions.

These revisions lead us to a second set of reader expectations. Each unit of discourse, no matter what the size, is expected to serve a single function, to make a single point. In the case of a sentence, the point is expected to appear in a specific place reserved for emphasis.

The Stress Position

It is a linguistic commonplace that readers naturally emphasize the material that arrives at the end of a sentence. We refer to that location as a "stress position." If a writer is consciously aware of this tendency, she can arrange for the emphatic information to appear at the moment the reader is naturally exerting the greatest reading emphasis. As a result, the chances greatly increase that reader and writer will perceive the same material as being worthy of primary emphasis. The very structure of the sentence thus helps persuade the reader of the relative values of the sentence's contents.

The inclination to direct more energy to that which arrives last in a sentence seems to correspond to the way we work at tasks through time. We tend to take something like a "mental breath" as we begin to read each new sentence, thereby summoning the tension with which we pay attention to the unfolding of the syntax. As we recognize that the sentence is drawing toward its conclusion, we begin to exhale that mental breath. The exhalation produces a sense of emphasis. Moreover, we delight in being rewarded at the end of a labor with something that makes the ongoing effort worthwhile. Beginning with the exciting material and ending with a lack of luster often leaves us disappointed and destroys our sense of momentum. We do not start with the strawberry shortcake and work our way up to the broccoli.

When the writer puts the emphatic material of a sentence in any place other than the stress position, one of two things can happen; both are bad. First, the reader might find the stress position occupied by material that clearly is not worthy of emphasis. In this case, the reader must discern, without any additional structural clue, what else in the sentence may be the most likely candidate for emphasis. There are no secondary structural indications to fall back upon. In sentences that are long, dense or sophisticated, chances soar that the reader will not interpret the prose precisely as the writer intended. The second possibility is even worse: The reader may find the stress position occupied by something that does appear capable of receiving emphasis, even though the writer did not intend to give it any stress. In that case, the reader is highly likely to emphasize this imposter material, and the writer will have lost an important opportunity to influence the reader's interpretive process.

The stress position can change in size from sentence to sentence. Sometimes it consists of a single word; sometimes it extends to several lines. The definitive factor is this: The stress position coincides with the moment of syntactic closure. A reader has reached the beginning of the stress position when she knows there is nothing left in the clause or sentence but the material presently being read. Thus a whole list, numbered and indented, can occupy the stress position of a sentence if it has been clearly announced as being all that remains of that sentence. Each member of that list, in turn, may have its own internal stress position, since each member may produce its own syntactic closure.

Within a sentence, secondary stress positions can be formed by the appearance of a properly used colon or semicolon; by grammatical convention, the material preceding these punctuation marks must be able to stand by itself as a complete sentence. Thus, sentences can be extended effortlessly to dozens of words, as long as there is a medial syntactic closure for every piece of new, stress-worthy information along the way. One of our revisions of the initial sentence can serve as an example:

The smallest of the URF's is URFA6L, a 207-nucleotide (nt) reading frame overlapping out of phase the NH2-terminal portion of the adenosinetriphosphatase (ATPase) subunit 6 gene; it has been identified as the animal equivalent of the recently discovered yeast H^+ — ATPase subunit 8 gene.

By using a semicolon, we created a second stress position to accommodate a second piece of information that seemed to require emphasis.

We now have three rhetorical principles based on reader expectations: First, grammatical subjects should be followed as soon as possible by their verbs; second, every unit of discourse, no matter the size, should serve a single function or make a single point; and, third, information intended to be emphasized should appear at points of syntactic closure. Using these principles, we can begin to unravel the problems of our example prose.

Note the subject-verb separation in the 62-word third sentence of the original passage:
Recently, however, immunoprecipitation experiments with antibodies to purified, rotenone-sensitive NADH-ubiquinone oxido-reductase [hereafter referred to as respiratory chain NADH dehydrogenase or complex I] from bovine heart, as well as enzyme fractionation studies, have indicated that six human URF's (that is, URF1, URF2, URF3, URF4, URF4L, and URF5, hereafter referred to as ND1, ND2, ND3, ND4, ND4L, and ND5) encode subunits of complex I.
After encountering the subject ("experiments"), the reader must wade through 27 words (including three hyphenated compound words, a parenthetical interruption and an "as well as" phrase) before alighting on the highly uninformative and disappointingly anticlimactic verb ("have indicated"). Without a moment to recover, the reader is handed a "that" clause in which the new subject ("six human URF's") is separated from its verb ("encode") by yet another 20 words.

If we applied the three principles we have developed to the rest of the sentences of the example, we could generate a great many revised versions of each. These revisions might differ significantly from one another in the way their structures indicate to the reader the various weights and balances to be given to the information. Had the author placed all stress-worthy material in stress positions, we as a reading community would have been far more likely to interpret these sentences uniformly.

6. *We cannot succeed in making even a single sentence mean one and only one thing; we can only increase the odds that a large majority of readers will tend to interpret our discourse according to our intentions.*

We couch this discussion in terms of "likelihood" because we believe that meaning is not inherent in discourse by itself; "meaning" requires the combined participation of text and reader. All sentences are infinitely interpretable, given an infinite number of interpreters. As communities of readers, however, we tend to work out tacit agreements as to what kinds of meaning are most likely to be extracted from certain articulations. We cannot succeed in making even a single sentence mean one and only one thing; we can only increase the odds that a large majority of readers will tend to interpret our discourse according to our intentions. Such success will follow from authors becoming more consciously aware of the various reader expectations presented here.

Here is one set of revisionary decisions we made for the example:

The smallest of the URF's, URFA6L, has been identified as the animal equivalent of the recently discovered yeast W-ATPase subunit 8 gene; but the functional significance of other URF's has been more elusive. Recently, however, several human URF's have been shown to encode subunits of rotenone-sensitive NADH-ubiquinone oxido-reductase. This is a large complex that also contains many subunits synthesized in the cytoplasm; it will be referred to hereafter as respiratory chain NADH dehydrogenase or complex I. Six subunits of Complex I were shown by enzyme fractionation studies and immunoprecipitation experiments to be encoded by six human URF's (URF1, URF2, URF3, URF4, URF4L, and URF5); these URF's will be referred to subsequently as ND1, ND2, ND3, ND4, ND4L, and ND5.

Sheer length was neither the problem nor the solution. The revised version is not noticeably shorter than the original; nevertheless, it is significantly easier to interpret. We have indeed deleted certain words, but not on the basis of wordiness or excess length. (See especially the last sentence of our revision.)

When is a sentence too long? The creators of readability formulas would have us believe there exists some fixed number of words (the favorite is 29) past which a sentence is too hard to read. We disagree. We have seen 10-word sentences that are virtually impenetrable and, as we mentioned above, 100-word sentences that flow effortlessly to their points of resolution. In place of the word-limit concept, we offer the following definition: A sentence is too long when it has more viable candidates for stress positions than there are stress positions available. Without the stress position's locational clue that its material is intended to be emphasized, readers are left too much to their own devices in deciding just what else in a sentence might be considered important.

In revising the example passage, we made certain decisions about what to omit and what to emphasize. We put subjects and verbs together to lessen the reader's syntactic burdens; we put the material we believed worthy of emphasis in stress positions; and we discarded material for which we could not discern significant connections. In doing so, we have produced a clearer passage—but not one that necessarily reflects the author's intentions; it reflects only our interpretation of the author's intentions. The more problematic the structure, the less likely it becomes that a grand majority of readers will perceive the discourse in exactly the way the author intended.

7. *The information that begins a sentence establishes for the reader a perspective for viewing the sentence as a unit.*

It is probable that many of our readers—and perhaps even the authors—will disagree our choices. If so, that disagreement underscores our point: The original failed to communicate its ideas and their connections clearly. If we happened to have interpreted the passage as you did, then we can make a different point: No one should have to work as hard as we did to unearth the content of a single passage of this length.

The Topic Position

To summarize the principles connected with the stress position, we have the proverbial wisdom, "Save the best for last." To summarize the principles connected with the other end of the sentence, which we will call the topic position, we have its proverbial contradiction, "First things first." In the stress position the reader needs and expects closure and fulfillment; in the topic position the reader needs and expects perspective and context. With so much of reading comprehension affected by what shows up in the topic position, it behooves a writer to control what appears at the beginning of sentences with great care.

The information that begins a sentence establishes for the reader a perspective for viewing the sentence as a unit: Readers expect a unit of discourse to be a story about whoever shows up first. "Bees disperse pollen" and "Pollen is dispersed by bees" are two different but equally respectable sentences about the same facts. The first tells us something about bees; the second tells us something about pollen. The passivity of the second sentence does not by itself impair its quality; in fact, "Pollen is dispersed by bees" is the superior sentence if it appears in a paragraph that in-tends to tell us a continuing story about pollen. Pollen's story at that moment is a passive one.

Readers also expect the material occupying the topic position to provide them with linkage (looking backward) and context (looking forward). The information in the topic position prepares the reader for upcoming material by connecting it backward to the previous discussion. Although linkage and context can derive from several sources, they stem primarily from material that the reader has already encountered within this particular piece of discourse. We refer to this familiar, previously introduced material as "old information." Conversely, material making its first appearance in a discourse is "new information." When new information is important enough to receive emphasis, it functions best in the stress position.

When old information consistently arrives in the topic position, it helps readers to construct the logical flow of the argument: It focuses attention on one particular strand of the discussion, both harkening backward and leaning forward. In contrast, if the topic position is constantly occupied by material that fails to establish linkage and context, readers will have difficulty perceiving both the connection to the previous sentence and the projected role of the new sentence in the development of the paragraph as a whole.

Here is a second example of scientific prose that we shall attempt to improve in subsequent discussion:

Large earthquakes along a given fault segment do not occur at random intervals because it takes time to accumulate the strain energy for the rupture. The rates at which tectonic plates move and accumulate strain at their boundaries are approximately uniform. Therefore, in first approximation, one may expect that large ruptures of the same fault segment will occur at approximately constant time intervals. If subsequent main-shocks have different amounts of slip across the fault, then the recurrence time may vary, and the basic idea of periodic mainshocks must be modified. For great plate boundary ruptures the length and slip often vary by a factor of 2. Along the southern segment of the San Andreas fault the recurrence interval is 145 years with variations of several decades. The smaller the standard deviation of the average recurrence interval, the more specific could be the long-term prediction of a future mainshock.

This is the kind of passage that in subtle ways can make readers feel badly about themselves. The individual sentences give the impression of being intelligently fashioned: They are not especially long or convoluted; their vocabulary is appropriately professional but not beyond the ken of educated general readers; and they are free of grammatical and dictional errors. On first reading, however, many of us arrive at the paragraph's end without a clear sense of where we have been or where we are going. When that happens, we tend to berate ourselves for not having paid close enough attention. In reality, the fault lies not with us, but with the author.

We can distill the problem by looking closely at the information in each sentence's topic position:

Large earthquakes
The rates
Therefore . . . one
subsequent mainshocks
great plate boundary ruptures
the southern segment of the San Andreas fault
the smaller the standard deviation . . .

Much of this information is making its first appearance in this paragraph—in precisely the spot where the reader looks for old, familiar information. As a result, the focus of the story constantly shifts. Given just the material in the topic positions, no two readers would be likely to construct exactly the same story for the paragraph as a whole.

If we try to piece together the relationship of each sentence to its neighbors, we notice that certain bits of old information keep reappearing. We hear a good deal about the recurrence time between earthquakes: The first sentence introduces the concept of nonrandom intervals between earthquakes; the second sentence tells us that recurrence rates due to the movement of tectonic plates are more or less uniform; the third sentence adds that the recurrence rate of major earthquakes should also be somewhat predictable; the fourth sentence adds that recurrence rates vary with some conditions; the fifth sentence adds information about one particular variation; the sixth sentence adds a recurrence-rate example from California; and the last sentence tells us something about how recurrence rates can be described statistically. This refrain of "recurrence intervals" constitutes the major string of old information in the paragraph. Unfortunately, it rarely appears at the beginning of sentences, where it would help us maintain our focus on its continuing story.

In reading, as in most experiences, we appreciate the opportunity to become familiar with a new environment before having to function in it. Writing that continually begins sentences with new information and ends with old information forbid both the sense of comfort and orientation at the start and the sense of fulfilling arrival at the end. It misleads the reader as to whose story is being told; it burdens the reader with new information that must be carried further into the sentence before it can be connected to the discussion; and it creates ambiguity as to which material the writer intended the reader to emphasize. All of these distractions require that readers expend a disproportionate amount of energy to unravel the structure of the prose, leaving less energy available for perceiving content.

We can begin to revise the example by ensuring the following for each sentence:

(1) *The backward-linking old information appears in the topic position.*
(2) *The person, thing or concept whose story it is appears in the topic position.*
(3) *The new, emphasis-worthy information appears in the stress position.*

Once again, if our decisions concerning the relative values of specific information differ from yours, we can all blame the author, who failed to make his intentions apparent. Here first is a list of what we perceived to be the new, emphatic material in each sentence:

time to accumulate strain energy along a fault
approximately uniform
large ruptures of the same fault
different amounts of slip
vary by a factor of 2
variations of several decades
predictions of future mainshock

Now, based on these assumptions about what deserves stress, here is our proposed revision:

Large earthquakes along a given fault segment do not occur at random intervals because it takes time to accumulate the strain energy for the rupture. The rates at which tectonic plates move and accumulate strain at their boundaries are roughly uniform. Therefore, nearly constant time intervals (at first approximation) would be expected between large ruptures of the same fault segment. [However?] *, the recurrence time may vary; the basic idea of periodic mainshocks may need to be modified if subsequent mainshocks have different amounts of slip across the fault.* [Indeed?] *, the length and slip of great plate boundary ruptures often vary by a factor of 2.* [For example?] *, the recurrence interval along the southern segment of the San Andreas fault is 145 years with variations of several decades. The smaller the standard deviation of the average recurrence interval, the more specific could be the long-term pre-*

diction of a future mainshock.

Many problems that had existed in the original have now surfaced for the first time. Is the reason earthquakes do not occur at random intervals stated in the first sentence or in the second? Are the suggested choices of "however", "indeed" and "for example" the right ones to express the connections at those points? (All these connections were left unarticulated in the original paragraph.) If "for example" is an inaccurate transitional phrase, then exactly how does the San Andreas fault example connect to ruptures that "vary by a factor of 2"? Is the author arguing that recurrence rates must vary because fault movements often vary? Or is the author preparing us for a discussion of how in spite of such variance we might still be able to predict earthquakes? This last question remains unanswered because the final sentence leaves behind earthquakes that recur at variable intervals and switches instead to earthquakes that recur regularly. Given that this is the first paragraph of the article, which type of earthquake will the article most likely proceed to discuss? In sum, we are now aware of how much the paragraph had not communicated to us on first reading. We can see that most of our difficulty was owing not to any deficiency in our reading skills but rather to the author's lack of comprehension of our structural needs as readers.

8. *In our experience, the misplacement of old and new information turns out to be the No. 1 problem in American professional writing today.*

In our experience, the misplacement of old and new information turns out to be the No. 1 problem in American professional writing today. The source of the problem is not hard to discover: Most writers produce prose linearly (from left to right) and through time. As they begin to formulate a sentence, often their primary anxiety is to capture the important new thought before it escapes. Quite naturally they rush to record that new information on paper, after which they can produce at their leisure the contextualizing material that links back to the previous discourse. Writers who do this consistently are attending more to their own need for unburdening themselves of their information than to the reader's need for receiving the material. The methodology of reader expectations articulates the reader's needs explicitly, thereby making writers consciously aware of structural problems and ways to solve them.

9. *Put in the topic position the old information that links backward; put in the stress position the new information you want the reader to emphasize.*

A note of clarification: Many people hearing this structural advice tend to oversimplify it to the following rule: "Put the old information in the topic position and the new information in the stress position." No such rule is possible. Since by definition all information is either old or new, the space between the topic position and the stress position must also be filled with old and new information. Therefore the principle (not rule) should be stated as follows: "Put in the topic position the old information that links backward; put in the stress position the new information you want the reader to emphasize."

Perceiving Logical Gaps

When old information does not appear at all in a sentence, whether in the topic position or elsewhere, readers are left to construct the logical linkage by themselves. Often this happens when the connections are so clear in the writer's mind that they seem unnecessary to state; at those moments, writers underestimate the difficulties and ambiguities inherent in the reading process. Our third example attempts to demonstrate how paying attention to the placement of old and new information can reveal where a writer has neglected to articulate essential connections.

> *The enthalpy of hydrogen bond formation between the nucleoside bases 2' deoxyguanosine (dG) and 2' deoxycvtidine (dC) has been determined by direct measurement. dG and dC were derivatized at the 5' and 3 hydroxyls with triisopropylsilyl groups to obtain solubility of the nucleosides in non-aqueous solvents and to prevent the ribose hydroxyls from forming hydrogen bonds. From isoperibolic titration measurements, the enthalpy of dC: dG base-pair formation is -6.65 ± 0.32 kcal/mol.*

Although part of the difficulty of reading this passage may stem from its abundance of specialized technical terms, a great deal more of the difficulty can be attributed to its structural problems. These problems are now familiar: We are not sure at all times whose story is being told; in the first sentence the subject and verb are widely separated; the second sentence has only one stress position but two or three pieces of information that are probably worthy of emphasis—"solubility...solvents," "prevent...from forming hydrogen bonds" and perhaps "triisopropylsilyl groups." These perceptions suggest the following revision tactics:

> (1) *Invert the first sentence, so that (a) the subject-verb-complement connection is unbroken, and (b) "dG" and are introduced in the stress position as new and interesting information. (Note that inverting the sentence requires stating who made the measurement; since the authors performed the first direct measurement, recognizing their agency in the topic position may well be appropriate.)*
> (2) *Since "dG" and "dC" become the old information in the second sentence, keep them up front in the topic position.*
> (3) *Since "triisopropylsilyl groups" is new and important information here, create for it a stress position.*
> (4) *"Triisopropylsilyl groups" then becomes the old information of the clause in which its effects are described; place it in the topic position of this clause.*

(5) Alert the reader to expect the arrival of two distinct effects by using the flag word "both." "Both" notifies the reader that two pieces of new information will arrive in a single stress position.

Here is a partial revision based on these decisions:

We have directly measured the enthalpy of hydrogen bond formation between the nucleoside bases 2'deoxyguanosine (dG) and 2'deoxycytidine (dC). dG and dC were derivatized at the 5' and 3' hydroxyls with triisopropylsilyl groups; these groups serve both to solubilize the nucleosides in non-aqueous solvents and to prevent the ribose hydroxyls from forming hydrogen bonds. From isoperibolic titration measurements, the enthalpy of dC: dG base-pair formation is -6.65 ± 0.32 kcal/mol.

The outlines of the experiment are now becoming visible, but there is still a major logical gap. After reading the second sentence, we expect to hear more about the two effects that were important enough to merit placement in its stress position. Our expectations are frustrated, however, when those effects are not mentioned in the next sentence: "From isoperibolic titration measurements, the enthalpy of dC: dG base-pair formation is -6.65 ± 0.32 kcal/mol." The authors have neglected to explain the relationship between the derivatization they performed (in the second sentence) and the measurements they made (in the third sentence). Ironically, that is the point they most wished to make here.

At this juncture, particularly astute readers who are chemists might draw upon their specialized knowledge, silently supplying the missing connection. Other readers are left in the dark. Here is one version of what we think the authors meant to say, with two additional sentences supplied from a knowledge of nucleic acid chemistry:

We have directly measured the enthalpy of hydrogen bond formation between the nucleoside bases 2'deoxyguanosine (dG) and 2'deoxycytidine (dC). dG and dC were derivatized at the 5' and 3' hydroxyls with triisopropylsilyl groups; these groups serve both to solubilize the nucleosides in non-aqueous solvents and to prevent the ribose hydroxyls from forming hydrogen bonds. Consequently, when the derivatized nucleosides are dissolved in non-aqueous solvents, hydrogen bonds form almost exclusively between the bases. Since the inter-base hydrogen bonds are the only bonds to form upon mixing, their enthalpy of formation can be determined directly by measuring the enthalpy of mixing. From our isoperibolic titration measurements, the enthalpy of dC: dG base-pair formation is -6.65 ± 0.32 kcal/mol.

Each sentence now proceeds logically from its predecessor. We never have to wander too far into a sentence without being told where we are and what former strands of discourse are being continued. And the "measurements" of the last sentence has now become old information, reaching back to the "measured directly" of the preceding sentence. (It also fulfills the promise of the "we have directly measured" with which the paragraph began.) By following our knowledge of reader expectations, we have been able to spot discontinuities, to suggest strategies for bridging gaps, and to rearrange the structure of the prose, thereby increasing the accessibility of the scientific content.

Locating the Action

Our final example adds another major reader expectation to the list.

Transcription of the 5S RNA genes in the egg extract is TFHIA-dependent. This is surprising, because the concentration of TFIIIA is the same as in the oocyte nuclear extract. The other transcription factors and RNA polymerase III are presumed to be in excess over available TFIIIA, because tRNA genes are transcribed in the egg extract. The addition of egg extract to the oocyte nuclear extract has two effects on transcription efficiency. First, there is a general inhibition of transcription that can be alleviated in part by supplementation with high concentrations of RNA polymerase III. Second, egg extract destabilizes transcription complexes formed with oocyte but not somatic 5S ENA genes.

The barriers to comprehension in this passage are so many that it may appear difficult to know where to start revising. Fortunately, it does not matter where we start, since attending to any one structural problem eventually leads us to all the others.

We can spot one source of difficulty by looking at the topic positions of the sentences: We cannot tell whose story the passage is. The story's focus (that is, the occupant of the topic position) changes in every sentence. If we search for repeated old information in hope of settling on a good candidate for several of the topic positions, we find all too much of it: egg extract, TFIIIA, oocyte extract, RNA polymerase III, 5S RNA, and transcription. All of these reappear at various points, but none announces itself clearly as our primary focus. It appears that the passage is trying to tell several stories simultaneously, allowing none to dominate.

We are unable to decide among these stories because the author has not told us what to do with all this information. We know who the players are, but we are ignorant of the actions they are presumed to perform. This violates yet another important reader expectation: Readers expect the action of a sentence to be articulated by the verb.

Here is a list of the verbs in the example paragraph:

is
is . . . is
are presumed to be
are transcribed
has
is. . . can be alleviated
destabilizes

The list gives us too few clues as to what actions actually take place in the passage. If the actions are not to be found in the verbs, then we as readers have no secondary structural clues for where to locate them. Each of us has to make a personal interpretive guess; the writer no longer controls the reader's interpretive act.

10. As critical scientific readers, we would like to concentrate our energy on whether the experiments prove the hypotheses.

Worse still, in this passage the important actions never appear. Based on our best understanding of this material, the verbs that connect these players are "limit" and "inhibit." If we express those actions as verbs and place the most frequently occurring information—"egg extract" and "TFIIIA"—in the topic position whenever possible, 2 we can generate the following revision:

In the egg extract, the availability of TFIIIA limits transcription of the 5S RNA genes. This is surprising because the same concentration of TFIIIA does not limit transcription in the oocyte nuclear extract. In the egg extract, transcription is not limited by RNA polymerase or other factors because transcription of tRNA genes indicates that these factors are in excess over available TFIIIA. When added to the nuclear extract, the egg extract affected the efficiency of transcription in two ways. First, it inhibited transcription generally; this inhibition could be alleviated in part by supplementing the mixture with high concentrations of RNA polymerase m. Second, the egg extracts destabilized transcription complexes formed by oocyte but not by somatic 5S genes.

As a story about "egg extract," this passage still leaves something to be desired. But at least now we can recognize that the author has not explained the connection between "limit" and "inhibit." This unarticulated connection seems to us to contain both of her hypotheses: First, that the limitation on transcription is caused by an inhibitor of TFIIIA present in the egg extract; and, second, that the action of that inhibitor can be detected by adding the egg extract to the oocyte extract and examining the effects on transcription. As critical scientific readers, we would like to concentrate our energy on whether the experiments prove the hypotheses. We cannot begin to do so if we are left in doubt as to what those hypotheses might be—and if we are using most of our energy to discern the structure of the prose rather than its substance.

Writing and the Scientific Process

We began this article by arguing that complex thoughts expressed in impenetrable prose can be rendered accessible and clear without minimizing any of their complexity. Our examples of scientific writing have ranged from the merely cloudy to the virtually opaque; yet all of them could be made significantly more comprehensible by observing the following structural principles:

(1) *Follow a grammatical subject as soon as possible with its verb.*
(2) *Place in the stress position the "new information" you want the reader to emphasize.*
(3) *Place the person or thing whose "story" a sentence is telling at the beginning of the sentence, in the topic position.*
(4) *Place appropriate "old information" (material already stated in the discourse) in the topic position for linkage backward and contextualization forward.*
(5) *Articulate the action of every clause or sentence in its verb.*
(6) *In general, provide context for your reader before asking that reader to consider anything new.*
(7) *In general, try to ensure that the relative emphases of the substance coincide with the relative expectations for emphasis raised by the structure.*

None of these reader-expectation principles should be considered "rules." Slavish adherence to them will succeed no better than has slavish adherence to avoiding split infinitives or to using the active voice instead of the passive. There can be no fixed algorithm for good writing, for two reasons. First, too many reader expectations are functioning at any given moment for structural decisions to remain clear and easily activated. Second, any reader expectation can be violated to good effect. Our best stylists turn out to be our most skillful violators; but in order to carry this off, they must fulfill expectations most of the time, causing the violations to be perceived as exceptional moments, worthy of note.

11. *It may seem obvious that a scientific document is incomplete without the interpretation of the writer; it may not be so obvious that the document cannot "exist" without the interpretation of each reader.*

A writer's personal style is the sum of all the structural choices that person tends to make when facing the challenges of creating discourse. Writers who fail to put new information in the stress position of many sentences in one document are likely to repeat that unhelpful structural pattern in all other documents. But for the very reason that writers tend to be consistent in making such choices, they can learn to improve their writing style; they can permanently reverse those habitual structural decisions that mislead or burden readers.

We have argued that the substance of thought and the expression of thought are so inextricably intertwined that changes in either will affect the quality of the other. Note that only the first of our examples (the paragraph about URF's) could be revised on the basis of the methodology to reveal a nearly finished passage. In all the other examples, revision revealed existing conceptual gaps and other problems that had been submerged in the originals by dysfunctional structures. Filling the gaps required the addition of extra material. In revising each of these examples, we arrived at a point where we could proceed no further without either supplying connections between ideas or eliminating some existing material altogether. (Writers who use reader-expectation principles on their own prose will not have to conjecture or infer; they know what the prose is intended to convey.) Having begun by analyzing the structure of the prose, we were led eventually to reinvestigate the substance of the science.

The substance of science comprises more than the discovery and recording of data; it extends crucially to include the act of interpretation. It may seem obvious that a scientific document is incomplete without the interpretation of the writer; it may not be so obvious that the document cannot "exist" without the interpretation of each reader. In other words, writers cannot "merely" record data, even if they try. In any recording or articulation, no matter how haphazard or confused, each word resides in one or more distinct structural locations. The resulting structure, even more than the meanings of individual words, significantly influences the reader during the act of interpretation. The question then becomes whether the structure created by the writer (intentionally or not) helps or hinders the reader in the process of interpreting the scientific writing.

The writing principles we have suggested here make conscious for the writer some of the interpretive clues readers derive from structures. Armed with this awareness, the writer can achieve far greater control (although never complete control) of the reader's interpretive process. As a concomitant function, the principles simultaneously offer the writer a fresh re-entry to the thought process that produced the science. In real and important ways, the structure of the prose becomes the structure of the scientific argument. Improving either one will improve the other.

Appendix 2

The Skills in Paper Writing

在英文科技写作过程中，存在一些简单的语言技巧，可以使得论文更好地符合英语的习惯，有利于论文的发表。

1. 如何评价前人的工作

自己的研究是建立在前人研究的基础上的，一定要注意绝对不能全面否定前人的成果，即使在你看来前人的结论完全不对。这是对前人工作最起码的尊重，英文叫作给别人的工作"credits"。所以文章不要出现非常"negative"的评价，比如

Their results are wrong, very questionable, have no commensence, 等等

遇到这类情况，可以婉转地提出：

Their studies may be more reasonable if they had considered this situation...

Their results could be better convinced if they... 或

Their conclusion may remain some uncertainties ....

2. 如何引出自己的观点或方法

如何指出当前研究的不足以及有目的地引导出自己的研究的重要性，通常在叙述了前人成果之后，用 However 来引导不足，比如：

... However, little information (little attention, little work, little data, little research)... 或

... or however few studies (few investigations, few researchers, few attempts, no, none of these studies)... 或

... However, ... has (have) been less done on (focused on, attempted to, conducted, investigated, studied, respected to) ... 或

... However, ... respected to...

对于提出一种新方法或者一种新方向，常用引导句型为：

Previous research (studies, records) has (have) ..., however, failed to consider (ignored, misinterpreted, neglected to, overestimated, underestimated, misleaded). Thus, these previus results are (inconclisive, misleading, unsatisfactory, questionable, controversial...). Uncertainties (discrepancies) still exist ...

如果研究的方法以及方向和前人一样，可以通过下面的方式强调自己工作的作用：

However, data is still scarce (rare, less accurate, there is still dearth of). We need to [aim to, have to, provide more documents (data, records, studies), increase the dataset] 或

However, data is still scarce (rare, less accurate, there is still dearth of). Further studies are still necessary (essential)

为了强调自己研究的重要性，一般还要在"However"之前介绍自己研究问题的反方面，另一方面等等比如：

(1) 时间问题：如果你研究的问题在时间上比较新，就可以大量提及对时间较老的问题的研究及重要性，然后说"However"，对时间尺度比较新的问题研究不足。

（2）物性及研究手段问题：如果你要应用一种新手段或者研究方向，可以提出当前比较流行的方法以及物质性质，然后说对你所研究的方向和方法，在现在的研究中甚少。

（3）研究区域问题：首先总结相邻区域或者其他区域的研究，然后强调这一区域研究的不足。

（4）不确定性：虽然前人对这一问题研究很多，但是目前有两种或者更多种的观点，这种"uncertainties, ambiguities"，值得进一步澄清。

（5）提出自己的假设来验证：如果自己的研究完全是新的，没有前人的工作进行对比，在这种情况下，你可以自信地说，根据提出的过程，存在这种可能的结果，本文就是要证实这种结果。

We aim to test the feasibility (reliability) of the ...

It is hoped that the question will be resolved (fall away) with our proposed method (approach).

如果是提出自己的观点，可用下述句型：

We aim to...

This paper reports on (provides results, extends the method, focus on) ...

The purpose of this paper is to...

Furthermore, Moreover, In addition, we will also discuss...

3. 如何圈定自己的研究范围

前言的另外一个作用就是告诉读者（包括"reviewer"）你的文章的主要研究内容。如果处理不好，"reviewer"会提出严厉的建议，比如你没有考虑某种可能性、某种研究手段等等。为了减少这种争论，在前言的结尾你就要明确提出本文研究的范围。

（1）时间尺度问题：如果你的问题涉及比较长的时序，可以明确地提出本文只关心这一时间范围的问题。可用句型为：

We preliminarily focus on the older (younger) ...

或者有两种时间尺度的问题（long-term and short term），你可以说两者都重要，但是本文只涉及其中一种。

（2）研究区域的问题：和时间问题一样，明确提出你只关心这一地区。

（3）最后的圆场：在前言的最后，还可以总结性地提出这一研究对其他研究的帮助。或者说，"further studies on ... will be summarized in our next"或"study (or elsewhere)"。

总之，其目的就是让读者把思路集中到你要讨论的问题上来，减少争论（arguments）。

4. 怎样提出自己的观点

在提出自己的观点时，采取什么样的策略很重要。不合适的句子通常会遭到"reviewer"的置疑。

（1）如果观点不是这篇文章最新提出的，通常要用"We confirm that..."。

（2）对于自己很自信的观点，可用"We believe that..."。

（3）在更通常的情况下，由数据推断出一定的结论，用"Results indicate, infer, suggest, imply that..."。

（4）在及其特别的情况才可以用"We put forward (discover, observe..)..for the first time". 来强调自己的创新。

(5) 如果自己对所提出的观点不完全肯定，可用 "We tentatively put forward (interrprete this to...)..." 或 "The results may be due to (caused by) attributed to (rsulted from)..." 或 "This is probably a consequence of..." "It seems that .. can account for (interpret) this..." 或 "It is pisible that it stem from..."。

5. 连接词与逻辑

写英文论文最常见的一个毛病就是文章的逻辑不清楚。解决的方法有：

(1) 句子上下要有连贯，不能让句子之间独立，常见的连接词语有：

However, also, in addition, consequently, afterwards, moreover,

Furthermore, further, although, unlike, in contrast;

Similarly,

Unfortunately, alternatively, parallel results,

In order to, despite,

For example,

Compared with other results, thus, therefore...

(2) 用好这些连接词，可使自己的观点表达得有层次，更加明确。比如，如果叙述有时间顺序的事件或者文献，最早的文献可用 "AA advocated it for the first time."；接下来，可用 "Then BB further demonstrated that..."；接下来，可用 "Afterwards, CC..."；如果还有，可用 "More recent studies by DD..."。如果叙述两种观点，要把它们截然分开：

"AA put forward that...; in contrast, BB believe..." 或者 "Unlike AA, BB suggest; on the contrary (表明前面的观点错误，如果只是表明两种对立的观点，用 in contrast) BB..."。

如果两种观点相近，可用："AA suggest..., similarily (alternatively), BB..." 或 "Also, BB...; BB also does..."

表示因果或者前后关系，可用 "Consequently, therefore, as a result" 表明递进关系，或者用 "furthermore, further, moreover, in addition"。当写完一段英文，最好首先检查一下是否较好地应用了这些连接词。

(3) 段落的整体逻辑。经常我们要叙述一个问题的几个方面。这种情况下，一定要注意逻辑结构。首先第一段要明确告诉读者你要讨论几个部分：

... Therefore, there are three aspects of this problem have to be addressed.

... The first question involves...; the second problem relates to...; 同 he thrid aspect deals with...

上面的例子可以清晰地把观点逐层叙述。或直接用 "First, Second, Third... Finally,.."；当然，Furthermore, in addition 等可以用来补充说明。

6. 讨论部分的整体结构

小标题是把要讨论的问题分为几个片段的比较好的方法。一般第一个片段指出文章最为重要的数据与结论。补充说明的部分可以放在最后一个片段。一定要明白文章的读者会分为多个档次。文章除了本专业的专业人士读懂以外，一定要想办法能让更多的外专业人读懂。所以可以把讨论分为两部分，一部分提出观点，另一部分详细介绍过程以及论述的依据。这样专业外的人士可以了解文章的主要观点。对于比较专业的讨论，外专业人士可以把它当成"黑箱子",而这一部分本专业人士可以进一步研究。

为了使文章清楚,第一次提出概念时最好加一个括号,给出较为详细的解释。如果文章用了很多"Abbreviation",有两种方法加以解决:在文章最好加上个"Appendix",把所有"Abreviation"列表;在不同的页面上,不时地给出"Abbreciation"的含义,用来提醒读者。

讨论部分主要包括主要数据特征的总结、主要结论以及与前人观点的对比和本文的不足。给出文章的不足恰恰是保护自己文章的重要手段。如果刻意隐藏文章的漏洞,觉得别人看不出来,是非常不明智的。所谓不足,包括以下内容:

(1) 研究的问题有点片面

讨论时一定要说,可选择的句型如下:

It should be noted that this study has examined only...

We concentrate (focus) on only...

We have to point out that we do not...

Some limitations of this study are...

(2) 结论有些不足

The results do not imply...

The results can not be used to determine...

The results be taken as evidence of...

Unfortunately, we can not determien this from this data...

Our results are lack of ...

但是,在指出这些不足之后,一定要再一次加强本文的重要性以及可能采取的手段来解决这些不足,为别人或者自己的下一步研究打下伏笔。

Notwithstading its limitation, this tudy does suggest...

However, these problems culd be solved if we consider...

Despite its preliminary character, this study can clearly indicate...

把审稿人想到的问题提前给一个交代,同时表明你已经在思考这些问题,但是由于文章长度、试验进度或者试验手段的制约,暂时不能回答这些问题。但是,通过你的一些建议,这些问题在将来的研究中有可能解决。

Appendix 3

Skills in Abstracts Writing

文章摘要是对所写文章主要内容的精炼概括。由于大多数检索系统只收录论文的摘要部分或其数据库中只有摘要部分免费提供，并且有些读者只阅读摘要而不读全文或常根据摘要来判断是否需要阅读全文。因此，摘要的清楚表达十分重要。美国人称摘要为"Abstract"，而英国人则喜欢称其为"Summary"。论文摘要的重点应放在所研究的成果和结论上，用简练的语言来表达论文的精华，做到简明扼要、切题、能独立成文，最好用第三人称的完整的陈述句。

1. 中英文摘要的差异

英文摘要的内容要求与中文摘要一样，包括研究目的、方法、结果和结论四部分。但是，英文摘要有其自身特点，最主要的是中译英时往往造成所占篇幅较长，同样内容的一段文字，若用英文来描述，其占用的版面可能比中文多一倍。因此，撰写英文摘要更应注意简洁明了，力争用最短的篇幅提供最主要的信息。第一，对所掌握的资料进行精心筛选，不属于上述"四部分"的内容不必写入摘要。第二，对属于"四部分"的内容，也应适当取舍，做到简明扼要，不能包罗万象。比如研究"目的"，在多数标题中就已初步阐明，若无更深一层的目的，摘要完全不必重复叙述；再如研究"方法"，有些在国外可能早已成为常规的方法，在撰写英文摘要时就可仅写出方法名称，而不必一一描述其操作步骤。

中英文摘要应该在内容方面保持一致性，但目前对这个问题的认识存在两个误区：一是认为两个摘要的内容"差不多就行"，因此在英文摘要中随意删去中文摘要的重点内容，或随意增补中文摘要所未提及的内容，这样很容易造成摘要的重心转移，甚至偏离主题；二是认为英文摘要是中文摘要的硬性对译，对中文摘要中的每一个字都不遗漏，这往往使英文摘要用词累赘、重复，显得拖沓、冗长。

英文摘要应严格、全面的表达中文摘要的内容，不能随意增删，但这并不意味着一个字也不能改动，具体撰写方式应遵循英文语法修辞规则，符合英文专业术语规范，并照顾到英文的表达习惯。

2. 英文摘要撰写中应注意事项

（1）为确保简洁而充分地表述论文的 IMRD (Introduction, Methods, Results and Discussion) 结构的写作模式，可适当强调研究中的创新、重要之处（但不要使用评价性语言）；尽量包括论文中的主要论点和重要细节（重要的论证或数据）。

（2）使用简短的句子，表达要准确、简洁、清楚；注意表述的逻辑性，尽量使用指示性的词语来表达论文的不同部分（层次），如使用"We found that..."表示结果；使用"We suggest that..."表示讨论结果的含义等。

（3）应尽量避免引用文献、图表，用词应被潜在的读者所熟悉。若无法回避使用引文，应在引文出现的位置将引文的书目信息标注在方括号内；如确有需要（如避免多次重复较长的术语）使用非同行熟知的缩写，应在缩写符号第一次出现时给出其全称。

（4）为方便检索系统转录，应尽量避免使用化学结构式、数学表达式、角标和希腊文等特殊符号。

(5) 查询拟投稿期刊的读者须知，以了解其对摘要的字数和形式的要求。如果是结构式摘要，应了解其分为几段，使用何种标识、时态，是否使用缩写或简写，等等。

3. 英文摘要写作的时态

选择适当的时态和语态，是使摘要符合英文语法修辞规则的前提。通常情况下，摘要中谓语动词的时态和语态都不是通篇一律的，而应根据具体内容而有所变化，否则容易造成理解上的混乱。

在叙述研究过程时，多采用一般过去时；在采用一般过去时叙述研究过程当中提及在此过程之前发生的事，宜采用过去完成时；说明某课题现已取得的成果，宜采用现在完成时。摘要开头表示本文所"报告"或"描述"的内容，以及摘要结尾表示作者所"认为"的观点和"建议"的做法时，可采用一般现在时。

(1) 介绍背景资料时，如果句子的内容为不受时间影响的普遍事实，应使用现在时；如果句子的内容是对某种研究趋势的概述，则使用现在完成时。例如：

The authors **review** risk and protective factors for drug abuse, **assess** a number of approaches for drug abuse prevention potential with high-risk groups, and **make** recommendations for research and practice. [Psychological Bulletin, 1992, 112 (1): 64-105]

Previous research **has confirmed** four dimensions of temperament: .... [Archives of General Psychiatry, 1993, 50 (12): 975-990]

(2) 在叙述研究目的或主要研究活动时，如果采用"论文导向"，多使用现在时（如：This paper presents...）；如果采用"研究导向"，则使用过去时（如：This study investigated...）。

This article **summarizes** research on self-initiated and professionally facilitated change of addictive behaviors using the key transtheoretical constructs of stages and processes of change. [American Psychologist, 1992, 47 (9): 1102-1114]

We **investigated** whether captopril could reduce morbidity and mortality in patients with left ventricular dysfunction after a myocardial infarction. [New England Journal of Medicine, 1992, 327 (10): 669-677]

(3) 概述实验程序、方法和主要结果时，通常用现在时。

We **describe** a new molecular approach to analyzing the genetic diversity of complex microbial populations. [Applied and Environmental Microbiology, 1993, 59 (3): 695-700]

Our **results indicate** that p21 may be a universal inhibitor of cyclin kinases. [Nature, 1993, 366 (6456): 701-704]

(4) 叙述结论或建议时，可使用现在时、臆测动词或 may, should, could 等助动词。

We **suggest** that climate instability in the early part of the last interglacial **may** have delayed the melting of the Saalean ice sheets in America and Eurasia, perhaps accounting for this discrepancy. [Nature, 1993, 364 (6434): 218-220]

4. 英文摘要写作的语态

在多数情况下可采用被动语态。但在某些情况下，特别是表达作者或有关专家的观点时，又常用主动语态，其优点是鲜明有力。但由于主动语态的表达更为准确，且更易阅读，因而目前大多数期刊都提倡使用主动态。国际知名科技期刊"Nature"，"Cell"等尤其如此，其中第一人称和主动语态的使用十分普遍。在撰写论文时应认真调查一下拟投稿期刊有关人称和语态的使用习惯。为简洁、清楚地表达研究成果，在论文摘要的撰写中不应刻意回避第一人称和主动语态。

5. 英文摘要写作的用词

掌握一定的遣词造句技巧的目的是便于简单、准确的表达作者的观点，减少读者的误解。用词力求简单，在表达同样意思时，尽量用短词代替长词，以常用词代替生僻词。但是当描述方法、步骤时，应该用狭义词代替广义词。例如，英文中有不少动词 do、run、get、take 等，虽简单常用，但其意义少则十几个，多则几十个，用这类词来描述研究过程，读者难免产生误解，甚至会不知所云，这就要求根据具体情况，选择意义相对明确的词，诸如 perform，achieve 等，以便于读者理解。

6. 英文摘要写作的造句

（1）熟悉英文摘要的常用句型

尽管英文的句型种类繁多，丰富多彩，但摘要的常用句型却很有限，而且形成了一定的规律。

表示研究目的，常用在摘要之首用：

In order to...

This paper describes...

The purpose of this study is...

表示研究的对象与方法用：

The [curative effect/sensitivity/function] of certain [drug/kit/organ...] was [observed/detected/studied...]

表示研究的结果用：

[The result showed/It proved/The authors found] that...

表示结论、观点或建议用：

The authors [suggest/conclude/consider] that...

（2）关于从句的用法

尽量采用 -ing 分词和 -ed 分词作定语，少用关系代词 which，who 等引导的定语从句。由于摘要的时态多采用一般过去时，使用关系代词引导的定语从句不但会使句式变得复杂，而且容易造成时态混乱（因为定语和它所修饰的主语、宾语之间有时存在一定的"时间差"，而过去完成时、过去将来时等往往难以准确判定）。采用 -ing 分词和 -ed 分词作定语，在简化语句的同时，还可以减少时态判定的失误。

7. 不同类型论文摘要的写作举例

（1）陈述型论文摘要

陈述型论文摘要（Descriptive Abstract）一般只说明论文的主题是什么，多半不介绍内容。

例1 ABSTRACT

This article extends Lehand and Pyle (1997) model to include the possibility that manager may exploit cooperate wealth through transactions with affiliated companies and/

or individuals. The results of our model show that the amounts of wealth exploitation are affected by several factors. They are inorderly managerial stock holding, the severity of penalty, the manager's risk attitude, firm's expected future cash flows, and the variance of future cash flows. What's more, the relation between managerial holding and the amount of wealth exploitation is not seperate. Wealth exploitation rises with management share holdings before the break point. When managerial holding exceeds the break point, any further increase in the management holdings will decrease the amount of wealth exploitation.

例 2   ABSTRACT

The effect of price limit on the stock return, volatility and the structural change is analyzed through a generalized autoregressive conditional heterscedasticity (GARCH) model. The interaction between stock returns and its volatility is permitted in each price limit regime. While the stock return does not go up when the price limit goes down from 5% to 7%. The stock volatility, on the other hand, is substantially different across three regimes. The higher the price limit, the larger is the volatility. In end the GARCH model does not suffer from the structural change when price limits change.

例 3   ABSTRACT

This issue presents a complete survey on the integrin, immunoglobulin, and selection families cellular expression patterns on endothelial, resident cells and graft infiltrating cells in human stomach, heart, and lung transplants. It describes the patterns of cellular expression and inducibility in different pathological conditions of the graft. It also discusses the implications for the organ specific appearance of inflammatory reactions in human stomach, heart and lung transplants as for immunosuppressive and therapeutic interventions.

例 4   ABSTRACT

We classfied firm's finacing decisions into four categories: internal financing banking, bank loans, convertible bonds and preferred stocks, and new common shares. This paper uses pooled cross-section and time series data. When the adjusted data is used, we find that B/M has explanatory power to stock returns for the new-issue category; C/F for the internal financing category. The size variable is also significant in all categories. However, the sign is not consistent. The convertible category has reverse size effect and others have size effects. As for the E/P variable, we don't find any additional explanatory power to stock returns for any category. The use of different definition of size hardly changes our results.

例 5   Contact Problems in the Classical Theory of Elasticity (by G. M. Gladwell)

This article gives an account of contact problems in the classical theory of elasticity. It begins from fundamental principles and aims to offer information on recent developments on this subject that the reader can take advantage to widen his horizon of contact problems.

例 6   Use of Engineering Metals

This article shows us the importance of metals in our daily life, especially in machine building and engineering construction. Metals that are used in industry are called engineering metals, of which the most widely-used is iron. Therefore, production of iron is very critical to the development of a nation.

例7 Pure and Applied Science

In this article, the author holds that pure science is fundermentally concerned with the development of theories establishing relationships between the phenomena of the universe. On the other hand applied science, is directly connected with the application of the working laws of pure science to the practical affairs in the life, and to man's increasing control over the environment, thus leading to the development of new techniques, processes and machines. It is evident that many branches of applied science are practical extensions of purely theoretical or experimental work. It shows to us that these two branches of science are mutually dependent and interacting.

（2）信息型论文摘要

信息型论文摘要（Informational Abstract）与陈述型论文摘要相比，略有不同。它除了介绍论文的主题外，还应介绍论文的主要观点以及各观点的核心内容。这种形式在实际中比较少见。显然，这种形式的摘要比陈述型摘要长得多。

例1 Probable Development of Agricultural Mechanization

This article reviews the past achievements in agricultural mechanization and aims to predict the developments in the immediate future. It is widely acknowledged that to cope with greater crop output, it is necessary to make use of more powerful equipment that is capable of more extensive work within the limited time. The main problems are discussed under the following headings: 1) General Trend of Machanization; 2) Mechanization of Crop Farming; 3) Mechanization of Animal Husbandry. The article concludes that further progress of agricultural mechanization depends on, to a great extent, the development of mechanical industry, which should be spurred and encouraged.

例2 Cultural Factors in Translation and Their Transfer

This thesis attempts to apply general knowledge of linguistics, anthropology and translation to the research of cultural factors in translation and their transfer. The thesis concludes a brief introduction and four chapters. The introduction presents the specific topic and raises several related issues. Chapter One is devoted to the relationship between language, culture and translation. Chapter Two focus on the categorization of cultural factors in translation. Chapter Three talks about three topics: translatability, major translation methods of cultural factors, and the transfer of some important cultural subcategories. The last chapter, gives an analysis of some inappropriate translations and draws a general conclusion for the whole thesis.

例3 ABSTRACT

Internationalization and liberalization of business activities have become two of the most significant phenomena for the successful operations of contemporary enterprises. At present, the prevalence of national protectionism and the establishment of regional economic communites, have further enhanced many firms to take part in overseas investments in order that the tariff and non-tarriff barriers of the products could be reduced or eliminated. Yet, companies operate in different countries with different political, legal, economic, and social cultural surroundings might have different management styles and strategic operating patterns. Traditionally, the Japanese and the US management patterns are two of the most important reference models for the operations of domestic enterprises. Thus, to promote the success of business internationalization activities, it would be very helpful for the domestic firms in Taiwan China to evaluate the similarites and differences of the characteristics of business operations for the US and the Japanese firms especially on management styles, organizational structures, and business performances.

Through a series of personal interview and mail survey, this study concluded the following findings:

1) Firms with different investment origins (i. e. , Taiwanese, American, and Japanese firms) tended to operate differently on constructs of cooperate objectives, organizational structures, competitive strategies, and management performances.

2) For cooperate objectives, Japanese firms tended to emphasize organizational development objectives through educating and activating human resources, while American firms favor financial objective such as improving cash flow and profitability. For organizational structures, American firms achieved higher levels of delegation authority and operation formalization than those of Japanese and Taiwanees firms. For competitive strategies, American firms tended to emphasize product differentiation and cost leadership strategies while Japanese firms favor process innovation and product differentiation strategies.

3) Firms selecting process innovation and product development strategies tended to emphasize both organizational development and sales growth objectives . These firms tended to be very sensitive on the potentials of industry development. In addition, firms adopting cost leadership strategies tended to emphasize sales growth objectives. In a word, these firms seemed to exercise higher levels of formalization and standardization of business activities.

Appendix 4
与环境科学相关的网站地址

United States Environmental Protection Agency (EPA)
http://www.epa.gov
United States Geological Survey (USGS)
http://www.usgs.gov
Endangered Species Home Page
http://endangered.fws.gov
Environment News Service (ENS)
http://www.ens-news.com
EnviroSources
http://www.envirosources.com
National Institute of Environmental Health Sciences (NIEHS)
http://www.niehs.nih.gov
Natural Resource Conservation Service (NRCS)
http://www.nrcs.usda.gov
Bureau of Reclamation
http://www.usbr.gov/
Hydrologic Engineering Center (HEC)
http://www.hec.usace.army.mil/
Iowa Institute of Hydraulic Research
http://www.iihr.uiowa.edu/
Bureau of Reclamation
http://www.usbr.gov
EPA's Surf Your Watershed
http://www.epa.gov/surf/
Hydrologic Conditions-Delaware River Basin
http://www.state.nj.us/drbc/hydro.htm
Hydrologic Engineering Center (HEC)
http://www.hec.usace.army.mil/
Hydrogeologists Home Page
http://www.thehydrogeologist.com/index.htm
Hydrology and Remote Sensing Laboratory
http://www.ars.usda.gov/main/site_main.htm?modecode=12650600
National Ground Water Association
http://www.ngwa.org
California Water Quality Program

http://www.omwq.water.ca.gov/
EPA Microbiology Home Page
http://www.epa.gov/nerlcwww/
Society of Environmental Toxicology and Chemistry
http://www.setac.org
Water Quality Information Center
http://www.nal.usda.gov/wqic/
Water Quality Monitoring
http://www.epa.gov/OWOW/monitoring
Clean Lakes Program
http://www.epa.gov/OWOW/lakescllkspgm.html
Delaware Nonpoint Source Management Plan
http://www.dnrec.state.de.us/dnrec2000/Library/NPS/NPSPlan.pdf
Index of Watershed Indicators (IWI)
http://www.epa.gov/iwi
National Wetlands Research Center (NWRC)
http://www.nwrc.usgs.gov/
National Wetlands Inventory Center (NWI)
http://www.nwi.fws.gov
Nonpoint Source Pollution Control Program
http://www.epa.gov/OWOW/NPS/
North American Lake Management Society (NALM)
http://www.nalms.org
Office of Wetlands, Oceans, and Watersheds
http://www.epa.gov/owowwtr1/oceans/
Guide for Teaching About Coastal Wetlands
http://www.nwrc.usgs.gov/fringe/ff_index.html
Wetlands Homepage
http://www.epa.gov/owow/wetlands
Watershed Assessment, Tracking and Environmental Results (WATERS)
http://www.epa.gov/waters
American Water Works Association
http://www.awwa.org
Current Drinking Water Standards
http://www.epa.gov/ogwdw/mcl.html
Office of Groundwater and Drinking Water
http://www.epa.gov/OGWDW/
Public Drinking Water Systems Programs
http://www.epa.gov/safewater/pws/pwss.html

Safe Drinking Water Overview
http：//www. epa. gov/enviro/html/sdwis/sdwis_ov. html
Canadian Pipeliner
http：//www. pipeline. ca/
Ductile Iron Pipe Research Association (DIPRA)
http：//www. dipra. org/
Plastic Pipe Institute
http：//www. plasticpipe. org/
Underground Construction
http：//www. undergroundinfo. com/
Water One
http：//www. waterone. org/
Water Transmission and Distribution
http：//www. cwlp. com/Water_division/T_DWater/t&dwater. htm
American Concrete Pipe Association
http：//www. concrete-pipe. org
Hopatcong Sanitary Sewer Project
http：//www. hopatcong. org/
Water Corporation
http：//www. watercorporation. com. au/
New Orleans Drainage System
http：//www. swbnola. org/
Story of Sewerage in Leeds
http：//www. dsellers. demon. co. uk/sewers/sew_ch1. htm
Greater Vancouver Regional District
http：//www. metrovancouver. org/services/Pages/default. aspx
Metropolitan Water Reclamation District of Greater Chicago
http：//www. mwrdgc. dst. il. us/
Controlling Nonpoint Source Pollution
http：//www. epa. gov/OWOW/NPS/roads. html
HydroCAD Software Sampler
http：//www. hydrocad. net
Department of Natural Resources & Environmental Control (DNREC)
http：//www. dnrec. delaware. gov/Pages/default. aspx
Managing Urban Runoff
http：//www. epa. gov/OWOW/NPS/facts/point7. htm
EPA's Office of Wastewater Management
http：//www. epa. gov/OWM/
Largest Wastewater Treatment Plant in the World

http://www.mwrdgc.dst.il.us/plants/stickney.htm
Environment Canterbury
http://www.ecan.govt.nz/home
Water Environment Federation (WEF)
http://www.wef.org/Home
Environmental Industry Interactive
http://www.envasns.org/
EPA's Wastes
http://www.epa.gov/epawaste/index.htm
Municipal Solid Waste Publications
http://www.epa.gov/epawaste/index.htm
Solid Waste Association of North America (SWANA)
http://www.swana.org/www/default.aspx
Solid Waste Disposal Facilities criteria
http://www.epa.gov/epaoswer/non-hw/muncpl/criteria.htm
Solid Waste Online
http://www.solidwaste.com/
EPA's Brownfields Web Page
http://www.epa.gov/swerosps/bf/
Groundwater Remediation Technology Analysis Center (GWRTAC)
http://www.gwrtac.org/
Hazardous Waste Cleanup Information
http://www.clu-in.com/
Office of Solid Waste-Hazardous Waste
http://www.epa.gov/epaoswer/osw/hazwaste.htm
Superfund Site
http://www.epa.gov/superfund/
Underground Injection Control (UIC)
http://www.epa.gov/ogwdw/uic.html
AIR Now
http://www.epa.gov/airnow/index.html
Air Releases Overview
http://www.epa.gov/enviro/html/airs/
Clean Air Market Programs
http://www.epa.gov/airmarkets/
Global Warming Site
http://www.epa.gov/climatechange/index.html
Indoor Air Quality
http://www.epa.gov/iaq

Office of Air and Radiation
http://www.epa.gov/oar/
Office of Air Quality Planning and Standards
http://www.epa.gov/oar/oaqps/
Ozone Depletion
http://www.epa.gov/ozone
Plan English Guide to the Clean Air Act
http://www.epa.gov/oar/oaqps/peg_caa/pegcaain.html
United Nations Convention on Climate Change
http://www.unfccc.de/
Highway Traffic Noise in the United States
http://www.nonoise.org/library/highway/probresp.htm
Noise Pollution Clearinghouse (NOC)
http://www.nonoise.org/
Quiet Communities Act of 2005
http://www.barkingdogs.net/actamqca2005.shtml
International Association for Impact Assessment (IAIA)
http://iaia.org
American Society for Engineering Education
http://www.engineeringk12.org/
American Society of Civil Engineers
http://www.asce.org/asce.cfm
Construction Metrication Newsletter
http://www.nibs.org/cmcnews.html
International System of Units
http://physics.nist.gov/cuu/Units/index.html
Toward A Metric America
http://ts.nist.gov/ts/htdocs/200/202/mpo_home.htm